I0066339

Agricultural Productivity Enhancement: Techniques and Technologies

Agricultural Productivity Enhancement: Techniques and Technologies

Edited by **Jordan Smith**

New York

Published by Syrawood Publishing House,
750 Third Avenue, 9th Floor,
New York, NY 10017, USA
www.syrawoodpublishinghouse.com

Agricultural Productivity Enhancement: Techniques and Technologies
Edited by Jordan Smith

International Standard Book Number: 978-1-68286-143-1 (Hardback)

Printed in the United States of America.

Contents

Preface

The world is advancing at a fast pace like never before. Therefore, the need is to keep up with the latest developments. This book was an idea that came to fruition when the specialists in the area realized the need to coordinate together and document essential themes in the subject. That's when I was requested to be the editor. Editing this book has been an honour as it brings together diverse authors researching on different streams of the field. The book collates essential materials contributed by veterans in the area which can be utilized by students and researchers alike.

The enhancement of agricultural productivity is important for the economic growth of any country. This book presents researches and studies performed by experts across the globe. Discussed herein are topics such as impact of fertilizers, postharvest crop management, irrigation techniques, etc. The text aims to equip students and professionals engaged in the field of agricultural science and similar fields of study with the advanced topics and upcoming concepts in this area. It aims to serve as a resource guide and contribute to the growth of the discipline.

Each chapter is a sole-standing publication that reflects each author´s interpretation. Thus, the book displays a multi-facetted picture of our current understanding of applications and diverse aspects of the field. I would like to thank the contributors of this book and my family for their endless support.

Editor

1

Clean production technology of integrated pretreatment for Lignocellulose

Chen Hongzhang[1]* and Zhao Junying[1,2]

[1]National Key Laboratory of Biochemical Engineering, Institute of Process Engineering, Chinese Academy of Sciences, Beijing 100190, China.
[2]Graduate University of Chinese Academy of Sciences, Beijing 100049, China.

Pretreatment is a key step for lignocellulose converting into universal raw material for industries as alternative of petroleum. This review discusses that single pretreatment lacks cost-competitiveness because only one product takes responsibility for all cost. Based on the inheterogeneity of lignocellulose and the advantages of steam explosion pretreatment, steam explosion integrated pretreatment is proposed as strategy to solve the problem above. Steam explosion integrates with different pretreatment technologies to deconstruct lignocelluloses selectively for multiple products. By comparison, steam explosion-washing-alkali extraction-mechanical carding integrated pretreatment is considered the best. Finally, industrial demonstration of this integrated pretreatment is given out. With this integrated process, the recovery yields of cellulose, hemicelluloses and lignin reached 70, 80 and 80%, respectively.

Key words: Clean production technology, Lignocellulose, Integrated pretreatment, steam explosion, industrialization.

INTRODUCTION

Making the process cost-competitive is critical now for the industrialization of lignocellulose refining. Researchers have mainly concentrated on ethanol production from cellulose which only accounts for 30% of lignocellulose. However, about 70% lignocellulose including lignin and hemicelluloses are wasted. So, no matter how high the yield of enzymatic hydrolysis is, the cost of raw material is all apportioned to single product, leading to its high price. Therefore, we propose the concept of fractional conversions (Chen and Qiu, 2010) that fractionates lignocelluloses into different fractions suitable for various products, leaving almost no residues (Chen and Qiu, 2010). So the cost of raw materials and refining process are apportioned to multiple products. However, during long time evolution process, lignin content in three tissues increases: vascular tissue to transport water, sclerenchyma tissue to support stem and epidermic tissue to avoid moisture evaporation. Moreover, cellulose connects lignin and hemicellulose with hydrogen bond, and hemicellulose combines lignin with covalent bond. So, cell wall with complex structure is recalcitrant to enzymatic hydrolysis and other physical or chemical effects. If separating cellulose, hemicelluloses and lignin completely and subsequently hydrolyzing big molecule into chemicals, it would involve high solvent and energy, not to mention synthesizing other products from chemicals. This process is contrary to the principle of energy utilization, leading to high cost. So we design the process of selective deconstruction for fractional conversion. By this way, structure characteristics would be considered and applied while converting lignocellulose to biofuel, biomaterial or biochemical. And economic feasibility for lignocellulose industrialization would be improved with the principle multiproducts orientation. To obtain multiple products with selective deconstruction, single pretreatment method could hardly fulfill the requirements (Agbor et al., 2011). So we establish

*Corresponding author. E-mail: hzchen@home.ipe.ac.cn.

integrated pretreatment technologies to make the advantages of each pretreatment technology complementary, bringing to multiple products through selective deconstruction.

This paper highlights the mechanism to integrate various pretreatment technologies based on the analysis of the heterogeneity of lignocellulose. Moreover, research advances are presented and industrial demonstration is provided.

INEVITABILITY OF INTEGRATED PRETREATMENT FOR LIGNOCELLULOSE

Pretreatment method for lignocellulose

The reason to use lignocellulose in the early time was to complement the shortage of food and fuel owing to congestion of population in many developing countries (Klyosov, 1986). Lignocellulose was mainly applied by three kinds of methods. The first kinds focused on single product with single pretreatment technology before 1990 (Kawamori et al., 1986; Taniguchi et al., 1982a; Zadra il, 1977; Wayman and Parekh, 1988). And then integrated pretreatment technology appeared for single product (Carrasco and Roy, 1992; Maekawa, 1996; Holtzapple et al., 1991) also. In the 21 century, human began to care about the problems of environment, energy and rural development. So, integrated pretreatment for multiproducts was developed in order to find the alternative of petroleum. However, back in 1987 (Klyosov, 1986), COALBAR Company in Brail had used lignocellulose for both ethanol and charcoal. So the division time above is not so strict, but research focuses are different in different stages.

Single pretreatment technology for single product

Before 1990, lignocellulose was only converted to ethanol with single pretreatment technology including chemical, physical and biological methods.

Chemical methods referred to dilute acid or alkali (Kawamori et al., 1986), ozone (Bono et al., 1985), zinc chloride (Cao et al., 1995). Dilute acid included dilute sulfuric acid; dilute hydrochloric acid, hypochlorite (David and Atarhouch, 1987), peracetic acid (Taniguchi et al., 1982b), sulfur dioxide (Wayman and Parekh, 1988). There were also single pretreatment nowadays for single product, such as fumaric acid, maleic (Kootstra et al., 2009) and ionic liquid (Liu and Chen, 2006). All the researches above were aimed to improve enzyme hydrolysis rate. And phosphoric acid was also applied to improve lignocellulose edibility (Deschamps et al., 1996). Physical pretreatment technology included mechanical comminution, steam explosion (Grous et al., 1986), hot water (van Walsum et al., 1996), gamma-ray (Bono et al., 1985), microwave (Ooshima et al., 1984), organic solvent (Lipinsky and Kresovich, 1982), ammonia freeze explosion (Holtzapple et al., 1991). With the hot water pretreatment, two stages hydrolysis at high temperature was applied to avoid the effect of furfural (Torget and Teh-An, 1994). Biological method was mainly used to hydrolyze lignin with microorganism (Hatakka, 1983), reducing inefficacious absorption and enhancing enzyme hydrolysis rate. And microorganism pretreatment was also researched to produce single cell protein (Taniguchi et al., 1982a) and feed (Zadra il, 1977).

Industrial application about single pretreatment for single product was backed to 1913 (Klyosov, 1986). In Georgetown, South Carolina, pin mill waste was pretreated with 2% sulfuric acid at 175°C in rotary steam heated digesters, producing 5000 gallon ethanol per day. Subsequently, a second plant was built in Fullerton, Louisiana. However, both plants became profitable until the middle 1920s because of extremely low cost sawdust waste and the lack of environmental controls. So, lignocellulose conversion into single product with single pretreatment method could hardly industrialize nowadays because of high feedstock price and strong environmental consciousness.

Integrated pretreatment for single product

As early as 1986 (Klyosov, 1986), steam explosion pretreatment was pointed out to be one of the most economical and effective technology. Pretreatment methods integrated with steam explosion included methanol, peroxide hydrogen, sodium hydroxide (Maekawa, 1996) and ammonia (Carrasco, 1992). Moreover, in order to enhance biogas production in anaerobic digestion, corn stover was pretreated with sodium hydroxide integrated with green oxygen or 1,4-dihydroxy anthraquinone (Ping et al., 2010), and rice straw was pretreated with grinding integrated with calcium hydroxide (Cui et al., 2011).

To prepare fiber, ultrasonic integrated chemical pretreatment was applied for poplar. During chemical pretreatment, there were two processes including delignification with sodium hypochlorite and removal of gel as well as hemicelluloses with sodium hydroxide (Chen et al., 2011b). Though the price of petroleum is rising and environmental consciousness is becoming stronger and stronger, technology to produce ethanol or biogas still lacks of competitiveness because of low yield. So, how to convert all components into products is a key step to reduce cost.

The inevitability of integrated pretreatment for multiproduct

During forty years after the Second World War, only one

biofuel company in Soviet Union succeeded as a commercial operation. The essential reason was that lignocellulose was considered as multifunctional raw material and converted into not only ethanol (cellulose) but also yeast, furfural (hemicellulose) and fuel (lignin). Up to 1986, this company consisted of 40 full-scale plants with a maximum capacity of 1000 tons of wood material per plant per day. The production from this technology was 1.5 million tons of fodder yeasts and 195 million L of ethanol. It proved that integrated pretreatment for multiproduct was the only way for the industrialization of lignocellulose refining.

As early as 1986 (Klyosov, 1986), integrated pretreatment for multiproduct was proposed as an effective way to apply lignocellulose which was considered as wastes of large number then. In 1993 (Wyman and Goodman, 1993), economic model revealed that multiproduct from lignocellulose could decrease cost of each product. However, effective pretreatment method had not reported.

It is an inevitable direction that bioproducts become alternative of those made from petroleum. However, lignocellulose refining technology could hardly been industrialized due to the lack of competitiveness, though related technology has been researched as long as 100 years and the price of petroleum has been rising nowadays. Related researches mainly focused on biofuels, but in the long term, lignocellulose would be the only raw materials for chemicals when petroleum is running out. Therefore, the importance of biochemicals should not be ignored. Even biochemicals could not take place of those from petroleum, multiproduct can make the industrialization of lignocelluloses refining beneficial.

DESIGN OF INTEGRATED PRETREATMENT FOR LIGNOCELLULOSE

Connotation of integrated pretreatment for lignocellulose

Integrated pretreatment is the integration of two or more physical, chemical and biological pretreatments instead of their simple combination. It is conceived according to the heterogeneity of lignocellulose, abiding by the principle of selective deconstruction and fractional conversion, obtaining economic, steady and environment friendly products.

According to the heterogeneity of lignocellulose

It has been found that the component and structure are heterogeneous for different kinds of lignocelluose such as corn stover, wheat straw and rice straw, and for different organ, tissue and cell from the same kind.

1. For different feedstock, components and structure are obviously different.
2. For different organs from the same feedstock, such as leaf, sheath, rind, pith and so on, components and structure are different because the percentage of various tissues are different, resulting in diversity of enzyme hydrolysis of different organs (Chen et al., 2011a).
3. The component and structure of different tissue are different. According to the classifying method proposed by Sachs (1875), the mature tissues of plant are classified as dermal tissue system, vascular tissue system and ground tissue system. The dermal tissue system includes epidermis and periderm. Vascular tissue system is composed of xylem and phloem. Ground tissue system consists of parenchyma (including secretory), collenchymas and sclerenchyma.

The surface of dermal tissue system often contains cuticle. Cuticle consists of cutin and wax, and cutin is polymer of C_{16} and C_{18}. The parenchyma under epidermis is hardened, leading to high lignin content. Vessel wall is secondary thickened and lignified, so, lignin content is high. However, in phloem, sieve tube mainly consists of primary wall. In the ground tissue system, cell wall of parenchyma is lightly lignified but sclerenchyma contains secondary wall which is high in lignin.

4. The component and structure of the different cell are different. As presented above, component of different cell in the same tissue is also different. There is also some specialized cell, such as silica cell which is located in the middle of long cell around vein. The wall of silica cell is silicidized.
5. The heterogeneity of different component is mainly demonstrated in lignin and hemicellulose. Phenylpropane is the basic framework of lignin mainly including syringyl, guaiacyl and hydroxy-phenyl. The heterogeneity of lignin from different cell, tissue or organ is caused by variations in the polymer composition, size, cross-linking and functional groups (Bonini et al., 2008).

Structure units of hemicellulose include D-xylose, D-glucose, D-galactose, D-allose, 4-O-methyl-D-glucuronic acid, D-galacturonic acid and D- glucuronic acid. There is also less L-rhamnose, L-fucose and other neutral sugar with methoxy and acetyl. Heterogeneous polysaccharide in Hemicellulose consists of 2 to 4 structure units above.

Therefore, conversion technology for these raw materials should be established on the basis of the heterogeneity in different levels. For example, vascular tissue is high in lignin, so it is suitable for lignin extraction instead of ethanol preparation. But for ground tissue system, especially parenchyma is easy to be hydrolyzed by cellulase because of high cellulose content.

Comply with principle of selective deconstruction and fractional conversion

Selective deconstruction means two aspects. On one hand, considering from raw materials, integrated pretreatment should be based on the heterogeneous property of raw materials instead of separating cellulose, hemicelluloses and lignin completely. On the other hand, considering from products, integrated pretreatment should be based on the requirements of products to make best of functional components or structure. That is to convert directly with selective deconstruction, avoiding hydrolyzing into small molecules to reduce energy consumption.

For example, ethanol from polysaccharide, especially cellulose is essential to obtain a long-term sustainable supply of energy. Cellulose content is high in parenchyma tissue and phloem which are suitable for enzyme hydrolysis. However, epidermal tissue, xylem and sclerenchyma, especially vessel in xylem are high in lignin. As the raw materials for ethanol production, high cellulose content and less lignin content are required. So parenchyma and phloem should be obtained and then hydrolyzed directly without further lignin removal. The residual epidermal tissue, xylem and sclerenchyma, which are high in lignin, can be converted into board or further refined for lignin product.

Fractional conversion means to produce mainly one or two products as well as other by-products based on the principle of eco-industry. Multiproduct got from fractional conversion includes biofuel, biomaterials and biochemicals. Biofuel includes ethanol, butanol, bio-oil, hydrogen, biogas and feedstock for power generation. Biomaterials include not only board and phenolic resin from lignin but also biomass composite materials made by adding other polymer, metal or inorganic salts. Biochemicals are products from lignocellulose with the conversion technology of thermochemistry or biochemistry.

Compared with single product, multiple products would apportion the cost of raw material and other consumption to reduce to cost of each product. For example, if the price of corn stover was 93.78$/t, ethanol would be only apportioned one third (31.26$) because ethanol is made from cellulose which accounts for about one third of corn stover. And other 62.52$ is apportioned to products of lignin (for example, phenolic resin) and hemicelluloses (for example, butanol and acetone).

Complementary integration of different pretreatment technologies

Complementary integration is to make different pretreatment technologies complementary, obtaining multiproduct with one or two main products.

Usually, in order to obtain one product with high yield, several pretreatment technologies are involved, which is lack of competitiveness because of high cost and large wastes. Complementary integration highlights multiple-product orientation to reduce the cost and remove wastes. By this way, different components including cellulose, hemicellulose, lignin, even ash are converted into final products almost without wastes.

Take steam explosion for example, hemicellulose is hydrolyzed by auto catalyzing effect at high temperature. And then hydrolysates could be washed out and converted into biogas or butanol through fermentation. However, lignin could not be separated by steam explosion.

Lignin could be dissolved by alkali. Therefore, steam explosion and alkali extraction technologies could be integrated to separated selectively hemicellulose, lignin and cellulose. Meantime, the strength of steam explosion reduces without the aim to separate lignin. And alkali used is less after steam explosion pretreatment because the tight structure of lignocellulose is destroyed. By this way, hydrolysate of hemicelluloses, cellulose and lignin could be converted into different products.

Advantages of integrated pretreatment for lignocellulose

Compared with single pretreatment technology, the advantages of integrated pretreatment for lignocellulose are as follows.

1. Integrated pretreatment is oriented by multiproduct to make best of each component for different products. Therefore, the economic value of lignocellulose increases and discharge reduces.
2. Integrated pretreatment is based on the heterogeneity of lignocellulose in different levels, especially on cell level. For example, cells of vessel and parenchyma could be separated by steam explosion integrated mechanical carding (Chen and Fu, 2012), which is regarded as selective separation. And the vessel, which is high in lignin, could be used for spinning, pulping and board. Meanwhile, parenchyma cell and phloem cell which are rich of cellulose could be converted into biofuels by enzymatic hydrolysis and fermentaion.
3. Integrated pretreatment is of wide adaptability because there are more parameters could be adjusted. Therefore, integrated pretreatment is also flexible and fit for various raw materials from different regions and planting manners.
4. According to the property of lignocellulose in terms of components and structure, many ideas and technologies could be integrated including those from paper making, pinning and board industry, especially petroleum refining which should be rational, economic, effective, clear and operable. For example, the idea of petroleum refining could be used in lignocellulose conversion to separate selectively for different products.
5. Integrated pretreatment makes different technologies complementary in advantages. By this way, the strength of each technology is reduced, leading to law pretreatment cost.

STATUS OF INTEGRATED PRETREATMENT FOR LIGNOCELLULOSE

As discussed previously, integrated pretreatment is to integrate pretreatment technologies to make best of the advantages of each single technology. That means to make best of one pretreatment method as well as to create advantages for next pretreatment methods. Steam explosion has been regarded as one potential pretreatment technology (Grous et al., 1986; Guo et al., 2011) and applied for various products from lignocelluloses (Biswas et al., 2011; Brugnago et al., 2011; Jacquet et al., 2011; Li et al., 2009; Rocha et al., 2012; Zimbardi et al., 2009). At present, integrated pretreatment oriented by multiproduct is mainly about steam explosion integrated pretreatment. The authors Zhang and Chen (2012), Chen and Liu (2007), Jin and Chen (2006) and Xu et al. (2001) have summarized advantages of steam explosion from decades of research and industrialization of steam explosion technology as follows:

1. Steam explosion hydrolyzes most hemicellulose, reducing the complexity of cell wall component (Chen and Qiu 2010).
2. At the instant of steam explosion, steam in and among the cells spurts out, destroying cell wall and the connection of tissue. As a result, the tight structure of tissue and cell wall is loosened. Plus the hydrolysis of hemicellulose, porous structure of lignocellulose is formed after steam explosion. Therefore, specific surface area increases, which could enhance mass transfer rate and improve the accessibility ((Zhang and Chen 2012).
3. Purity of products is enhanced after separating hydrolysate of hemicellulose and lignin, as well as other water soluble components (Peng and Chen, 2012).
4. Steam explosion pretreatment is clear, effective, efficient and none chemical additives (Chen and Qiu 2010).
5. Steam explosion is easy to operate and spread.

Therefore, lignocellulose after steam explosion is suitable for further physical, chemical and biological pretreatment, which is the very reason for steam explosion to become research focus. Meantime, steam explosion integrated pretreatment explains the connotation of integrated pretreatment technology in this paper.

Steam explosion-washing-solvent extraction-Bauer screening integrated pretreatment (SE-W-S-BS)

Acetyl attached in hemicellulose is hydrolyzed during steam explosion and become acetic acid. Then, acetic acid catalyzes the degradation of hemicellulose at high temperature. Monosaccharide from hemicellulose is water soluble. Therefore, hydrolysates from hemi-celluloses could be washed out first and then converted into final product. After steam explosion, lignin is more soluble in ethanol and glycerin because phenolic hydroxyl buried in lignin is exposed out. To obtain hemicelluloses and lignin fractions, washing and solvent extraction has been applied after steam explosion. It was revealed that (Chen and Liu, 2007) hemicellulose recovery yield reached 80% by washing steam exploded wheat straw with water for four times.

The solution obtained from washing mainly contained xylose. The residual solid after washing was extracted by ethanol. Ethanol was distilled then at low temperature and the recovery yield reached 88.04%. The extraction liquid after distilling was glue. Lignin in the glue could be precipitated by adding 0.3 mol/L hydrochloric acid. And the precipitated lignin was purified later with recovery yield of 75%. Enzymatic hydrolysis test revealed that, after extraction, reduced sugar yield reached 31.41% which was 5.13 times higher than that of grinding wheat straw.By comparison to methanoic acid, propionic acid, ethylene glycol, butanediol, glycerin was the best one to extract lignin. Cellulose of wheat straw was reserved by 92% (Sun and Chen, 2008) and the yield of enzymatic hydrolysis reached 54% after steam explosion-washing-glycerin extraction integrated pretreatment.

The enzymatic hydrolysis yield of cellulose was also low after steam explosion-washing-solvent extraction pretreatment because of lignin residuals. In plant tissue, cells which are high in lignin are longer including cells from epidermis tissue, sclerenchyma and vessel tissue. However, cells which are low in lignin are shorter such as parenchyma cells. Short cells would be fit for enzymatic hydrolysis if they could be separated from long cells. So, Bauer screening pretreatment was applied.Two grades were obtained after Bauer screening. Scanning electronic microscope analysis showed that grade 1 was mainly fiber tissue cell (from epidemic tissue and mechanical tissue) and grade 2 was mainly miscellaneous cell (non-fiber tissue cell) with percentage of 47.6% and 19.9% respectively.

The average length and width of the long fiber were 1.067 mm and 13.893 μm which was similar to that reported (1.39 mm and 13.0 μm) in Papermaking Raw Materials of China an Atlas of Micrographs and the Characteristics of Fibers. The ratio of length to width was 76.81, which was among that of conifer and broadleaf but higher than the requirement of paper making materials (Fang and Liu, 1996).Steam explosion -Bauer screening integrated pretreatment was also used for corn stover. It revealed that product smaller than 75 μm was easier to be hydrolyzed by cellulase than the product bigger than 600 μm. Therefore, process was designed to produce the main product of butanol as shown in Figure 1.

Steam explosion-washing-alkali extraction-mechanical carding integrated pretreatment (SE-W-A-MC)

Though hydrolyzed lignin was extracted by steam explosion integrated solvent extraction pretreatment, lignin

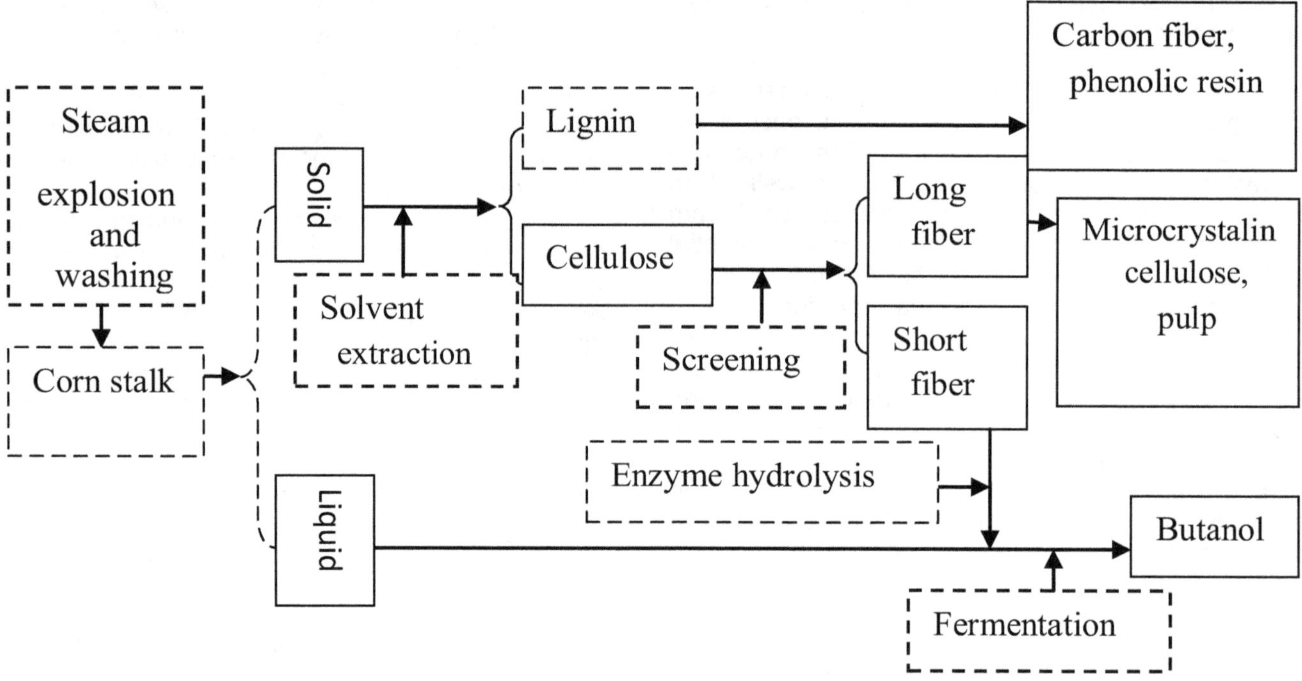

Figure 1. Flow chart of steam explosion-washing-solvent extraction-Bauer screening integrated pretreatment for corn stalk with the main product of butanol.

lignin residuals was also recalcitrant to cellulase. Moreover, lignin residuals were inhibitor in fermentation for butanol. So, methods were further explored to hydrolyze lignin.

Research revealed that lignin could be separated by alkali oxygenation. Wang and Chen (2011) and Cara et al. (2006) also reported that lignin in lignocellulose could be removed by pretreatment to improve the yield of enzymatic hydrolysis and fermentation property. Tissues consist of different cells and components could be fractionated by mechanical carding equipment (Chen and Fu, 2012). To enhance the yield of lignin, enzymatic hydrolysis and fermentation, steam explosion-washing-alkali extraction-mechanical carding integrated pretreatment was researched.

With this process, the recovery yields of cellulose and lignin reach 70 and 80%, respectively.

Normally, for butanol fermentation, sugar concentration in culture medium should be lower than 60 g/L. The total sugar content was 45.24 g/L in the enzyme hydrolysate of corn stover pretreated with steam explosion combined 1% NaOH+4% H_2O_2 (Wang and Chen, 2011). So, hydrolysate from enzymatic hydrolysis could satisfy the requirements for butanol fermentation. Further research revealed that soluble lignin content in enzyme hydrolysis was 1.2 g/L which would not inhibit butanol fermentation. As a result, the total solvent content was 12.10 g/L with the yield of 0.27 and productivity of 0.17 g/(L·h).

Therefore, process was designed with butanol as main product (Figure 2).

Steam explosion integrated super grinding pretreatment (SE-SG)

Superfine-grinding has been applied widely in the chemical industry, and its energy consumption is not as high as traditional grinding methods (Yang, 1988).

Therefore, superfine-grinding technology is considered to be imported into steam explosion integrated pretreatment to satisfy lignocellulose conversion requirements (Jin and Chen, 2006). Water-holding capacity and toughness is different between fiber tissue and non-fiber tissue including ground tissue and epidemic tissue (Yang, 2001), leading to different grinding property.

Therefore, steam explosion integrated wet grinding pretreatment was studied by taking rice straw as material.Fiber part accounted for 70.4% of dry weight after steam explosion integrated wet grinding pretreatment (Jin and Chen, 2006). In the fiber part, fiber cell accounted for 63.1% which was higher than that of raw rice straw by 37.8%. Fiber component accounted for 65.6% in fiber part which was higher than that of raw rice straw by 74.9%, and the ground tissue cell accounted for 33.5%.

This indicated that steam explosion-wet superfine grinding pretreatment could separate fiber cell as well as fiber component to some extent.

Enzymatic hydrolysis test revealed that hydrolysis rate and reducd sugar yield of fiber tissue part was the highest compared with that of raw rice straw and steam exploded rice straw. Especially, reduced sugar yield was

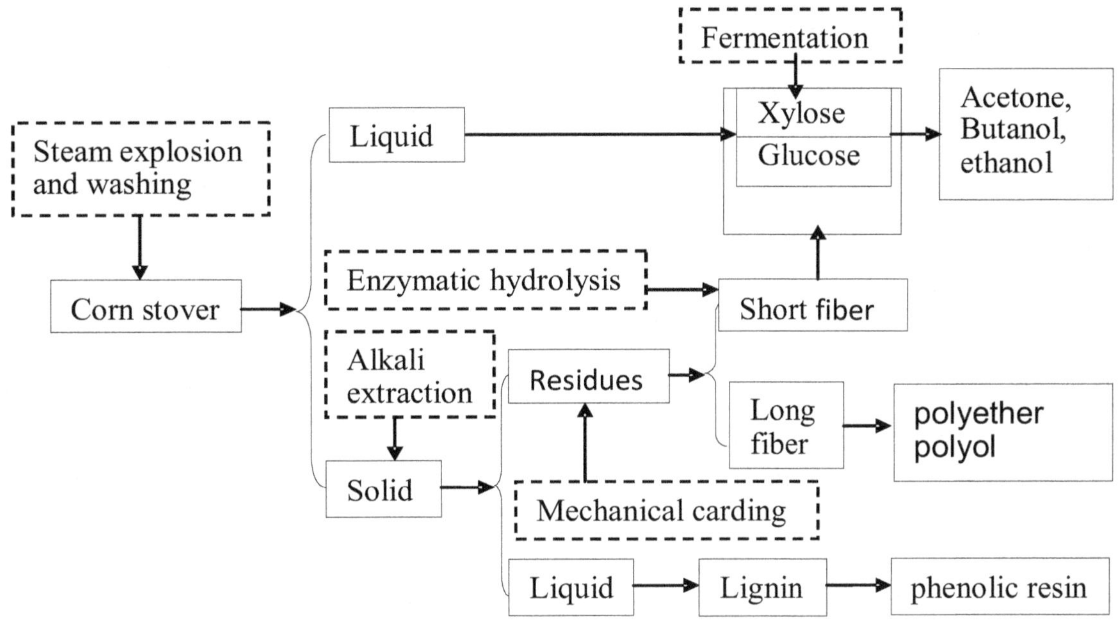

Figure 2. Flow chart of steam explosion-washing-alkali extraction-mechanical carding integrated pretreatment for corn stover with butanol as main product.

higher than that of raw rice straw by 60.0%. Therefore, fiber tissue separated by steam explosion combined wet superfine grinding was suitable as enzymatic hydrolysis material.

Simultaneous saccharification and fermentation test showed that ethanol yield of fiber tissue part pretreated by steam explosion integrated wet superfine grinding method reached 8.6%, but that of steam exploded rice straw and raw rice straw were 7.3% and 6.7%, respectively after 48 h fermentation. Therefore, steam explosion combined wet superfine grinding pretreatment increases the ethanol yield as well as its productivity.

Other combined pretreatment waiting for integrating steam explosion combined ionic liquid pretreatment (SE-IL)

Compared with the organic solvent and electrolyte, there are some advantages for ionic liquid and it is regarded as one of the three green solvents including supercritical carbon dioxide, aqueous two-phase and ionic liquid. At first, it is nearly no steam pressure, so, the pollution problem of other volatile organic solvent is avoided. Then, it is chemically and physically stable among broad range temperature adjustability (from lower than 25 to 300°C) and electrochemical window. Finally, it can be adjusted by the design of cation and anion, so the solubility to inorganics, organics and water is changed and its acidity can be adjusted to super acid.

Previous experiment (Liu and Chen, 2006) showed that the contents of cellulose, hemicellulose and lignin of wheat straw decreased after steam explosion combined ionic liquid pretreatment. The ionic liquid was self-prepared. To protect cellulose, acid or alkali was inputted into ionic liquid. It was revealed that sodium hydroxide could improve solubility and protect cellulose. However, 1% sulfuric acid reduced the content of hemicellulose and cellulose, but improved the lignin content. Further research showed that, enzymatic hydrolysis of wheat straw reached 102.49% (the hydrolysate of hemicelluloses was included) after steam explosion combined ionic liquid pretreatment.

Steam explosion combined laccase pretreatment (SE-L)

Lignin-carbohydrate complex (LCC) inhibits enzymatic hydrolysis. So it is important to separate LCC before hydrolysis. Enzyme pretreatment is mild and specific. Therefore, steam explosion combined enzyme pretreatment was researched to remove LCC (Qiu and Chen, 2012). The enzyme used included laccase and feruloyl esterase or xylanase. Steam exploded material was pressed by screwing to separate the solid and liquid. And then, enzyme combination was input into the solid part. When laccase and feruloyl esterase were used, the yield of cellulose, hemicellulose and lignin hydrolysis of corn stover were 5.65%, 10.97% and 33.93%, respectively compared with the content of steam explosion corn stover. Hydrolysis products of lignin and hemicellulose could be converted to other products of high value. This pretreatment was high-rate, efficient,mild and specificity.

Moreover, water wastage was decreased by screwing press instead of washing. Importantly, the increased hydrolysis rate could provide subtract with high sugar content for subsequent fermentation.

Steam explosion combined electricity catalyzing pretreatment (SE-EC)

Electricity catalyzing is being applied in waste water pretreatment because it is without chemical reagent and second pollution, and easy to operate. Pulp could be bleached by redox effect of electricity catalyzing. Redox effect catalyzes lignin to destroy its structure, consequently, lignin is removed and pulp is bleached (Ma et al., 2003).

Therefore, steam explosion combined electricity catalyzing pretreatment was researched (Ding, 2010). With this pretreatment, hemicellulose was hydrolyzed in steam explosion process and lignin was degraded in electricity catalyzing process. As a result, hemicellulose, lignin and cellulose were separated and could be converted into high value products separately.

Compared with that of steam exploded corn stover, the contents of lignin of core straw pretreated by steam explosion combined electricity catalyzing at the voltage 1.5, 2.5, and 5 decreased 6.9 12.7 and 20.0%, respectively. And the content of cellulose was reduced 1.0 2.8 and 4.3%. The content of hemicellulose decreased 10.1, 14.6 and 16.9%. However the soluble content increased 50.0, 84.2 and 111.8% respectively at voltage 1.5, 2.5 and 5 V. This revealed that electricity catalyzing pretreatment degraded lignin as well as hemicellulose and cellulose, leading to soluble component improvement. Glucose content in corn stover hydrolysate increased by 25.4% at the voltage 1.5 V after steam explosion combined electricity catalyzing pretreatment.

Steam explosion combined photochemistry pretreatment (SE-P)

Steam explosion combined photochemistry pretreatment was researched (Bonini et al., 2008) to convert lignin into by product of pulp and bleaching process. Pine and corn stover were impregnated in dilute acid, steam exploded and washed by hot water firstly. Then, lignin was extracted by 1.5% sodium hydroxide solution and precipitated at pH 2 with 20% H_2SO_4. Lignin from pine was put into a tube and saturated with bubbling oxygen for 1 h, and the irradiation was performed by using a 50 W tungsten-halogen lamp. Lignin from corn was dissolved in 1:1 acetonitrile-methanol (20 ml) and flushed with ozone generated by using Ozone Generator 500 M. Characterization with NMR revealed that ozonolysis of steam exploded lignin did not recovery fine chemicals with the exception of oxalic acid. This was caused by acid impregnation which made lignin destroyed. And an initial polymerization process was observed, which was induced by the formation of radicals in lignin components obtained from steam explosion pine. It was found that steam explosion was a method affording the highest destroying of corn stover lignin, making lignin easily oxidized by ozone.

Steam explosion combined acid pretreatment (SE-A)

Dilute-acid can make hemicelluloses soluble, therefore it has been widely applied in lignocellulose pretreatment. It was found (Chen et al., 2011b) that rice straw pretreated by dilute-acid/steam explosion had a higher xylose yield, a lower level of inhibitor in hydrolysate and a greater degree of enzymatic hydrolysis, which resulted in a 1.5-fold increase in the overall sugar yield while compared to acid-catalyzed steam explosion.

Alexandre Emmel (Emmel et al., 2003) investigated steam explosion of Eucalyptus grandis under various pretreatment conditions (200 to 210°C, 2 to 5 min) after impregnation of the wood chips with 0.087% and 0.175% (w/w) H_2SO_4. It was found that the yield of alkali-soluble lignin increased with higher acid impregnation at high pretreatment temperature.

Non-cellulosic constituents such as hemicelluloses and lignin were removed by steam explosion combined bleaching or acid treatments (Deepa et al., 2011). And fiber diameter reduced after steam explosion followed by acid treatments. Percentage yield and aspect ratio of nanofiber obtained by this technique was found to be higher than other conventional methods.

Comparison of different integrated pretreatment

Steam explosion integrated pretreatment technologies are compared in terms of recovery yield, purity of fractions, cost-competitiveness and environment-friendliness (Table 1). It shows that SE-W-A-MC is both cost-competitive and environment-friendly because there is no chemical additions and easy to operate. Compared with that of SE-W-S-BS, the purity of fractions is enhanced by the process of SE-W-A-M. SE-SG, SE-EC, SE-P require pretreatment of raw materials, so they are not cost-competitive. For SE-IL and SE-L, the cost of ion liquid and laccase is high. And it is obvious that SE-A involves chemical additions which are not acceptable for environment.

DEMONSTRATION OF STEAM EXPLOSION INTEGRATED PRETREATMENT FOR LIGNOCELLULOSE

Lignocellulose is heterogeneous, the same as its

Table 1. Comparison of different integrated pretreatment.

Pretreatment	Recovery yield (%)[a]	Purity of fractions (%)[b]	Cost-Competiveness[c]	Environment-friendliness[d]	Total[e]
SE-W-S-BS	70-85	75-85	Y	N	62.0-N
SE-W-A-MC	70-90	80-95	Y	Y	70.0-Y
SE-SG	85-95	80-90	N	Y	76.5-N
SE-IL	80-90	85-95	N	Y	76.5-N
SE-L	80-95	75-85	N	Y	72.0-N
SE-EC	80-95	70-85	N	Y	72.0-N
SE-P	80-90	70-85	N	N	68.0-N
SE-A	80-95	70-85	Y	N	70.0-N

SE, Steam explosion; W, washing; S, solvent extraction; BS, Bauer screening; A, alkali extraction; MC, mechanical carding; SG, super grinding; IL, ion liquid; L, laccase; EC, electricity catalyzing; P, photochemistry; A, acid. [a]Recovery yield means to the weight percentage of all fractions to total raw materials. [b]Purity of fractions means to the weight percentage of ideal fraction to actual fraction. [c]Cost includes equipment depreciation, additions, labor and power. [d]Environment-friendliness means that there is no chemical additions, no wastes left, less electricity and water consumption. [e]For total; it is calculated as follows: $Percentage(\%) = recovery\ yield \times purity\ of\ fraction$; $N \times N = N; N \times Y = N; Y \times N = N; Y \times Y = Y$.

conversion property. The authors have established three platforms for lignocelluloe refining including law pressure, non-pollution steam explosion platform (Zhang and Chen, 2012), solid-state enzymatic hydrolysis and solid-state fermentation platform, fermentation and separation coupling platform (Li and Chen, 2007). For each platform, pilot test apparatus have been designed, as well as industrial equipment. By process integration, straw refining systems of multi-layer have been established and industrialized in plant about lignocellulose ethanol, butanol, board, spinning and so on. Especially, in September, 2010, Laihe Chemcial Co., Lid. located in Jilin Province, China introduced SE-W-A-MC (Wang and Chen, 2011; Chen and Fu, 2012). Its capacity is 300 thousand tons corn stover annually.

The products from this technology were 50 thousand tons butanol, acetone and ethanol, 30 thousand tons pure lignin which could be converted into 20 thousand tons phenol formaldehyde resin adhesive, 120 thousand tons cellulose which could be converted into 50 thousand tons biological polyether polyol. The cost of solvent reduces more than 50% after cost apportioning by lignin and cellulose. Other lignocellulose refining plants which adopt integrated pretreatment are also on the way.

CONCLUSION AND PERSPECTIVES

Research and industrial demonstration above reveal that integrated pretreatment is an effective way to convert different components into multiproduct of high value through selective deconstruction. Reduced sugar from the process of enzymatic hydrolysis become cost-competitive though cost apportioning by other products. Moreover, little waste leaves by converting cellulose, hemicellulose, lignin and other components into relative products separately. To provide valid theory for the process engineering of lignocellulose refining, more understanding are needed about the properties of various lignocellulose materials, as well as changes of molecular structure resulted from different single pretreatment technologies. And then, mechanism to integrate different pretreatment technologies in this paper could be basic references for designing processes for other raw materials. Moreover, equipment for industrialization should be designed and relative theory for industrial scale-up should be established to overcome problems in application.

ACKNOWLEDGEMENTS

This work was financially supported by the National Basic Research Program of China (973 Project, No. 2011CB707401) and the National High Technology Research and Development Program of China (863 Program, SS2012AA022502).

REFERENCES

Agbor VB, Cicek N, Sparling R, Berlin A, Levin DB (2011). Biomass pretreatment: fundamentals toward application. Biotechnol. Adv. 29(6):675-685.

Biswas AK, Yang W,Blasiak W (2011). Steam pretreatment of Salix to upgrade biomass fuel for wood pellet production. Fuel Proc. Technol. 92(9):1711-1717.

Bonini C, D'Auria M, Di Maggio P,Ferri R (2008). Characterization and degradation of lignin from steam explosion of pine and corn stalk of lignin: the role of superoxide ion and ozone. Ind. Crops Products 27(2):182-188.

Bono JJ, Gas G,Boudet AM (1985). Pretreatment of poplar lignocellulose by gamma-ray or ozone for subsequent fungal biodegradation. Appl. Microbiol. Biotechnol. 22(3):227-234.

Brugnago RJ, Satyanarayana KG, Wypych F,Ramos LP (2011). The effect of steam explosion on the production of sugarcane bagasse/polyester composites. Composites Part A: Appl. Sci. Manufact. 42(4):364-370.

Cao N, Xu Q, Chen L (1995). Xylan hydrolysis in zinc chloride solution. Appl. Biochem. Biotechnol. 51(1):97-104.

Cara C, Ruiz E, Ballesteros I, Negro MJ,Castro E (2006). Enhanced enzymatic hydrolysis of olive tree wood by steam explosion and alkaline peroxide delignification. Proc. Biochem. 41(2):423-429.

Carrasco F (1992). Thermo-mechano-chemical pretreatment of wood in a process development unit. Wood Sci. Technol. 26(6):413-428.

Carrasco F, Roy C (1992). Kinetic study of dilute-acid prehydrolysis of xylan-containing biomass. Wood Sci. Technol. 26(3):189-208.

Chen HZ, Fu XG (2012). Method and equipment for fractionating long and short fiber of biomass with day carding. CN102357465A.

Chen HZ, Li HQ,Liu L (2011a). The inhomogeneity of corn stover and its effects on bioconversion. Biomass and Bioenergy 35(5): 1940-1945

Chen WS, Yu HP, Liu YX, Chen P, Zhang MX, Hai YF (2011b). Individualization of cellulose nanofibers from wood using high-intensity ultrasonication combined with chemical pretreatments. Carbohydr. Polym. 83(4):1804-1811.

Chen HZ,Qiu WH (2010). Key technologies for bioethanol production from lignocellulose. Biotechnol. Adv. 28(5):556-562.

Chen HZ,Liu LY (2007). Unpolluted fractionation of wheat straw by steam explosion and ethanol extraction. Bioresource Technology 98(3): 666-676.

Cui QJ, Zhu HG, Wang DY,Xiong FL (2011). Effect on biogas yield of straw with twin-screw extruder physical-chemical combination pretreatment. Transactions of the Chinese Soc. Agric. Eng. 27(1): 280-285.

David C,Atarhouch T (1987). Utilization of waste cellulose. Appl. Biochem. Biotechnol. 16(1):51-59.

Deepa B, Abraham E, Cherian BM, Bismarck A, Blaker JJ, Pothan LA, Leao AL, de Souza SF,Kottaisamy M (2011). Structure, morphology and thermal characteristics of banana nano fibers obtained by steam explosion. Bioresour. Technol.102(2):1988-1997.

Deschamps FC, Ramos LP, Fontana JD (1996). Pretreatment of sugar cane bagasse for enhanced ruminal digestion. Appl. Biochem. Biotechnol.57(1):171-182.

Ding Wy (2010).Synergistic enzymatic hydrolysis and ethanol production of steam-exploded straw by nonisothermal simultaneous saccharification and fermentation. Vol. DoctorBeijing: Process Engineering Institute of Chinese Academy of Sciences.

Emmel A, Mathias AL, Wypych F,Ramos LP (2003). Fractionation of Eucalyptus grandis chips by dilute acid-catalysed steam explosion. Bioresou. Technol. 86(2):105-115.

Fang H,Liu SH (1996). Apraisal about cellulose mateerials for paper making. Beijing Wood Industry 16:19-22.

Grous WR, Converse AO,Grethlein HE (1986). Effect of steam explosion pretreatment on pore size and enzymatic hydrolysis of poplar. Enzyme Microb. Technol. 8(5):274-280.

Guo P, Mochidzuki K, Zhang D, Wang H, Zheng D, Wang X,Cui Z (2011). Effects of different pretreatment strategies on corn stalk acidogenic fermentation using a microbial consortium. Bioresour. Technol. 102(16):7526-7531.

Hatakka AI (1983). Pretreatment of wheat straw by white-rot fungi for enzymic saccharification of cellulose. Appl. Microbiol. Biotechnol. 18(6):350-357.

Holtzapple MT, Jun JH, Ashok G, Patibandla SL, Dale BE (1991). The ammonia freeze explosion (AFEX) process. Appl. Biochem. Biotechnol. 28(1):59-74.

Jacquet N, Quiévy N, Vanderghem C, Janas S, Blecker C, Wathelet B, Devaux J, Paquot M (2011). Influence of steam explosion on the thermal stability of cellulose fibres. Polym. Degradation Stab. 96(9):1582-1588.

Jin SY, Chen HZ (2006). Superfine grinding of steam-exploded rice straw and its enzymatic hydrolysis. Biochem. Eng. J. 30(3):225-230.

Kawamori M, Morikawa Y, Ado Y, Takasawa S (1986). Production of cellulases from alkali-treated bagasse in Trichoderma reesei. Appl. Microbiol. Biotechnol.24(6):454-458.

Klyosov A (1986). Enzymatic conversion of cellulosic materials to sugars and alcohol. Appl. Biochem. Biotechnol. 12(3):249-300.

Kootstra AMJ, Beeftink HH, Scott EL,Sanders JPM (2009). Comparison of dilute mineral and organic acid pretreatment for enzymatic hydrolysis of wheat straw. Biochem. Eng. J. 46(2):126-131.

Li DM, Chen HZ (2007). Fermentation of Acetone and Butanol Coupled with Enzymatic Hydrolysis of Steam Exploded Cornstalk Stover in a Membrane Reactor. Chin. J. Proc. Eng. 7(6):1212-1216.

Li J, Gellerstedt G, Toven K (2009). Steam explosion lignins; their extraction, structure and potential as feedstock for biodiesel and chemicals. Bioresour. Technol. 100(9):2556-2561.

Lipinsky E, Kresovich S (1982). Sugar crops as a solar energy converter. Cell. Mol. Life Sci. 38(1):13-18.

Liu LY, Chen HZ (2006). Enzymatic hydrolysis of cellulose materials treated with ionic liquid [BMIM] Cl. Chin. Sci. Bull. 51(20):2432-2436.

Ma Z, Ou Y, Huang Q (2003). Advances in electrochemical bleaching research. Celluluose Sci. Technol. 11:56-61.

Maekawa E (1996). On an available pretreatment for the enzymatic saccharification of lignocellulosic materials. Wood Sci. Technol. 30(2):133-139.

Ooshima H, Aso K, Harano Y,Yamamoto T (1984). Microwave treatment of cellulosic materials for their enzymatic hydrolysis. Biotechnol. Lett. 6(5):289-294.

Peng XW, Chen HZ (2012). Hemicellulose sugar recovery from steam-exploded wheat straw for microbial oil production. Process Biochem. 47(2):209-215.

Ping W, Xiujin L, Hairong Y,Yunzhi P (2010). Anaerobic biogasification performance of corn stalk pretreated by a combination of green oxygen and NaOH. J. Be ijing University of Chem ical Technology (Natural Science) 37(003):115-118.

Qiu WH,Chen HZ (2012). Enhanced the enzymatic hydrolysis efficiency of wheat straw after combined steam explosion and laccase pretreatment. Bioresour. Technol. 118:8-12.

Rocha GJM, Gonçalves AR, Oliveira BR, Olivares EG,Rossell CEV (2012). Steam explosion pretreatment reproduction and alkaline delignification reactions performed on a pilot scale with sugarcane bagasse for bioethanol production. Industrial Crops Products 35(1):274-279.

Sachs J (1875). Text-book of botany: morphological and physiological. Clarendon press. P.3

Sun FB, Chen HZ (2008). Organosolv pretreatment by crude glycerol from oleochemicals industry for enzymatic hydrolysis of wheat straw. Bioresour. Technol. 99(13):5474-5479.

Taniguchi M, Kometani Y, Tanaka M, Matsuno R,Kamikubo T (1982a). Production of single-cell protein from enzymatic hydrolyzate of rice straw. Appl. Microbiol. Biotechnol. 14(2):74-80.

Taniguchi M, Tanaka M, Matsuno R,Kamikubo T (1982b). Evaluation of chemical pretreatment for enzymatic solubilization of rice straw. Appl. Microbiol. Biotechnol. 14(1):35-39.

Torget R,Teh-An H (1994). Two-temperature dilute-acid prehydrolysis of hardwood xylan using a percolation process. Appl. Biochem. Biotechnol. 45(1):5-22.

Van Walsum GP, Allen SG, Spencer MJ, Laser MS, Antal MJ,Lynd LR (1996). Conversion of lignocellulosics pretreated with liquid hot water to ethanol. Appl. Biochem. Biotechnol. 57(1):157-170.

Wang L,Chen HZ (2011). Increased fermentability of enzymatically hydrolyzed steam-exploded corn stover for butanol production by removal of fermentation inhibitors. Proc. Biochem. 46(2):604-607.

Wayman M,Parekh SR (1988). SO 2 prehydrolysis for high yield ethanol production from biomass. Appl. Biochem. Biotechnol. 17(1): 33-43.

Wyman CE, Goodman BJ (1993). Biotechnology for production of fuels, chemicals, and materials from biomass. Appl. Biochem. Biotechnol. 39(1):41-59.

Xu FJ, Chen HZ, Li ZH (2001). Solid-state production of lignin peroxidase (LiP) and manganese peroxidase (MnP) by Phanerochaete chrysosporium using steam-exploded straw as substrate. Bioresour. Technol. 80(2):149-151.

Yang S (2001). Plant Cellulose Chemistry. Beijing: Chinese Light Industry Press. pp. 33-38.

Yang Z (1988). Super Micron Grinding: Principle, Equipment and Application. Beijing: Chem. Indust. Press. pp.289-291.

Zadra il F (1977). The conversion of straw into feed by basidiomycetes. Appl. Microbiol. Biotechnol. 4(4):273-281.

Zhang YZ, Chen HZ (2012). Multiscale modeling of biomass pretreatment for optimization of steam explosion conditions. Chem. Eng. Sci. 75:177-182.

Zimbardi F, Viola E, Nanna F, Cardinale G, Villone A, Valerio V, Braccio G (2009). Lignocellulosic biomass as carbon source by steam explosion pretreatment. New Biotechnol. 25(0):S275.

2

A review on postharvest handling of avocado fruit

A. Kassim, T. S. Workneh and C. N. Bezuidenhout

School of Engineering, Bioresources Engineering, University of KwaZulu-Natal, Pietermaritzburg, Private Bag X01, Scottsville, 3209, South Africa.

This paper reviews the literature on the effect of pre- and postharvest treatment effects on the postharvest quality of avocado fruit. It is evident from literature that pre-harvest factors such as ambient field temperature and water stress affects the postharvest physiology of avocado fruit. The postharvest handling treatments and environmental conditions identified were pre-packaging treatments, different density packaging films, storage conditions and cyclic temperature storage environments during the avocado fruit cold chain. Temperature was found to have the greatest influence on the physical, sensory and chemical quality of avocado fruit after harvest. Maintenance of an optimal temperature regime from harvest to final market destination is, therefore, required to maintain fruit quality. The influence of varying temperature and relative humidity with time during the avocado fruit cold chain in South Africa requires further study.

Key words: Pre-packaging, packaging, cyclic storage conditions, quality, cold chain, South Africa.

INTRODUCTION

Shelf life can be defined as the period in which a product should maintain a predetermined level of quality under specified storage conditions (Perez et al., 2004). Avocado (*Persea americana* Miller) is a highly perishable commodity (Yahia and Gonzalez-Aguilar, 1998; Jeong et al., 2002; Yahia, 2002; Perez et al., 2004; Gamble et al., 2010) and yet valued for export. The leading exporter of avocados to Europe is Israel, supplying 29% of imports, followed by South Africa contributing 21% and Spain with 17% (Van Zyl and Ferreira, 1995). In 2008 and 2009, approximately 64% of South African produced avocados were exported (Department of Agriculture Forestry and Fisheries, unpublished, 2010), however, in 2009, South African exported avocados were considered to be of an inferior quality (Nelson, 2010).

Avocado quality at its final destination is a major concern during exportation. Thus, the development of valuable postharvest technologies could improve the quality and consequently extend the shelf life of avocados locally and during export to distant markets. Avocados continue respiring even after harvest, commencing the ripening process almost immediately due to their climacteric characteristic of high respiration rates (Wu et al., 2011). Villa-Rodriguez et al., (2011) states the complete ripening process to be five to seven days at 25°C. Numerous studies have been conducted to exhibit the effect of pre-packaging treatments, packaging materials and storage conditions on the effect of avocados (Meir et al., 1997; Hofman et al., 2003; Woolf et al., 2003; Perez et al., 2004; Wu et al., 2011). Pre-packaging methods such as hot and cold treatments, waxes and 1-Methylcyclopropene (1-MCP) were shown to reduce chilling injury and improve avocado quality. Polyethylene (PE) and biodegradable packaging films can extend the avocado shelf life. Optimal temperature and relative humidity conditions have also proven beneficial in maintaining high quality avocados. Studies by Tefera et al. (2007) demonstrated positive effects on

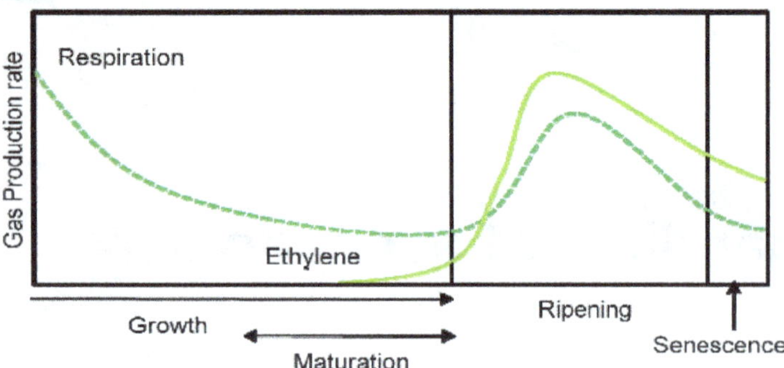

Figure 1. Physiological developmental stages of avocados (Blakey, 2011).

the quality of mangos in Ethiopia by integrating suitable pre-packaging, packaging and storage conditions. Cold chain management is a crucial factor regarding the quality of avocados from harvest till final market destination (Milne, 1998; Blakey and Bower, 2009; Kok et al., 2010). The interplay of time and temperature under cyclic storage conditions are important in maintaining a superior quality product deemed acceptable by both national and international standards.

An overview of the pertinent postharvest handling conditions adopted to enhance the quality of avocados and prolonging their shelf life are presented. An outline of the significant physical, sensory and chemical quality parameters associated with evaluating avocado maturity are provided. Subsequently, the effects of pre-packaging treatments, packaging materials and storage conditions on the quality parameters of avocados are reviewed. Focus was also given to pre-harvest conditions such as exposure of avocados to ambient temperature and water stress as these factors were proven to affect the postharvest behaviour of avocados. Finally the avocado cold chain, regulations, guidelines and recommendations during postharvest handling are identified.

PRE-HARVEST AND HARVESTING FACTORS that AFFECT AVOCADO QUALITY

Understanding avocado postharvest behaviour requires an understanding of the basic physiological principles. Here, some vital pre-harvest and harvesting factors affecting postharvest behaviour of avocados are presented.

Respiration and ripening

Respiration is described as a natural process occurring within all living organisms whereby organic materials are broken down into simpler end products. During respiration oxygen is expended and carbon dioxide liberated accompanied by the production of energy in the form of heat (Workneh and Osthoff, 2010). This process is predominantly responsible for the ripening of avocados. Starrett and Laties (1991), Jeong et al. (2002), Yahia (2002), Jeong et al. (2003) and Wu et al. (2011) described avocados as being climacteric, characterised by a surge in ethylene production and respiration at the start of the climacteric ripening (Figure 1).

The respiration of avocados follows three characteristic climacteric stages viz. preclimacteric minimum of least respiration, climacteric maximum of highest respiration and a postclimacteric stage synonymous with a decline in respiration. It is during the preclimacteric and climacteric stages where much of the changes associated with ripening occur (Perez et al., 2004). The shelf life of fresh commodities is inversely related to respiration and ethylene rates as stated by Perez et al. (2004). An increase in the respiration rate hastens senescence contributing to poor fruit quality (Maftoonazad and Ramaswamy, 2008). Therefore, to improve handling of avocados once harvested it is essential to lower the respiration rates by reducing the temperature, increasing carbon dioxide and reducing oxygen concentrations within limits.

Pre-harvest factors

Here, the significant interaction and impact of pre-harvest factors on postharvest requirements, development and inherent quality characteristics of avocados are highlighted.

Ambient heat exposure

Ferguson et al. (1999) observed the predominant pre-harvest factor influencing postharvest quality of avocados to be ambient temperature during growth. This was confirmed by Woolf et al. (1999) who demonstrated that the side of the avocado exposed directly to sunlight while still on the tree was able to withstand higher temperatures during postharvest treatments of approximately 50°C compared to the shaded side. Avocados exposed to high ambient temperature on the field also demonstrated a tolerance to low postharvest temperatures and external chilling injury (Woolf and Ferguson, 2000). This, however, does not apply to already harvested avocados as it is imperative that they be protected against direct sunlight to elude overheating and subsequent early postharvest deterioration (Milne, 1998; Yahia, 2002).

Results obtained by Woolf et al. (2000), further, confirmed that avocados exposed to direct sunlight were capable of tolerating postharvest hot water treatments of 50 and 55°C and were found to be firmer. Sun exposed fruit showed a higher endurance to chilling injury when stored at 0°C for periods between three to six weeks. The ethylene peak of sun exposed fruit was delayed by two to five days during ripening at 20°C. Postharvest heat treatments are based on a similar principle of exposing avocados to high temperatures, however, for predetermined periods of time in order to stimulate the production of heat-shock proteins (Florissen et al., 1996; Woolf et al., 1999; Woolf and Ferguson, 2000; Woolf et al., 2000; Fallik, 2004; Wu et al., 2011).

Water stress

Adato and Gazit (1974) found that pre-harvest water stress resulted in premature fruit abscission and an increase in ethylene production leading to accelerated ripening by 40 and 25% depending on the degree of water stress. Both Kaluwa (2010) and Blakey (2011) confirm that water stress decreased the normal avocado ripening time hence reducing the shelf life accompanied by an increased risk of physiological disorders. The effect of water stress on avocados can further be linked to temperature. When a plant is water stressed, the temperature of the fruit rises. Cooling is as a result of evaporative water loss from the plant tissue (Woolf and Ferguson, 2000). Once the fruit is harvested this cooling effect is halted and the rise in temperature is further exacerbated by exposure to the sun. Water stressed 'Hass' avocado trees were found to bear more elongated fruit (Yahia, 2002). The reason for this is unknown and further studies are required to investigate this phenomenon. Water stress reduces the internal quality of avocados due to the increased activity in polyphenol oxidase leading to browning of the flesh. Lower concentrations of calcium were found in water stressed fruit resulting in high incidence of physiological disorders.

Harvesting techniques

Avocados unlike other fruit do not ripen on the tree, but only once harvested (Lee et al., 1983; Hopkirk et al., 1994; Baryeh, 2000; Ozdemir and Topuz, 2004; Perez et al., 2004; Gamble et al., 2010; Osuna-Garcia et al., 2010). The time at which avocados are harvested plays an important role in maturation and the expected shelf life. Harvesting too early in the season contributes to low pulp dry matter. This is associated with irregular ripening, watery texture, flavourless, shriveled, blackened fruit (Gamble et al., 2010; Osuna-Garcia et al., 2010) and a low oil concentration (Blakey, 2011). Perez et al. (2004) reported that harvesting prior to physiological maturity results in irregular softening, a poor taste and higher susceptibility to decay. Generally if the fruit are not harvested at the correct time the quality becomes compromised and the shelf life shortened (Wu et al., 2011).

The time of harvesting among other factors depends on the avocado cultivar. Whiley et al. (1996a) showed that early harvesting at 21 and 24% dry matter lead to a higher cumulative and average yield on the early-maturing 'Fuerte'. However, delaying harvesting till a value of 30% dry matter was attained, reduced yields by 26% and lead to alternate bearing. Similar studies by Whiley et al. (1996b) indicated that early harvesting of late-maturing 'Hass' at 25 to 30% dry matter resulted in high productivity whereas delayed harvesting at 35% dry matter reduced yields also leading to alternate bearing. Chen et al. (2009) observed that late season 'Sharwil' were smaller in size, had higher oil content and dry matter and demonstrated a shorter shelf life than early and mid-season fruit.

Hofman et al. (2002a) recommended that picking avocados when wet should be avoided as this increases the incidence of cold injury, pulp spot and lenticel damage. Fruit harvested in the morning or late afternoon tend to have less field heat. Colour, size or oil content, generally serve as indications as to the most appropriate time for harvesting (Ozdemir and Topuz, 2004).

Harvesting is mainly accomplished by manual techniques such as clipping and snapping. Eaks (1973) investigated the influence of these two methods on the postharvest development of avocados and found no significant difference in terms of weight loss and the rate of ripening. Yahia (2002) found that clipping helped reduce bruising and puncturing of adjacent fruit while in containers as well as reduce the onset of stem end rot. Hofshi and Witney (2002) referring to studies that indicated snapped avocados ripened at a faster rate than those that were clipped. Further studies are required to establish the effect of snapping and clipping on the consequent quality of avocados.

AVOCADO QUALITY CHARACTERISTICS

The quality of horticultural produce are composed of sensory attributes, nutritive attributes, chemical constituents, mechanical properties, functional properties and defects (Abbott, 1999; Yahia, 2002; Forero, 2007). Colour, texture, flavour and aroma have been found to be essential in determining the eating quality of avocados and are the main characteristics to which consumers refer to when purchasing (Lee et al., 1983; Forero, 2007). In order to investigate and maintain the quality of avocados it is essential to be aware of the quality related attributes which are outlined here.

Physical properties

Physical properties are mainly related to the appearance and aesthetic appeal of avocados to which consumers are initially exposed influencing their decision to purchase. The physical quality parameters of avocados include skin colour, firmness, texture, and physiological disorders.

Skin colour

Avocado skin colour is an important indication of the stage of ripening for industry and consumers (Cox et al., 2004; Arzate-Vázquez et al., 2011). Skin colour can be measured either objectively, commonly using a chroma meter or colorimeter or alternatively using subjective means by experienced sensory panellists using eye colour rating. Skin colour is found to vary among different avocado cultivars. The 'Hass' variety is characteristic of a colour change from green to purple and eventually black (Cox et al., 2004; Forero, 2007; Arzate-Vazquez et al., 2011). However, Chen et al. (2009) revealed that the skin colour of the 'Sharwil' variety does not darken with maturity, therefore, other methods must be utilised to distinguish the various stages of maturity. The parameters relating to colour measurement are (Maftoonazad and Ramaswamy, 2008):

L = Lightness or brightness,
a* = Redness or greenness and
b* = Yellowness or blueness.

From these parameters the chroma (C) and hue angle can be determined as follows (Maftoonazad and Ramaswamy, 2008):

$$C = \sqrt{a^2 + b^2} \quad (1)$$

$$hue\ angle = \tan^{-1}(\tfrac{a}{b}) \quad (2)$$

where a and b are as described previously. Cox et al. (2004) found that 'Hass' stored at 15°C did not exhibit a fully black colouration as compared to when stored at temperatures of 20 and 25°C despite the fruit being considered as ripe. This demonstrated the dependence of colour change on storage temperature. The change in colour influenced by the ripening process in 'Hass' was attributed to a decrease in chlorophyll, L, C and hue of the skin and an increase in cyanidin 3-O-glucoside. Ashton et al. (2006) found comparable results by observing a decline in the chlorophyll of the skin during ripening but at a faster rate as compared to Cox et al. (2004).

Firmness

The firmness of avocados is a vital determinant in assessing the degree of ripening (Mizrach and Flitsanov, 1999; Flitsanov et al., 2000; Arzate-Vazquez et al., 2011). Firmness can be described as the resistance to penetration (Mizrach and Flitsanov, 1999) determined by employing invasive, such as hand tactile methods, destructive methods such as the Magness-Taylor puncture test (M-T), or non-destructive methods such as impulse response and ultrasonic methods. Destructive techniques do not allow for continuity in monitoring on a commercial basis but is, rather, well suited for laboratory analysis. Mizrach and Flitsanov (1999); Flitsanov et al. (2000) and Mizrach et al. (2000) employed ultrasonic techniques to evaluate the firmness in a non-destructive manner which rendered comparable results to that of destructive methods. Gomez et al. (2005) found that impulse response techniques were more sensitive to firmness changes in pear fruit as compared to the M-T tests and has the potential to replace destructive testing methods in determining fruit firmness and shelf life. A stiffness coefficient to determine the firmness of spherical fruit can be calculated using Equation 1 (Gomez et al., 2005).

$$S = f^2\, m^{2/3} \quad (3)$$

where, S = stiffness coefficient ($kg^{2/3}\ s^{-2}$); f = dominant frequency where response magnitude is the greatest (Hz), and m = fruit mass (g).

Previous studies by Mizrach et al. (2000), demonstrated a strong correlation between fruit firmness, maturity stage and expected storage time. Storage temperature is fundamental in the diminution of firmness as avocados subjected to low temperatures exhibited reduced rates of softening (Paull, 1999; Flitsanov et al., 2000; Mizrach et al., 2000). Villa-Rodriguez et al. (2011) discovered that at storage day 0, 4, 8 and 12 the firmness diminished from approximately 130.51 N to 54.62 N, 19.92 N and 7.37 N, respectively, when stored at 15°C. Arzate-Vazquez et al. (2011) also observed a reduction in the firmness from 75.43 to 2.63 N over a period of 12 days at 20°C and 75% relative humidity.

Texture

Texture is a significant indicator of avocado quality and of concern to the consumer (Maftoonazad and Ramaswamy, 2008; Toivonen and Brummell, 2008; Landahl et al., 2009). Avocados undergo drastic changes in texture (Landahl et al., 2009; Li et al., 2010). Chen et al. (2009) stated that the oil content is a key component in the texture of avocados and which Hofman et al. (2002a) identified as contributing to the 'smoothness'. Despite the relation between texture and oil content, it was discovered by Chen et al (2009) that an increase in the oil content over the harvest period did not manifest into any change in the texture. Storage temperature, oxygen and carbon dioxide concentrations and wounding directly affect the texture (Maftoonazad and Ramaswamy, 2008). The relationship between texture and firmness can be extended to the strength of avocados and the ability of the fruit to withstand loading during storage. Firmness can be described as the force necessary to attain a previously defined deformation during textural evaluation (Landahl et al., 2009). It was found that as the avocados ripened the texture, firmness and strength were reduced.

Physiological disorders

Every biological system operates optimally within specific limits. If these limits are significantly reduced or increased, physiological disorders are likely to ensue. Storage at low temperature is commonly used to extend the shelf life of fresh commodities, however, this results in chilling injury of avocados (Eaks, 1976; Florissen et al., 1996; Yahia and Gonzalez-Aguilar, 1998; Woolf et al., 2003; Hershkovitz et al., 2005; Woolf et al., 2005; Adams and Brown, 2007). The main symptoms associated with chilling injury are black spots on the peel or gray or dark-brown discolouration of the mesocarp (Pesis et al., 1994, 2002; Meir et al., 1995; Florissen et al., 1996; Hershkovitz et al., 2005). Florissen et al. (1996), Hofman et al. (2002b) and Hofman et al. (2003) found that employing hot water treatments prior to storage were effective in reducing the effects of chilling injury. Exposing avocados to low temperature storage conditions just above those at which chilling injury is likely to occur prior to storage have been proven to alleviate the effects of chilling injury (Woolf et al., 2003). The optimum low temperature was found to be 6 or 8°C for three to five days. However, Sanxter et al. (1994) and Woolf et al. (2003) found the minimum temperature to be 4°C. Hopkirk et al. (1994) experimented on the 'Hass' variety to demonstrate that postharvest disorders increase with the increase in storage time and temperature.

Other disorders include sunburn leading to development of symptoms which emerge as yellowing or bleaching or a roughened skin (Woolf and Ferguson, 2000). On the other hand controlled or modified atmospheres that expose avocados to too low of oxygen or too high carbon dioxide concentrations can lead to disorders (Ferguson et al., 1999). Application of exogenous ethylene was also found to contribute to mesocarp discolouration (Pesis et al., 2002).

Sensory properties

One of the main sensory properties of avocados is flavour which encompasses both aroma and taste and forms an important component of the eating quality of the fruit (Abbott, 1999).

Flavour

Paull (1999), Workneh and Osthoff (2010) and Paull and Duarte (2011) defined flavor as the ratio of sugar to acid influenced by temperature as in the case of grapefruit held at 8ºC resulting in a sugar and acid decline as compared to those stored at 12ºC. Premature harvesting can lead to an undesirable taste (Brown, 1972; Perez et al., 2004 and Wu et al., 2011) or lack of flavour thereof (Gamble et al., 2010; Osuna-Garcia et al., 2010). The off flavour can be ascribed to increased levels of ethanol and acetaldehyde (Thompson, 2010; Paull and Duarte, 2011). Treatment with 0.25% oxygen and 80% carbon dioxide causes an increase in ethanol and acetaldehyde (Ke et al., 1995). Burdon et al. (2007) observed that exposure of 'Hass' to oxygen and carbon dioxide concentrations of less than 0.5% and up to 20%, respectively, resulted in increased levels of acetaldehyde and ethanol which is in accordance with that found by Ke et al. (1995). As with texture, the oil content also forms a key component of flavour (Chen et al., 2009) and, hence, it can be deduced that a positive correlation exists between texture and taste.

Chemical properties

Identification of horticultural maturity is often difficult to determine in avocados as changes in external appearance are sometimes not easily distinguishable (Lee et al., 1983). Other techniques of determining maturity and that employ chemical properties are therefore required. The chemical properties of avocados discussed here are total soluble sugar, pH, total titratable acid, moisture content, dry matter and oil content.

Total soluble sugars

Carbohydrates are an essential source of energy for growth, development and maintenance in avocados (Liu et al., 1999a, b; Tesfay et al., 2012). Five major soluble sugars were identified within the avocado include the rare seven carbon (C7) reducing sugar mannoheptulose, its reduced polyol form, perseitol, the common disaccharide sucrose, and its component hexoses, fructose and glucose (Liu et al., 1999b). These constituted approximately 98% of the total soluble sugars (TSS). Liu et al. (1999b) demonstrated that ripening of avocados at 20°C resulted in a considerable decline in the TSS in the peel and flesh, particularly the C7 sugars, and that a decrease in the TSS was concomitant with an increase in the oil content. During storage at 1 and 5°C a decrease in the TSS was observed but at a slower rate. Similarly Liu et al. (2002) found a decrease in the C7 sugars during the progression of the ripening process. During avocado growth, carbohydrates are stored, however, once the fruit is harvested these carbohydrates are consumed for postharvest physiological processes such as respiration via enzymatic mechanisms that metabolize the C7 sugars (Liu et al., 1999b). This suggests that the C7 sugars play an important role in the respiration of the avocado during the ripening process.

pH

Avocado pH lies in the range of 6 to 6.5 (Soliva-Fortuny et al., 2004). Maftoonazad and Ramaswamy (2008) observed

Table 1. Maximum moisture content of avocados.

Cultivar	Fuerte	Pinkerton	Ryan	Hass	Lamb Hass	Maluma Hass	Nature's Hass	Other
Moisture content (%)	80	80	80	77	73	78	77	75

that the pH decreased with time during storage. Avocados treated with pectin based coatings illustrated a slower rate of decrease in pH values compared to untreated fruit and those exposed to higher temperatures. Exposure to low oxygen and/or high carbon dioxide for short periods of time has been used as a pretreatment to alleviate physiological disorders and for enhanced storage atmospheres. These conditions can also lead to a decrease in the intracellular pH thereby altering the various physiological processes that are dependent upon pH (Ke et al., 1995). Avocados subjected to (a) 0.25% oxygen, (b) 20% oxygen in combination with 80% carbon dioxide or (c) 0.25% oxygen and 80% carbon dioxide reduced the pH value from 6.9 to 6.7, 6.3, and 6.3, respectively, at 20°C (Ke et al., 1995). Similarly Lange and Kader (1997) stored avocados at 20°C in atmospheres of varying concentrations of oxygen and carbon dioxide and found that concentrations of (a) 21% oxygen, (b) 20% carbon dioxide (17% oxygen and the remainder Nitrogen) and (c) 40% carbon dioxide (13% oxygen and the remainder Nitrogen) produced pH values of 7.0, 6.6 and 6.4 respectively. These show that the lowest concentration of oxygen and highest concentration of carbon dioxide results in a reduction of pH to form an acidic medium.

Total titratable acid

Acidity is associated with both sweetness and sourness of fruit. The method used to measure acidity is titratable acidity (Lobit et al., 2002). Maftoonazad and Ramaswamy (2008) observed an increase in the titratable acid at higher storage temperatures in both pectin-based coated and non-coated avocados. Holcroft and Kader (1999) showed that strawberries exposed to higher concentrations of carbon dioxide at 5°C exhibited increased pH and decreased levels of titratable acidity. Mangoes subjected to postharvest treatments, packaging and storage for 28 days resulted in a decrease in the titratable acidity from 3.42 to 0.2% (Tefera et al., 2007). It can thus be deduced that the postharvest changes in strawberries and mangos differ to that of avocados in terms of total titratable acidity.

Moisture content

Moisture content is the preferred indicator of maturity in South Africa with the recommended moisture content in the range of 69 to 75% depending on the cultivar (Mans et al., 1995). Export of early season 'Fuerte' commences once the moisture content has reached 78 to 80% equivalent to an oil content of 9 to 11 standards of moisture percent (Dodd et al., 2008). The current determination (Table 1) and the specified limits of avocado moisture content in South Africa are according to the Standards and Requirements Regarding Control of the Export of Avocados (1998-1999). Studies are, however, ongoing to establish more accurate means of measurement (Retief, 2012).

Oil content

Avocado is considered to be an important oil fruit and oil content serves as a significant indicator of fruit maturity (Hofman et al., 2002a; Ozdemir and Topuz, 2004; Gamble et al., 2010; Blakey, 2011). As the fruit matures, the concentration of oil within the mesocarp increases as described by Hofman et al. (2002a), Ozdemir and Topuz (2004) and Chen et al. (2009). This increase in oil results in a reduction in the water by the same amount within the fruit implying that the percentage of total water plus oil remains constant throughout the avocado life (Hofman et al., 2002a; Ozdemir and Topuz, 2004). Lee et al. (1983) and Chen et al. (2009) observed a close correlation between the percentage oil content and percentage dry matter. The maturity index could then be calculated by either the oil content or dry matter.

Dry matter

Hofman et al. (2002a) and Gamble et al. (2010) referred to percentage dry matter determination as an alternative to oil content determination in assessing the maturity. Extending the maturation stage of the avocado allows for more oil accumulation and dry matter however the risk of increased disease is introduced. Maturity standards are being used by avocado producing countries to avoid marketing of low quality immature fruit. The standards adopted are the Californian minimum dry matter of 20.8% for 'Hass' or a slightly higher minimum dry matter content of approximately 25% to decrease disorders during storage (Gamble et al., 2010). An oil content of 8% has been reported by Ozdemir and Topuz (2004) to be acceptable for marketing of avocados.

Villa-Rodriguez et al. (2011) found that the dry matter had increased from 31.65 to 36.52% over eight days at 15°C and thereafter decreased to 32.91 on the day 12. Zauberman and Jobin-Decor (1995) found that as the dry matter increased the time to ripen decreased implying that the more mature the fruit the less time required to ripen.

Table 2. Heat treatment regimes of avocados.

Temperature (°C)	Heating media	Exposure time	Effects	Reference
37 - 38	Air	17-18 h	Reduced chilling injury	Sanxter et al. (1994)
38	Air	6-12 h	Reduced chilling injury and reduction in ripening time and flesh injury after ripening	Florissen et al. (1996)
38	Water	2 h	Reduced chilling injury	Woolf and Lay-Yee (1997)
40 and 41	Water	30 min	Reduced body rots and decreased vascular browning	Hofman et al. (2002b)
38	Water	30 min	Good appearance and internal quality	Wu et al. (2011)
38	Air	6 h	Reduced chilling injury	Wu et al. (2011)

Hofman et al. (2000) suggested that the percentage oil content and dry matter are not suitable indicators of avocado maturity in late-harvested 'Hass' due to late harvested fruit having inconsistent changes. No distinct correlation could be drawn between the effect of varying temperature and relative humidity on the percentage dry matter and oil content of avocados during storage, thus, motivating research to be conducted in this field.

PRE-PACKAGING TREATMENTS

Treatments prior to packaging have the added benefit of prolonging the shelf life and enhancing the quality of avocados when combined with suitable packaging and storage conditions (Tefera et al., 2007). Here, some common pre-packaging techniques and technologies applied to avocados are presented.

Heat treatments

In Kenya, Ouma (2001) observed that heating avocados at 38°C for periods of 24, 48 and 72 h improved the appearance and reduced the effects of chilling injury as opposed to untreated fruit. The maximum ethylene evolution was delayed; however, the rate of respiration was unchanged. Furthermore, weight loss was reduced as the number of days of heating increased leading to an improved shelf life.

To address the attack of insect pests in avocados, cold disinfestation is often used (Florissen et al., 1996; Hofman et al., 2002b; Hofman et al., 2003; Wu et al., 2011). This treatment requires exposing the fruit to a temperature of 1°C for 16 days but this induces chilling injury. To alleviate the onset of chilling injury, heating the avocados for various duration and temperature regimes are applied (Table 2).

Water is the preferred medium for most thermal applications as it is more efficient in transferring heat than air (Fallik, 2004). Heat treatment using air requires a longer heating time than with water. The variation in temperature and associated treatment times differs among the studies depending on the cultivar as well as pre-harvest environmental conditions such as sun exposure (Woolf et al., 1999). Such heat treatments are associated with the induction of heat shock proteins responsible for protecting the fruit against heat injury (Florissen et al., 1996).

Low temperature conditioning

Low temperature conditioning is based on the principal of holding the avocado at temperatures just above those at which it is susceptible to induce tolerance to low temperatures (Woolf et al., 2003). Low temperature conditioning between 4 to 8°C for a period of four days provided substantial protection against chilling injury (Hofman et al. (2003); similar to the results found by Woolf et al. (2003) of 6 or 8°C for three to five days. Hard skin, tissue breakdown and incidence of rot were reduced and even eliminated (Woolf et al., 2003); skin damage and internal quality were improved (Hofman et al., 2003). In both these studies, the researchers deduced that hot water treatments do not prove to be as successful in alleviating external chilling injury and improving the overall quality of the avocado when compared to low temperature conditioning.

Heat shock proteins can also be stimulated by near chilling temperatures to reduce chilling injury (Florissen et al., 1996). This can be attributed to heat shock proteins inhibiting the ethylene production rates associated with increased levels of chilling injury.

Surface coating and wax treatments

Postharvest water loss has a detrimental effect on avocados. Weight (water) loss leads to accelerated ripening and a higher degree of physiological disorders (Johnston and Banks, 1998). Waxes were found to address the challenge of water loss due to their impermeability characteristic. This allows for water retention, increasing turgidity and maintaining fruit weight for longer periods. Hagenmaier and Shaw (1992), Banks et al. (1997), Johnston and Banks (1998) and Maftoonazad and Ranaswamy (2008) describe waxes as providing a surface barrier, thus hindering the movement of gases. This creates an internal modified atmosphere resulting in lowered rates of respiration and delayed ripening. A PE based wax of concentration 11% visibly improved the exterior sheen and reduced mass loss (Johnston and Banks, 1998). Waxes are able to contribute to both the physiological characteristics of avocados and in enhancing the exterior aesthetic appeal by imparting a sheen and gloss to the exterior of the fruit (Hagenmaier and Shaw, 1992; Johnston and Banks 1998; Yahia, 2002; Maffoonazad and Ramaswamy 2008). The use of pectin-based waxes by Maftoonazad and Ramaswamy (2008) demonstrated improved results when compared to their earlier work of using methyl cellulose coatings in reducing respiration rates. In South Africa, the use of waxes is recommended prior to packaging provided that no more than 140 mg of the compound adheres per kg of avocados (Standards and Requirements Regarding Control of the Export of Avocados, 1998-9). 'StaFresh', a natural wax emulsion produced equivalent and enhanced results in physiological disorders, external appearance and shelf life when compared to a PE wax on South African avocados stored at 5.5°C for four weeks and ripened at 18°C (Kremer-Kohne and Duvenhage, 1997).

Low oxygen atmosphere/ hypoxic acclimation

Exposure of fruit to low levels of oxygen has been used as an alternative to chemical fumigation however fruit tissue are sensitive to low oxygen concentrations. Pesis et al. (1994) demonstrated through an experiment using 'Fuerte' that pre-treatment with low oxygen atmospheres inhibited chilling injury and fruit softening. Optimum results were found at 3% oxygen and 97% nitrogen for a period of 24 h at 17°C prior to storage at 2°C with relative humidity of 90% for a period of three weeks. El Mir et al. (2001) investigated the effect of exposing 'Hass' to low oxygen levels in order to acclimatize the fruit tissue to withstand insecticidal low oxygen atmospheres. It was concluded that treating avocados with 3% oxygen for 24 h and thereafter exposing the fruit to insecticidal treatment of 1 and 0.25% oxygen for one to three days at 20ºC produced greater fruit firmness.

1-Methylcyclopropene (1-MCP)

1-MCP is a synthetic cyclopropene used as an ethylene action inhibitor in many perishable fruit (Feng et al., 2000; Jeong et al., 2002, 2003; Hershkovtiz et al., 2005; Zhang et al., 2011). Earlier work by Jeong et al. (2002) indicated that 'Simmonds' avocado treated with 0.45 µl L^{-1} 1-MCP for 24 h at 20°C delayed ripening by four days. However, when compared to the later work by Jeong et al. (2003) which incorporated the use of wax and 1-MCP at a lower storage temperature of 13°C, less weight loss and a greener colour was observed. Weight loss of fruit treated with 1-MCP and wax was found to be 3.8% after 19 days of storage (Jeong et al., 2003) whereas after 8 days fruit treated with only 1-MCP was found to be 3.9% (Jeong et al., 2002). Avocados treated for 24 h with 1 µl L^{-1} 1-MCP at 20°C, followed by 21 days of storage at 4°C and 3.5% oxygen and thereafter transferred to 20°C for 14 days delayed the onset of respiration and climacteric peaks. Oxidation of lipids within the fruit pulp was also effectively controlled resulting in an extended shelf life. Zhang et al. (2011) found comparable results. Mid-climacteric avocados exposed to hypoxia conditions followed by treatment with 18.6 m mol m^{-3} of aqueous 1-MCP (1000 µg L^{-1}) exhibited reduced fruit softening and delays in peak ethylene production and respiration in comparison to those treated with 1-MCP only. The use of 1-MCP has been used with success in South Africa (Lemmer et al., 2002; Vorster, 2004-05). Lemmer et al. (2002) recommended that 500 ppb of 1-MCP be applied to avocados for 12 h at either 5 or 10°C, however he further stated that additional studies may be beneficial to refine this application dosage and exposure time.

PACKAGING AND STORAGE METHODS

The basic functions of food packaging are for storage, preservation and protection for prolonged periods of time (Garlic et al., 2011). A review on the past and current trends related to packaging and storage of avocados is presented here. The two most recognized techniques for avocados are modified atmosphere packaging (MAP) and controlled atmosphere storage (CAS). These methods have been proven to extend the shelf life and maintain the quality of avocados and other fresh fruit (Yahia and Gonzalez-Aguilar, 1998 and Berrios, 2002).

Packaging films

The plastic materials primarily used for MAP of whole fruit and vegetables are, low density polyethylene (LDPE), high density polyethylene (HDPE), polypropylene (PP), polyvinylchloride (PVC), polystyrene (PS), ethylene vinyl acetate (VA), ionomer, rubber hydrochloride (pliofilm) and polyvinylidine chloride (PVDC) (Workneh and Osthoff, 2010) (Table 3).

Storage of avocados in PE bags reduced the effects of chilling injury (Pesis et al., 1994; Meir et al., 1997). Thompson (2010) revealed that individually sealed 'Fuerte' in PE bags of 0.025 mm thickness for 23 days at 14 to 17°C ripened normally once removed from the bags. The atmosphere within the bags after storage was found to be 8% carbon dioxide and 5% oxygen. Similarly individually sealed 'Hass' stored at 10°C resulted in

Table 3. Packaging film permeabilities (Workneh and Osthoff, 2010).

Film type	Transmission rate		
	Oxygen*	Carbon dioxide*	Water vapour**
Low density polyethylene (LDPE)	3900 - 13000	7700 - 77000	6 - 23.2
Linear low density polyethylene (LL DPE)	7000 - 9300	-	16 - 31
Medium density polyethylene (MDPE)	2600 - 8293	7700 - 38750	8 - 15
High density polyethylene (HDPE)	52 - 4000	3900 - 10000	4 - 10
Polypropylene (PP)	1300 - 6400	7700 - 21000	4 - 10.8
Polyvinylchloride (PVC)	620 - 2248	4263 - 8138	> 8
Polyvinylchloride (PVC), plasticized	77 - 7750	770 - 55000	> 8
Polystyrene (PS)	2000 - 7700	10000 - 26000	108.5 - 155
Ethylene vinyl acetate copolymer (12% VA)	8000 - 13000	35000 - 53000	60
Ionomer	3500 - 7500	9700 - 17800	22 - 30
Rubber hydrochloride (Pliofilm)	130 - 1300	520 - 5200	> 8
Polyvinylidine chloride (PVDC)	8 - 26	59	1.5 - 5

*Measured in units of $cm^3m^{-2}day^{-1}$ at 1 atm. **Measured in units of g $m^{-2}day^{-1}$ at 37.8°C and 90% relative humidity.

increased storage life (Oudit and Scott, 1973). West Indian avocados stored in PE bags at 13°C exhibited delayed softening and increased shelf life (Thompson et al., 1971). This study further demonstrated perforated bags and unwrapped avocados have a similar effect on the storage life. Low density polyethylene (LDPE) packages displayed more suitable MAP conditions of low oxygen and high carbon dioxide in retaining avocado, papaya and mango freshness as compared to oriented PP and oriented PS films (Xiao and Kiyoto, 2001).

Biodegradable films and coatings are becoming more popular from an environmental perspective as they are easily recyclable (Aguilar-Mendez et al., 2008). The compositions of biodegradable films are essential in determining the postharvest behaviour of avocados and in the performance of the packaging itself. Gelatin-starch films and coatings are used on avocados with positive outcomes of firmer fruit pulps, skin colour retention and lower weight loss. Higher starch concentrations and pH of the biodegradable film causes greater carbon dioxide permeability while lower levels of starch lead to higher film puncture strength (Aguilar-Mendez et al., 2008). Gelatin and starch based films offer the benefit of being inexpensive and manufacturing is possible on a large scale.

Modified atmosphere packaging

There exists a misconception that MAP and CAS is one in the same. However, MAP incorporates a lower degree of control over the concentration of gases as it depends on the interaction between the commodity and the packaging (De Reuck et al., 2010). The aim of MAP is to create a micro environment within the package specific to the avocado requirements to delay ripening and maintain the quality. According to Meir et al. (1997); Mangaraj et al. (2009) and Sandhya (2010), ideally equilibrium is established between the avocado and the packaging based on the following factors:

1. Maturity stage and respiration rate of the commodity,
2. Storage temperature,
3. Film surface area to fruit volume or weight ratio, and
4. Type of film (thickness and permeability to oxygen, carbon dioxide and water vapour).

Equilibrium is assumed to be established once the quantity of gas exchanged through the avocado is equivalent to that through the film (Mangaraj et al., 2009; De Reuck, 2010). MAP is based on the principle of modifying the atmosphere within the package to lower oxygen concentrations and raise carbon dioxide concentrations (Meir et al., 1997; Yahia and Gonzalez-Aguilar, 1998; Berrios, 2002; Hertog et al., 2003; Valle-Guadarrama et al, 2004; Mangaraj et al., 2009). This modified atmosphere suppresses respiration and ethylene formation thereby promoting a longer avocado shelf life. Due to the numerous variations in the factors required to establish equilibrium, it is not possible to reach an ideal equilibrium, however, a reasonable average can be attained.

Gas concentrations for MAP were found to be 2 to 6% oxygen and 3 to 10% carbon dioxide at 5°C and 7°C (Meir et al., 1997). This combination inhibits avocado softening and decreases the effect of chilling injury. Meir et al. (1997) investigated the effect of MAP on the storage of 'Hass'. Optimum results were found when storing 3.2 kg of the avocados in 30 µm PE bags (40 × 70 cm) at 5°C. The concentration of oxygen and carbon dioxide attained values of approximately 4 and 5%, respectively, at 5 and 7°C. At 5°C lower ethylene evolution was detected with firmer fruit. These concentrations are in accordance with those prescribed by Sandhya (2010) of 2 to 5% oxygen, 3-

10% carbon dioxide and 85 to 95% nitrogen. Berrios (2002) recommends similar CAS and MAP conditions of 2 to 5% oxygen and 3 to 10% carbon dioxide at 5 to 13°C for transportation and storage of avocados. Temperature variation of 7 to 14°C resulted in varying oxygen and carbon dioxide concentrations between 2 to 6 and 3 to 7%, respectively (Meir et al., 1997). The avocados retained a good quality for up to seven weeks. Softening became evident within four weeks of storage as oxygen levels exceeded 9%.

A web-based software tool was developed by the University College Cork in Ireland, 'PACK-in-MAP', assisting in designing optimal modified atmosphere conditions of fresh commodities. This is achieved by user input of the commodity type. The software is then able to define the optimum temperature, range of oxygen and carbon dioxide concentrations and permeability of various packaging materials (Mangaraj et al., 2009).

Modification of the storage environment can be accomplished either through respiration of the commodity identified as natural or passive MAP or by intentionally introducing a gas mixture into the packaging identified as artificial or active MAP (Yahia and Gonzalez-Aguilar, 1998; Mangaraj et al., 2009; De Reuck et al., 2010).

Controlled atmosphere storage

In CAS the headspace gas is more precisely monitored and controlled on a continuous basis to suit the requirements of the fruit (Berrios, 2002; Sandhya, 2010; Workneh and Osthoff, 2010). Oxygen and carbon dioxide concentrations of 2 to 6 and 3 to 10% respectively were recommended by Meir et al. (1997) at 5 and 7°C to reduce chilling injury and inhibit softening of avocados. The most effective results were attained during storage at temperatures between 5 and 7°C accompanied by oxygen and carbon dioxide concentrations of 2 to 3 and 8 to 10%, respectively (Meir et al., 1997). Reduction in mesocarp discolouration in 'Fuerte' was achieved with 2% oxygen and 10% carbon dioxide at 5.5°C for 28 days (Pesis et al., 2002).

CAS is a capital intensive operation as stated by Workneh and Osthoff (2010) and is more suited for bulk storage of commodities and for prolonged storage periods (Sandhya, 2010).

Semi-active and modified atmosphere storage system

Related to MAP and CAS are alternative systems of packaging such as semi-active and modified atmosphere storage system (MASS). Semi-active atmosphere modification is the initial removal or addition of a specified volume of gas from/ to the package (Yahia and Gonzalez-Aguilar, 1998). This technique of packaging proved to reduce the accumulation of ethylene; however, no additional benefits in terms of fruit softening and weight loss were identified over passive MAP.

Berrios (2002) described MASS as a means of eliminating atmospheric air through a gas impermeable packaging containing the commodity in an intermittent mode. The system is composed of a gas impermeable packaging fitted with a positive or negative pressure release valve which enables movement of gases to either enter or exit the system. Avocados subjected to MASS demonstrated a longer shelf life with reduced weight loss as compared to those under MAP conditions. Berrios (2002) stated that MASS could possibly be converted into an inexpensive alternative to CAS in countries which do not have the means of such technology.

STORAGE PARAMETERS

The conditions prevailing within storage facilities are essential in attaining an extended shelf life and enhanced quality of avocados. The essential storage parameters and the subsequent effect on avocado quality is focused on here.

Temperature

Temperature is the single most important factor to consider in storage of fruit due to its involvement in biological processes (Workneh et al., 2011). Low temperature storage reduces the rate of respiration and ethylene production resulting in retarded metabolic rates and an extended shelf life (Hofman et al., 2002a; Perez et al., 2004; Workneh and Osthoff, 2010; Getinet et al., 2011). Theoretically, for every 10°C increase in temperature a resultant doubling in the rate of respiration occurs (Workneh and Osthoff, 2010). Equation 2 demonstrates this concept by determining the temperature quotient, Q_{10} which represents the rate of deterioration for each 10°C rise in temperature above the optimum (Perez et al., 2004).

$$Q_{10} = \frac{SL\ (T-10)}{SL\ x\ T} \quad (4)$$

Where, Q_{10} = temperature quotient, SL = Shelf life, T = storage temperature (°C), andShelf Life is expressed in days. Equation 3 can be used to predict the shelf life of avocados at different temperatures (Perez et al., 2004).

$$Q_{10}^{\Delta/10} = \frac{Shelf\ Life\ (T_1)}{Shelf\ Life\ (T_2)} \quad (5)$$

Where, Δ = Difference between $T_1$and T_2 (°C)

Too low a temperature can result in chilling injury of avocados. Zauberman and Jobin-Decor (1995) found that storage at 5 and 8°C resulted in early ripening and mesocarp discolouration. Perez et al. (2004) reported the

Table 4. Optimum temperature and relative humidity of avocado fruit (Yahia, 2002).

Cultivar	Temperature (°C)	Relative humidity (%)	Postharvest life (weeks)
'Hass'	3 - 7	85 - 90	2 - 4
'Fuerte'	3 - 7	85 - 90	2 - 4
'Fuchs'	13	85 - 90	2
'Pollock'	13	85 - 90	2
'Lula'	4	90 - 95	4 - 8
'Booth I'	4	90 - 95	4 - 8

optimum storage temperature for unripe avocados to be 5 to 13°C and for mature avocados 2 to 4°C resulting in a two to four week shelf life, depending on the cultivar. If, however, mature avocados were to be stored at 5 to 8°C the shelf life would be reduced to one to two weeks. According to Hopkirk et al. (1994) cool stored avocados at 6°C thereafter ripened at 15°C was the most effective in enhancing the fruit quality. This compares with Meir et al. (1995) who reported that temperatures of between 5 and 7°C yielded successful results in prolonging the shelf life of 'Hass' avocados by five to nine weeks. Combining a temperature of 7°C with 2% oxygen and >4% carbon dioxide extended the storage time to 9 weeks (Zauberman and Jobin-Decor, 1995).

Van Rooyen and Bower (2006) discovered that storage of 'Pinkerton' at below the recommended temperature of 5.5°C reduced the severity of mesocarp discolouration which was thought to be due to chilling injury while storage at temperatures above the recommended temperature intensified the disorder. However, cold storage increased the occurrence of mesocarp discolouration in 'Fuerte' and became more pronounced with increasing maturity (Cutting et al., 1992). These two papers demonstrate that the avocado cultivar and time of harvest contribute to the onset of chilling at specified temperatures. Too high temperatures are also undesirable with fruit failing to ripen adequately and proliferation of postharvest disorders at 30 and 25°C as compared to a ripening temperature of 20°C (Eaks, 1978; Hopkirk et al., 1994). Flitsanov et al. (2000) demonstrated the effect of temperature on the firmness of the 'Ettinger' variety. During the first four weeks of storage at 2, 4, 6 and 8°C the firmness decreased to 89.2, 79.2, 12.5 and 10.9 N, respectively indicating that the higher temperature accelerated the ripening process as measured by firmness. Results obtained by Mizrach et al. (2000) are in accordance with those found by Flitsanov et al. (2000).

Relative humidity

Most fresh commodities require high relative humidity conditions during storage (Hofman et al., 2002a; Getinet et al., 2011). By increasing the relative humidity, the vapour pressure deficit is reduced, resulting in less water loss (Blakey, 2011). The negative effect of low relative humidity on texture and appearance can be attributed to water loss (Paull, 1999). Adato and Gazit (1974) demonstrated that avocados at 10 to 20% relative humidity lost water three times faster than those stored at 90 to 95% relative humidity and 21 to 22°C. The ripening process was also hastened by 3.3 days. However, Hofman and Jobin-Decor (1999) discovered that holding avocados at 60% relative humidity or less for four days resulted in an increase in the dry mass by 1.5% and reduced the days to ripen as compared to a 98% relative humidity. Storage conditions of mature avocados at 5°C and a relative humidity of 85 to 90% could result in a shelf life of two to three weeks as compared to a shelf life of one to two weeks at 5 to 8°C (Perez et al., 2004) (Table 4).

Gas concentration

Gases significantly affect the storage of fresh commodities particularly oxygen, carbon dioxide, ethylene and nitrogen (Berrios, 2002). The combination of gas concentrations depend largely on the cultivar and intended use of the avocado.

Oxygen and carbon dioxide

Meir et al. (1995) describes oxygen and carbon dioxide as having a synergistic role in inhibiting the ripening process of avocados through the increase in carbon dioxide and decrease in oxygen. Previous studies have demonstrated that most successful atmospheres were created containing 2% oxygen and 10% carbon dioxide. The study undertaken by Meir et al. (1995) showed that a carbon dioxide concentration in combination with an oxygen concentration of 8 and 3%, respectively, yielded a storage time of nine weeks with marketable fruit and no chilling injury at 5°C. Similarly carbon dioxide concentrations of 5 or 10% delayed the respiratory rise and decreased the respiration rate contributing to a prolonged shelf life (Kosiyachinda and Young, 1976). Exposing 'Fuerte' to 25% carbon dioxide for three days prior to storage at 5°C for 28 days resulted in decreased disorders and lower levels of total phenols (Pesis et al., 1994).

Subjecting avocados to excessively high carbon dioxide and too low oxygen concentrations induced exocarp and mesocarp injury (Ke et al., 1995; Lange and Kader, 1997). Oxygen concentrations of less than 1% are likely to result in anaerobic respiration (Forero, 2007). Oxygen levels below 3% for prolonged periods are not recommended (Valle-Guadarrema et al., 2004). Lange and Kader (1995) showed that avocados stored in 40% carbon dioxide and 12.6% oxygen demonstrated increased respiration rates when compared to 20% carbon dioxide and 16.8% oxygen. Meir et al. (1995) related peel injury with low concentrations of oxygen and slower softening rates of avocados to be associated with higher carbon dioxide levels.

Ethylene

Climacteric fruit produce ethylene just before and during the climacteric rise. Ethylene has the potential to induce over-ripening, accelerate quality loss and increase susceptibility to pathogens during storage of fresh commodities (Martinez-Romero et al., 2007). The effect of ethylene on avocados can be identified as flesh softening, colour change and development of distinct aromas (Gerard and Gouble, 2005; Martinez-Romero et al., 2007). Zauberman and Fuchs (1973) found that treatment of avocados with ethylene at a storage temperature of 6°C contributed to accelerated respiration rates and softening. Fruit treated continuously with exogenous ethylene produced the least amount of ethylene compared to untreated fruit and those treated for 24 h. It is suspected that the ethylene evolved is merely due to the diffusion of the exogenous ethylene that had initially been absorbed rather than production of ethylene by the fruit. Findings by Zauberman and Fuchs (1973) concur with those of Hatton and Reeder (1972) which indicate that the removal of ethylene from storage atmospheres reduced the rate of softening. Eaks (1978) showed that avocados held at 35°C displayed the climacteric pattern and ripened with minute amounts of ethylene being evolved as compared to temperatures of 20 and 25°C. Ethylene formation in avocados have, thus, appeared to be independent at high temperatures of 35°C while at 40°C this process seems to be inhibited. These studies indicate that the storage temperatures and application of exogenous ethylene to the avocados play a vital role in the formation of ethylene. Pesis et al. (2002) suggests absorbent sachets to remove ethylene from the packaging after five weeks storage at 5°C to reduce mesocarp discolouration and decay in 'Hass'.

Nitrogen

Nitrogen is a tasteless, colourless, odourless gas and relatively un-reactive (Sandya, 2010). Nitrogen is commonly used as a filler gas in the gas mixture to prevent collapsing of packages due to its low solubility in food as demonstrated by Ke et al. (1995) and Lange and Kader (1997). Storage of avocados in anoxia conditions of 100% nitrogen resulted in irreparable damage (Moriguchi and Romani, 1995). Gouble et al. (1995) demonstrated that continuous treatment with 80% of the nitrogen composite, nitrous oxide and 20% oxygen inhibited the ethylene production in avocado fruit.

A summary of the pertinent postharvest conditions of avocados that were reviewed are presented in Table 5 including the essential storage parameters as discussed within the scope of this document.

POSTHARVEST MANAGEMENT SUPPLY CHAIN IN SOUTH AFRICA

The South African avocado industry is predominantly export based (Bower and Cutting, 1987) which necessitates the need to ensure that the avocado quality is capable of meeting international standards.

Cold chain

Transportation of avocados from the growing regions in South Africa to the port in Cape Town and eventually, to European supermarkets requires extensive logistical management (Bower and Cutting, 1987). Maintaining the cold chain is essential in avoiding soft fruit with physiological disorders (Nelson, 2006). The Perishable Products Export Control Board works in alliance with the South African Avocado Growers' Association providing recommendations and guidelines for handling of avocados during export (Eksteen, 1995, 1999).

Studies by Blakey and Bower (2009) and Kok et al. (2010) demonstrated a break in the avocado cold chain was detrimental to the quality of the fruit. These investigations indicated that storage of avocados at 1°C for 28 days to simulate shipping regimes reduced the rate of softening and mass loss. However, Kok et al. (2010) states that additional studies are required to confirm these findings. Milne (1998) described the vital role played by the combination of time and temperature during cold chain management of avocados by reporting that a break later in the cold chain lead to greater fruit softening. Bezuidenhout, (1992) conducted an analysis to address the softening of avocados experienced during export to Europe. It was found that an increase in temperature by 1°C during a transit time of 28 days resulted in increased softening.

Step down temperature is a technique adopted which gradually reduces the storage temperature of avocados (Milne, 1998). This was shown to reduce chilling injury and pulp spot symptoms. Early season 'Fuerte' stored at 7.5°C for week one, 5.5°C for weeks two and three followed by 3.5°C for week four resulted in reduced chilling injury compared to 5.5°C for the total four week period. However, Sekhune (2012) advised not to subject avocados to temperatures below 5°C and greater than 10°C after harvest (Table 6). Milne (1998), however,

Table 5. Summary of avocado storage conditions recommended by different authors.

Cultivar/ type	Storage temperature (°C)	Ripening temperature (°C)	Storage/ ripening time	Relative humidity (%)	O_2 (%)	CO_2 (%)	Additional information	Reference
'Hass'	6	15	10 days				Best quality fruit and reduced postharvest rots	Hopkirk et al. (1994)
	5	20	<1 week 1-3 days		0.25	80	Increase in ethanol and acetaldehyde, reduction in pH values from 6.9 to 6.3	Ke et al. (1995)
	5 & 7						Prolong the shelf life to 5-9 weeks	Meir et al. (1995)
	5	20		90-95	3	8	Remained green after 9 weeks and retarding chilling injury	Meir et al. (1995)
	2	22	4 weeks				Remained firm and green for 5 weeks, ripening was delayed	Zauberman and Jobin-Decor (1995)
	5 & 8	22	4 weeks				Development of mesocarp discolouration and vascular browning, fruit ripening commenced during storage	Zauberman and Jobin-Decor (1995)
	7				2	>4	Shelf life of 9 weeks	Zauberman and Jobin-Decor (1995)
					12.6	40	Increased respiration rates	Lang and Kader (1997)
	5 & 7				2-6	3-10	CA and MA - Inhibit softening and chilling injury	Meir et al. (1997)
	7-14				2-6	3-7	MA - Fruit retained good quality for 7 weeks	Meir et al. (1997)
	5 or 7				2-3	8-10	Recommended storage conditions	Meir et al. (1997)
Hass' avocado tree on clonal 'Duke 7' rootstock		20	12 days	85-90			Decline in the TSS content	Liu et al. (2002)
'Hass'	5-13				2-5	3-10	MA - For transport and storage	Berrios (2002)
'Fuerte'	5.5		28 days		2	10	Reduced mesocarp discolouration	Pesis et al. () 2002
'Hass'	4-8		4 days				Reduced chilling injury	Hofman et al. (2003)
'Hass'		16					Reduced tissue breakdown, hard skin and rot	Woolf et al. (2003)
Unripe	6 or 8 5-13		3-5 days				2-4 weeks shelf life	Perez et al. (2004)
Mature	2-4						2-4 weeks shelf life	Perez et al. (2004)
Cultivar/ type	**Storage temperature (°C)**	**Ripening temperature (°C)**	**Storage/ ripening time**	**Relative humidity (%)**	**O_2* (%)**	**CO_2** (%)**	**Additional Information**	**Reference**
Mature	5			85-90			2-3 weeks shelf life	Perez et al. () 2004
Mature	5-8						1-2 weeks shelf life	Perez et al. (2004)
'Pinkerton'	<5.5						Reduced mesocarp discolouration	Van Rooyen and Bower (2006)
'Hass'	6		17 days		0.5	20	Increase in ethanol and acetaldehyde	Burdon et al. (2007)

Table 5. Contd.

			85-95	2-5	3-10	MA - Recommended storage conditions	Sandhya (2010)
'Hass'	20	12 days	75			Reduction in firmness from 75.43 N to 2.63 N	Arzate-Vazquez et al. (2011)
'Hass'	15	12 days				Firmness reduced from 130.51 N to 7.37 N and increase in dry matter from 31.65% to 36.52% and thereafter decrease to 32.91%	Villa-Rodriguez et al. (2011)

Table 6. Moisture content and temperature guidelines for export of avocados.

Moisture content (%)	Cold room (°C)	Road transport (°C)	Port storage (°C)	Vessel - 1st week (°C)	Vessel - last week (°C)
78.5 - 80.0	7.5	7.5	7.5	7.5	5.5
77.5 - 78.4	7.5	7.5	7.5	6.5	5.5
76.5 - 77.4	7.0	7.0	7.0	6.1	5.5
75.5 - 76.4	6.5	6.5	6.5	6.0	5.5
74.5 - 75.4	6.5	6.5	6.5	5.5	5.5
73.5 - 74.4	6.0	6.0	6.0	5.5	5.5
72.5 - 73.4	6.0	6.0	6.0	5.5	5.5
71.5 - 72.4	5.5	5.5	5.5	5.5	5.5
69.5 - 71.4	5.5	5.5	5.5	5.5	4.5
69.4 and less	5.5	5.5	5.5	5.5	3.5

reports that this step down regime was not necessary for 'Fuerte' grown in KwaZulu Natal as a continuous storage temperature of 5.5°C was sufficient for both internal and external quality.

A basic avocado cold chain is depicted by Figure 2 where T_X represents the temperature at each stage X in °C. For example T_H represents the temperature at harvest.

DISCUSSION AND CONCLUSION

Pre-harvest factors play a significant role in the postharvest development of avocado fruit. Harvesting methods and time of harvest contribute to the final quality as too early harvesting can lead to low dry matter and flavourless fruit. Exposure to sunlight whilst still on the tree has been found to reduce chilling injury and improving the avocado fruit quality as opposed to shaded fruit. Temperature has the greatest influence on avocado quality both pre-harvest and throughout the postharvest stages. Therefore, suitable temperatures at all stages of avocado handling are essential. The minimum optimum storage temperature was found to be approximately 5°C (Meir et al., 1995; Zauberman and Jobin-Decor, 1995; Meir et al., 1997; Berrios, 2002; Perez et al., 2004) with a maximum of about 13°C (Berrios, 2002; Perez et al., 2004). Atmospheres containing 85 to 90% relative humidity were favourable for avocado quality depending on the cultivar (Yahia, 2002). In South Africa, the use of waxes and 1-MCP pre-treatments and non-perforated PE and biodegradable packaging were used with positive results for avocado quality.

The avocado cold chain is composed of different processes required for minimising quality loss. By maintaining optimum conditions at each stage specifically at suitable optimum temperature and time, the quality and shelf life of the fruit can be of a superior standard. In reality, avocados are subject to conditions that are not optimum due to breaks in the cold chain as a result of logistical problems that might be encountered. Therefore,

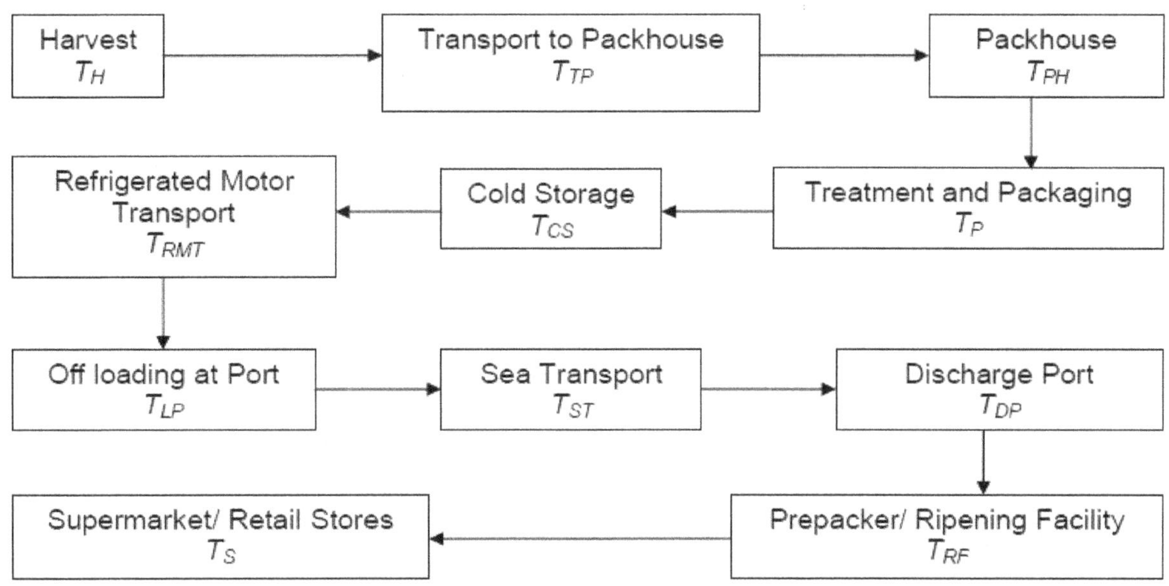

Figure 2. Avocado cold chain (Eksteen, 1999).

studies are needed to evaluate the effect of cyclic storage conditions on the quality of avocados.

Changes in skin colour, firmness, pH, total titratable acid, percent dry matter or oil content, weight loss, flavour and marketable quality are among the most common methods associated with avocado quality assessment.

REFERENCES

Abbott JA (1999). Quality measurement of fruits and vegetable. Posth. Biol. Technol. 15(3):207-225.

Adams JB, Brown HM (2007). Discoloration in raw and processed fruits and vegetables. Crit. Rev. Food Sci. Nutr. 47(3):319-333.

Adato I, Gazit S (1974). Water-deficit stress, ethylene production and ripening in avocado fruits. Plant Physiol. 53(1):45-46.

Aguilar-Mendez MA, Martin-Martinez ES, Tomas SA, Cruz-Orea A, Jaime-Fonseca MR (2008). Gelatine-starch films: Physicochemical properties and their application in extending the post-harvest shelf life of avocado (Persea americana). J. Sci. Food Agric. 88(2):185-193.

Arzate-Vazquez I, Chanona-Perez JJ, de Jesus Perea-Flores M, Calderón-Domínguez G, Moreno-Armendáriz MA, Calvo H, Godoy-Calderón S, Quevedo R, Gutiérrez-López G (2011). Image processing applied to classification of avocado variety Hass (Persea americana Mill.) during the ripening process. Food Bioproc. Technol. 4(7):1307-1313.

Ashton OBO, Wong M, McGhie TK, Vather R, Wang Y, Requejo-Jackman C, Ramankutty P, Woolf AB (2006). Pigments in avocado tissue and oil. J. Agric. Food Chem. 54(26):10151-10158.

Banks NH, Cutting JGM, Nicholson SE (1997). Approaches to optimising surface coatings for fruits. New Zeal. J. Crop Hort. Sci. 25(3):261-272.

Baryeh EA (2000). Strength properties of avocado pear. J. Agric. Eng. Res. 76(4):389-397.

Berrios JD (2002). Development of a dynamically modified atmosphere storage system applied to avocados. Food Sci. Technol. Int. 8(3):155-162.

Bezuidenhout JJ (1992). Analysis of transit temperature and fruit condition of South African export avocados. South African Avocado Growers' Association Yearbook 1992 15:39-40.

Blakey RJ (2011). Management of avocado postharvest physiology. Unpublished PhD Dissertation, School of Horticultural Science, University of Natal, Piertmaritzburg, RSA.

Blakey RJ, Bower JP (2009). The importance of maintaining the cold chain for avocado ripening quality. South African Avocado Growers' Association Yearbook 32:48-52.

Bower JP, Cutting JGM (1987). Some factors affecting post-harvest quality in avocado fruit. South African Avocado Growers' Association Yearbook. 10:143-146.

Brown BI (1972). Isolation of unpleasant flavor compounds in avocado (Persea-americana). J. Agric. Food Chem. 20(4):753-757.

Burdon J, Lallu,N, Yearsley,C, Burmeister D, Billing D (2007). The kinetics of acetaldehyde and ethanol accumulation in 'Hass' avocado fruit during induction and recovery from low oxygen and high carbon dioxide conditions. Posth. Biol. Technol. 43(2):207-214.

Chen NJ, Wall MM, Paull RE, Follett PA (2009). Variation in 'Sharwil' avocado maturity during the harvest season and resistance to fruit fly infestation. Hortscience 44(6):1655-1661.

Cox KA, McGhie TK, White A, Woolf AB (2004). Skin colour and pigment changes during ripening of 'Hass' avocado fruit. Posth. Biol. Technol. 31(3):287-294.

Cutting JGM, Wolstenholme BN, Hardy J (1992). Increasing relative maturity alters the base mineral-composition and phenolic concentration of avocado fruit. J. Hort. Sci. 67(6):761-768.

De Reuck K, Sivakumar D, Korsten L (2010). Effect of passive and active modified atmosphere packaging on quality retention of two cultivars of litchi (Litchi Chinensis Sonn.). J. Food Qual. 33(1):337-351.

Dodd M, Cronje P, Taylor M, Huysamer M, Fruger F, Lotz E, van de Merwe K (2008). A review of the postharvest handling of fruits in South Africa over the past twenty five years. South Afr. J. Plant Soil. 27(1):97-116.

Eaks IL (1973). Effects of clip vs. snap harvest of avocados on ripening and weight loss. J. Am. Soc. Horticult. Sci. 98(1):106-108.

Eaks IL (1976). Ripening, chilling injury, and respiratory response of 'Hass' and 'Fuerte' avocado fruits at 20° C following chilling. J. Amer. Soc. Hort. Sci. 101(5):538-540.

Eaks IL (1978). Ripening, respiration, and ethylene production of 'Hass' avocado fruits at 20 to 40°C. J. Am. Soc. Horticultural Sci. 103(5):576-578.

Eksteen GJ (1995). Handling Guidelines for Avocado -1995 Season. South African Avocado Growers' Association Yearbook 18:111-113.

Eksteen GJ (1999). Handling Procedures for Avocados 1999 Season. South African Avocado Growers' Association Yearbook 22:76-82.

El-Mir M, Gerasopoulos D, Metzidakis I, Kanellis AK (2001). Hypoxic acclimation prevents avocado mesocarp injury caused by subsequent

exposure to extreme low oxygen atmospheres. Posth. Biol. Technol 23(3):215-226.
Fallik E (2004). Prestorage hot water treatments (immersion, rinsing and brushing). Posth. Biol. Technol. 32(2):125-134.
Feng X, Apelbaum A, Sisler EC, Goren R (2000) .Control of ethylene responses in avocado fruit with 1-methylcyclopropene. Posth. Biol. Technol. 20(2):143-150.
Ferguson I, Volz R, Woolf A (1999). Preharvest factors affecting physiology disorders of fruit. Posth. Biol. Technol. 15(3):255-262.
Flitsanov U, Mizrach A, Liberzon A, Akerman M, Zauberman G (2000). Measurement of avocado softening at various temperatures using ultrasound. Posth. Biol. Technol. 20(3):279-286.
Florissen P, Ekman JS, Blumenthal C, McGlasson B, Conroy J, Holford P (1996). The effects of short heat-treatments on the induction of chilling injury in avocado fruit (Persea Americana Mill). Posth. Biol. Technol. 8(2):129-141.
Forero MP (2007). Storage life enhancement of avocado fruit. Unpublished MSc (space)Eng Dissertation, Department of Bioresources Engineering, McGill University, Ste-Anne De Bellevue, Canada.
Gamble J, Harker FR, Jaeger SR, White A, Bava C, Beresford M, Stubbings B, Wohlers M, Hofman PJ, Marques R, Woolf A (2010). The impact of dry matter, ripeness and internal defects on consumer perceptions of avocado quality and intentions to purchase. Posth. Biol. Technol. 57(1):35-43.
Garlic K, Scetar M, Kurek M (2011). The benefits of processing and packaging. Tren. Food Sci. Technol. 22(2-3):127-137.
Gerard M, Gouble B (2005). ETHY. A theory of fruit climacteric ethylene emission. Plant Physiol. 139(1):531-545.
Getinet H, Workneh TS, Woldetsadik K (2011). Effect of maturity stages, variety and storage environment on sugar content of tomato stored in multiple pads evaporative cooler. Afr. J. Biotechnol. 10(80):18481-18492.
Gomez AH, Wang J, Pereira AJ (2005). Impulse response of pear fruit and its relation to Magness-Taylor firmness during storage. Posth. Biol. Technol. 35(2):209–215.
Gouble B, Fath D, Soudain, P (1995). Nitrous oxide inhibition of ethylene production in ripening and senescing climacteric fruits. Posth. Biol. Technol. 5(4):311-321.
Hagenmaier RD, Shaw PE (1992). Gas permeability of fruit coating waxes. J. Amer. Soc. Hort. Sci. 117(1):105-109.
Hatton Jr. TT, Reeder WF (1972). Quality of 'Lula' avocados stored in controlled atmospheres with or without ethylene. J. Am. Soc. Hort. Sci. 97(3):339-341.
Hershkovitz V, Saguy SI, Pesis E (2005). Postharvest application of 1-MCP to improve thequality of various avocado cultivars. Posth. Biol. Technol. 37(3):252-264.
Hertog MLATM, Nicholson SE, Whitmore K (2003). The effect of modified atmospheres on the rate of quality change in 'Hass' avocado. Posth. Biol. Technol. 29(1):41-53.
Hofman PJ, Jobin-Decor M (1999). Effect of fruit sampling and handling procedures on the percentage dry matter, fruit mass, ripening and skin colour of 'Hass' avocado. J. Hort. Sci. Biotechnol. 74(3):277-282.
Hofman PJ, Jobin-Decor M, Giles J (2000). Percentage of dry matter and oil content are not reliable indicators of fruit maturity or quality in late-harvested 'Hass' avocado. Hortscience 35(4):694–695.
Hofman PJ, Fuchs Y, Milne DL (2002a). Harvesting, packaging, postharvest technology, transport and processing. In: ed. Whiley, AW, Schaffer, B and Wolstenholme, BN, The Avocado: Botany, Production and Uses, Ch. 14,363-401.CABI Publishing, Wallingford, Oxon.
Hofman PJ, Stubbings BA, Adkins MF, Meiburg GF, Woolf AB (2002b). Hot water treatments improve 'Hass' avocado fruit quality after cold disinfestations. Posth. Biol. Technol. 24(2):183-192.
Hofman PJ, Stubbings BA, Adkins MF, Corcoran RJ, White A, Woolf AB (2003). Low temperature conditioning before cold disinfestations improves 'Hass' avocado fruit quality. Posth. Biol. Technol. 28(1):123-133.
Hofshi R, Witney GW (2002). Should the California avocado industry consider "snap" harvesting? An overview of the South African avocado industry and planting trees on clonal rootstocks. AvoResearch 2(2):1-12.
Holcroft DM, Kader AA (1999). Controlled atmosphere-induced changes in pH and organic acid metabolism may affect color of stored strawberry fruit. Posth. Biol. Technol. 17(1):19-32.
Hopkirk G, White A, Beever DJ, Forbes SK (1994). Influence of postharvest temperatures and the rate of fruit ripening on internal postharvest rots and disorders of New Zealand 'Hass' avocado fruit. New Zeal. J. Crop Hort. Sci. 22(3):305-311.
Jeong J, Huber DJ, Sargent SA (2002). Influence of 1 - methylcyclopropene (1-MCP) on ripening and cell-wall matrix polysaccharides of avocado (Persea americana) fruit. Posth. Biol. Technol. 25(3):241-256.
Jeong J, Huber DJ, Sargent SA (2003). Delay of avocado (Persea americana) fruit ripening by 1-methylcyclopropene and wax treatments. Posth. Biol. Technol. 28(2):247-257.
Johnston JW, Banks NH (1998). Selection of a surface coating and optimization of its concentration for use on 'Hass' avocado (Persea Americana Mill.) fruit. Crop Hort. Sci. 26(2):143-151.
Kaluwa K (2010). Effect of postharvest silicon application on 'Hass' avocado (Persea Americana MILL) fruit quality. Unpublished MSc Dissertation, School of Agricultural Sciences and Agribusiness, University of Natal, Pietermaritzburg, RSA.
Ke D, Yahia E, Hess B, Zhou L, Kader AA (1995). Regulation of fermentative metabolism in avocado fruit under oxygen and carbon dioxide stresses. J. Am. Soc. Hort. Sci. 120(3):481-490.
Kok RD, Bower JP, Bertling I (2010). Low temperature shipping and cold chain management of 'Hass' avocados: An opportunity to reduce shipping costs. South African Avocado Growers' Association Yearbook 33:33-37.
Kosiyachinda S, Young RE (1976). Chilling Sensitivity of Avocado Fruit at Different Stages of the Respiratory Climacteric. J. Am. Soc. Hort. Sci. 101(6):665-667.
Kremer KS, Duvenhage JA (1997). Alternatives to polyethylene wax as post-harvest treatment for avocados. South African Avocado Growers' Association Yearbook. 20:97-98.
Landahl S, Meyer MD, Terry LA (2009). Spatial and temporal analysis of textural and biochemical changes of imported avocado cv. Hass during fruit ripening. J. Agric. Food Chem. 57(15):7039-7047.
Lange DD, Kader AA (1995). Respiration of 'Hass' avocados in response to elevated CO-2 levels. HortScience 30(4):809.
Lange DL, Kader AA (1997). Effects of elevated carbon dioxide on key mitochondrial respiratory enzymes in 'Hass' avocado fruit and fruit disks. J. Am. Soc. Hort. Sci. 122(2):238-244.
Lee SK, Young RE, Schiffman PM, Coggins Jr. CW (1983). Maturity studies of avocado fruit based on picking dates and dry weight. J. Amer. Soc. Hort. Sci. 108(3):390-394.
Lemmer D, Kruger FJ, Malumane TR, Nxudu KY (2002). 1-Methyl cyclopropene (1-MCP): An alternative for controlled atmosphere storage of South African export avocados. South African Avocado Growers' Association Yearbook. 25:25-34.
Li Xian, Xu Changjie, Korban SS, Chen K (2010). Regulatory mechanisms of textural changes in ripening fruits. Crit. Rev. Plant Sci. 29(4):222-243.
Liu X, Robinson PW, Madore MA, Witney GW, Arpaia, ML (1999a). 'Hass' avocado carbohydrate fluctuations. I. Growth and phenology. J. Am. Soc. Hort. Sci. 124(6):671-675.
Liu X, Robinson PW, Madore MA, Witney GW, Arpaia ML (1999b). 'Hass' avocado carbohydrate fluctuations. II. Fruit growth and ripening. J. Am. Soc. Hort. Sci. 124(6):679-681.
Liu X, Sievert J, Arpaia ML, Modore MA (2002). Postulated physiological roles of the seven-carbon sugars, mannoheptulose, and perseitol in avocado. J. Am. Soc. Hort. Sci. 127(1):108-114.
Lobit P, Soing P, Genard M, Habib R (2002). Theoretical analysis of relationships between composition, pH, and titratable acidity of peach fruit. J. Plant Nutr. 25(12):2775-2792.
Maftoonazad N, Ramaswamy HS (2008). Effect of pectin-based coating on the kinetics of quality change associated with stored avocados. J. Food Proc. Preserv. 32(4):621-643.
Mangaraj S, Goswami TK, Mahajan PV (2009). Applications of plastic films for modified atmosphere packaging of fruits and vegetables: a review. Food Eng. Rev. 1(2):133-158.
Mans CC, Donkin DJ, Boshoff M (1995). Maturity and storage temperature regimes for KwaZulu Natal avocados. South African Avocado Growers' Association Yearbook 18:102-105.

Martinez-Romero D, Bailin G, Serrano M, Guillen F, Valverde JM, Zapata P, Castillo S, Valero D (2007). Tools to maintain postharvest fruit and vegetable quality through the inhibition of ethylene action: A review. Crit. Rev. Sci. Nutr. 47(6):543-560.
Meir S, Akerman M, Fuchs Y, Zauberman G (1995). Further studies on the controlled atmosphere storage of avocados. Posth. Biol. Technol. 5(4):323-330.
Meir S, Naiman D, Akerman M, Hyman JY, Zauberman G, Fuchs Y (1997). Prolonged storage of 'Hass' avocado fruit using modified atmosphere packaging. Posth. Biol. Technol. 12(1):51-60.
Milne DL (1998). Avocado quality assurance: who? where? when? how? South African Avocado Growers' Association Yearbook. 21:39-47.
Mizrach A, Flitsanov U (1999). Nondestructive ultrasonic determination of avocado softening process. J. Food Eng. 40(3):139-144.
Mizrach A, Flitsanov U, Akerman M, Zauberman G (2000). Monitoring avocado softening in low-temperature storage using ultrasonic measurements. Comp. Electron. Agric. 26(2):199-207.
Moriguchi T, Romani RJ (1995). Mitochondrial self-restoration as an index to the capacity of avocado fruit to sustain atmospheric stress at two climacteric states. J. Am. Soc. Hort. Sci. 120(4):643-649.
Nelson RM (2006). Is the quality of South African avocados improving? South African Avocado Growers' Association Yearbook. 29:14-19.
Nelson, RM (2010). Quality challenges facing the South African avocado industry - an overview of the 2009 South African avocado season. South African Avocado Growers' Association Yearbook 33:7-13.
Osuna-Garcia JA, Doyon G, Salazar-Garcia S, Goenaga R, Gonzalz-Duran IJL (2010). Effect of harvest date and ripening degree on quality and shelf life of Hass avocado in Mexico. Fruits 65(6):367-375.
Oudit DD, Scott KJ (1973). Storage of 'Hass' avocados in polyethylene bags. Trop. Agric. 50(3):241-243.
Ouma G (2001). Heat treatments affect postharvest quality of Avocado. Hort. Sci. 36(3):504.
Ozdemir F, Topuz A (2004). Changes in dry matter, oil content and fatty acids composition of avocado during harvesting time and post-harvesting ripening period. Food Chem. 86(1):79-83.
Paull RE (1999). Effect of temperature and relative humidity on fresh commodity quality. Posth. Biol. Technol. 15(3):263-277.
Paull RE, Duarte O (2011). Postharvest Technology. In: ed. Paull, RE and Duarte, O, Tropical Fruits – Volume 1, Ch 5, 101-122. CABI, Wallingford, Oxfordshire.
Perez K, Mercado J, Soto-Valdez H (2004). Note. Effect of Storage Temperature on the Shelf Life of Hass Avocado (Persea americana).Food Sci. Technol. Inter. 10(2):73-77.
Pesis E, Marinansky R, Zauberman G, Fuchs Y (1994). Prestorage low-oxygen atmosphere treatment reduces chilling injury symptoms in 'Fuerte' avocado fruit. HortScience 29(9):1042-1046.
Pesis E, Ackerman M, Ben-Arie R, Feygenberg O, Feng X, Apelbaum A, Goren R, Prusky D (2002). Ethylene involvement in chilling injury symptoms of avocado during cold storage. Posth. Biol. Technol. 24(2):171–181.
Retief K (2012). Personal communication, Perishable Products Export Control Board, Pretoria, RSA 16 March 2012.
Sandhya (2010). Modified atmosphere packaging of fresh produce: current status and future needs. LWT – Food Sci. Technol. 43(4):381-392.
Sanxter SS, Nishijima KA, Chan Jr. HT (1994). Heat-treating 'Sharwil' avocado for cold tolerance in quarantine cold treatments. HortScience 29(10):1166-1168.
Sekhune S (2012). Personal communication, Everdon Estate, KZN, RSA 26 March 2012.
Soliva-Fortuny RC, Elez-Martinez P, Sebastian-Caldero M and Martin-Belloso O (2004). Effect of combined methods of preservation on the naturally occurring microflora of avocado puree. Food Control. 15(1):11-17.
Standards and Requirements Regarding Control of the Export of Avocados (1998-9). RSA Government Gazette No. R. 1983 of 1991: 23 August 1991, No 186. Pretoria, RSA. [Internet]. Available from: http://www.daff.gov.za/. [Accessed: 16 March 2012].
Starrett DA, Laties GG (1991). Involvement of wound and climacteric ethylene in ripening avocado disks. Plant Physiol. 97(2):720-729.
Tefera A, Seyoum T, Woldetsadlik (2007). Effect of Disinfection, Packaging, and Storage Environment on the Shelf Life of Mango. Biosys. Eng. 96(2):201–212.
Tesfay SZ, Bertling I, Bower JP, Lovatt C (2012). The quest for the function of 'Hass' avocado carbohydrates: clues from fruit and seed development as well as seed germination. Aust. J. Bot. 60(1):79-86.
Thompson AK, Mason GF, Halkon WS (1971). Storage of West Indian seedling avocado fruits. J. Am. Soc. Hort. Sci. 46(1):83-88.
Thompson AK (2010). Modified atmosphere packaging. In: ed. Thompson, AK, Controlled Atmosphere Storage of Fruits and Vegetables, CABI, Wallingford, Oxfordshire. Ch. 8:81-115.
Toivonen PMA, Brummell DA (2008). Biochemical bases of appearance and texture changes in fresh-cut fruit and vegetables. Posth. Biol. Technol. 48(1):1-14.
Valle-Guadarrama S, Saucedo-Veloz C, Pena-Valdivia CB, Corrales-Garcia JJE, Chavez-Franco SH (2004). Aerobic–anaerobic metabolic transition in 'Hass' avocado fruits. Food Sci. Technol. Int. 10(6):391-398.
Van Rooyen Z, Bower JP (2006). Effects of storage temperature, harvest date and fruit origin on post-harvest physiology and the severity of mesocarp discolouration in 'Pinkerton' avocado (Persea americana Mill.). J. Hort. Sci. Biotechnol. 81(1):89-98.
Van Zyl JL, Ferreira SG (1995). An overview of the avocado industry in South Africa as requested by: Development Bank of Southern Africa. South African Avocado Growers' Association Yearbook 18:23-30.
Villa-Rodriguez JA, Molina-Corral FJ, Ayala-Zavala JF, Olivas GI, González-Aguilar GA (2011). Effect of maturity stage on the content of fatty acids and antioxidant activity of 'Hass' avocado. Food Res. Int. 44(5):1231-1237.
Vorster LL (2004-05). The Avocado Industry in South Africa. California Avocado Society Yearbook. 87:59-62.
Whiley AW, Rasmussen TS, Saranah JB, Wolstenholme BN (1996a). Delayed harvest effects on yield, fruit size and starch cycling in avocado (Persea americana Mill.) in subtropical environments. I. the early-maturing cv. Fuerte. ScientiaHorticulturae 66(1-2):23-34.
Whiley AW, Rasmussen TS, Saranah JB, Wolstenholme BN (1996b). Delayed harvest effects on yield, fruit size and starch cycling in avocado (Persea americana Mill.) in subtropical environments. II. The late-maturing cv. Hass. ScientiaHorticulturae 66(1-2):35-49.
Woolf AB, Lay-Yee M (1997). Pretreatments at 38 °C of 'Hass' avocado confer thermotolerance to 50°C hot water treatments. Hort. Sci. 32(4):705–708.
Woolf AB, Bowen JH, Ferguson IB (1999). Preharvest exposure to the sun influences postharvest responses of 'Hass' avocado fruit. Posth. Biol. Technol. 15(2):143-153.
Woolf AB, Fergusen IB (2000). Postharvest responses to high fruit temperatures in the field. Posth. Biol. Technol. 21(1):7-20.
Woolf AB, Wexler A, Prusky D, Kobiler E, Lurie S (2000). Direct sunlight influences postharvest temperature responses and ripening of five avocado cultivars. J Am. Soc. Hort. Sci. 125(3):370-376.
Woolf AB, Cox KA, White A, Ferguson IB (2003). Low temperature conditioning treatments reduce external chilling injury of 'Hass' avocados. Posth. Biol. Technol. 28(1):113-122.
Woolf AB, Requejo-Tapia C, Cox KA, Jackman RC, Gunson A, Arpaia ML, White A (2005). 1-MCP reduces physiological storage disorders of 'Hass' avocados. Posth. Biol. Technol. 35(1):43-60.
Workneh TS, Osthoff G (2010). A review on integrated agro-technology of vegetables. Afr. J. Biotechnol. 9(54):9307-9327.
Workneh TS, Osthoff G, Steyn, MS (2011). Influence of preharvest and postharvest treatments onstored tomato quality. Afr. J. Agric. Res. 6(12):2725-2736.
Wu CT, Roan SF, Hsiung TC, Chen IZ, Shyr JJ, Wakana A (2011). Effect of harvest maturity and heat pretreatment on the quality of low temperature storage avocados in Taiwan. J Facul. Agric. Kyus. Univ. 56(2):255-262.
Xiao L, Kiyoto M (2001). Effects of modified atmosphere packages using films with different permeability characteristics on retaining freshness of avocado, papaya and mango fruits at normal temperature. Environ. Con. Biol. 39(3):183-189.
Yahia EM, Gonzalez-Aguilar G (1998). Use of passive and semi-active atmospheres to prolong the postharvest life of avocado fruit. Food Sci. Technol. 31(7-8):602-606.
Yahia EM (2002). Avocado. In: ed. Rees, D, Farrell, G and Orchard, J, Crop Postharv. Sci. Technol. 3(8):159-180. Jon Wiley and Sons,

Chichester, West Sussex.
Zauberman G, Fuchs Y (1973). Ripening processes in avocados stored in ethylene atmosphere in cold storage. J. Amer. Soc. Hort. Sci. 98(5):477-480.
Zauberman G, Jobin-Decor MP (1995). Avocado (Per-sea americana Mill.) quality changes in response to low-temperature storage. Posth. Biol. Technol. 5(3):234-243.
Zhang Z, Huber DJ, Rao J (2011). Ripening delay of mid-climacteric avocado fruit in response to elevated doses of 1-methylcyclopropene and hypoxia-mediated reduction in internal ethylene concentration. Posth. Biol. Technol. 60(2):83-91.

Methods for early evaluation for resistance to bacterial blight of coffee

Ithiru J. M.[1], Gichuru E. K.[1], Gitonga P. N.[2], Cheserek J. J.[1] and Gichimu B. M.[1]

[1]Coffee Research Foundation, P.O. Box 4 – 00232, Ruiru, Kenya.
[2]Kenya Methodist University, P.O. Box 267 – 60200, Meru, Kenya.

Bacterial Blight of Coffee (BBC) caused by *Pseudomonas syringae* pv garcae has become of major concern in Kenya due to its increasing incidence and severity. For decades, the disease was confined within and to the west of the Great Rift Valley, but recently it has spread to reach other coffee growing areas. In order to minimize the chemical input in its management, which apart from polluting the environment have high cost implications, development of resistant/tolerant cultivars is highly recommended. This study aimed at developing an effective method(s) for early evaluation of resistance to BBC and to use the method(s) to evaluate the reaction of selected coffee genotypes to different isolates of *P. syringae* pv *garcae.* Three isolates from different coffee growing areas in Kenya were used to inoculate thirteen coffee genotypes using injection and cut methods. The two inoculation methods were found to be effective and can be recommended with slight modifications. However, it was not possible to clearly authenticate the reaction of the different genotypes to BBC since the genotypes responded differently to different isolates and inoculation methods.

Key words: Coffee, *Pseudomonas syringae* pv *garcae*, inoculation method, Kenya.

INTRODUCTION

Although the genus *Coffea* is diverse and reported to comprise about 130 species (Davis et al., 2006), only two species namely Arabica (*Coffea arabica* L.) and Robusta (*Coffea canephora* Pierre) are under commercial cultivation (Lashermes et al., 1999; Anthony et al., 2002; Pearl et al., 2004). *Coffea arabica* L. is the most important species of the *Coffea* genus, followed by *C. canephora* (Silveira et al., 2003). Coffee production is fundamental for over 50 developing countries, for which it is the main foreign currency earner (Gichimu and Omondi, 2010). Its production is, however, constrained by a number of major diseases, including Coffee Leaf Rust (CLR) caused by *Hemileia vastatrix*, Coffee Berry disease (CBD) caused by *Colletotrichum kahawae* and Bacterial Blight of Coffee (BBC) caused by *Pseudomonas syringae pv. garcae* (Mugiira et al., 2011).

BBC has been described in Brazil, Kenya, Uganda and China where it is becoming of some concern due to its higher incidence and severity (Silva et al., 2006). In Kenya, the disease has been reported since the establishment of coffee plantations in 1893 but it was confined within and to the west of the Great Rift Valley. The symptoms include dark, water-soaked necrotic lesions on leaves, tips and nodes of vegetative and cropping branches culminating in a die-back (Mugiira et al., 2011). It can be a serious problem in high altitudes, where plants are injured from heavy winds (Jansen, 2005) and have a protracted bimodal pattern of rainfall and often experience storms accompanied by hail (Kairu et al., 1985). The disease was previously known as

Table 1. Details of isolates used.

Isolates	Date sampled	Source	Host Genotype	Altitude (masl)
Kap-1/012	25-01-2012	Kapsabet	Batian	1983
Kap-2/012	16-04-2012	Kapsabet	Ruiru 11	1983
Nak-1/012	31-05-2012	Nakuru	SL28	2099

"Elgon dieback" and "Solai dieback"; names derived from the areas where the disease occurred (Kairu et al., 1985). The inherent growth and flowering rhythm of *C. arabica* trees governed by the annual rainfall pattern, greatly influence seasonal periodicity of BBC (Ramos and Kamid 1981). Although the disease does not affect more than 5% of the crop in Kenya, it can cause total crop loss in some areas and severely affected trees sometimes have to be destroyed.

Over the years, coffee growers have relied greatly on copper-based formulations to control BBC. However, excessive use of copper sprays has certain drawbacks which include environmental pollution and high costs of chemicals. Besides, increased soil concentration of available copper may have phytotoxic effects on coffee trees which cause shortening and hardening of internodes of young shoots, chlorotic and diminished leaf area with consequent yield reduction (Kairu et al., 1985). In addition, chemical control accounts for up to 30% of the total cost of production and is a major constraint to economic coffee production especially to the small holders who find the use of pesticides beyond their financial and technical capabilities (Gichuru et al., 2008). There is a strong consensus that growing genetically resistant varieties is the most appropriate cost effective means of managing plant diseases and is one of the key components of crop improvement. It has also been recognized that a better knowledge of both the pathogen and the plant defense mechanisms will allow the development of novel approaches to enhance the durability of resistance (Silva et al., 2006). There is therefore, need to develop a breeding programme for the control of bacterial blight.

Crop improvement depends on the availability of adequate amounts of genetic diversity. It is recognized that the cultivated varieties, in particular *C. arabica,* have a very narrow genetic base (Van der Vossen, 1985; Anthony et al., 2002) that greatly limits the breeding programs especially for improvement of pest and disease resistance (Van der Vossen, 1985). Considerable success has been obtained in the use of classical breeding to control economically important plant diseases, such as the Coffee Leaf Rust and the Coffee Berry Disease (Silva et al., 2006). However, sources of resistance to Bacterial Blight of Coffee are not known. As a prerequisite to development of a breeding programme for BBC in *C. arabica*, there is need to develop a method for early selection of resistance to the disease and subsequently use the method to screen available accessions and developed hybrids for resistance/tolerance to the disease. These requirements formed the objectives of this study.

MATERIALS AND METHODS

Survey and sample collection

A BBC survey on occurrence of the bacterium was conducted between January and May 2012 and suspected diseased coffee samples collected for isolation and identification in the laboratory. Infected twigs or shoots were cut using sterilized pair of secateurs. The samples were put in well labeled paper bags and stored in an ice box. The samples were collected from diverse coffee agro-ecological zones in Kenya including Kisii, Kipkelion, Kapsabet, Nakuru, Ruiru and Nyeri. Nine out of twenty four isolates collected confirmed positive bacterial growth but only three isolates Kap-1/012, Kap-2/012, Nak-1/012 were used for inoculation (Table 1).

Test materials

A total of 13 coffee genotypes comprising of 1 Robusta accession and 12 Arabica varieties were used in this study. The Arabica genotypes included seven Kenyan commercial cultivars (Batian 1, Batian 2, Batian 3, Ruiru 11, K7, SL28 and SL34), two Indian commercial cultivars (Selection 5A and Selection 6) and three museum accessions (Rumesudan, Bourbon and Catimor 134). The reaction of all these genotypes to CBD and CLR is known but their reaction to BBC has not been documented.

Experimental layout and design

Four months old pre-germinated seedlings (with 2 pair of leaves) of the test genotypes were transplanted in black polythene bags measuring 9" x 5" with a potting media comprising of soil, river sand and well decomposed farm yard manure at a ratio of 3:2:1. Triple Super Phosphate (TSP) fertilizer (125 g/15 kg of potting mixture) was added in the media. The experiment was laid out in a temperature controlled room in an inoculation chamber and covered with a polythene sheet to ensure high relative humidity. They were arranged in a completely randomized design with three replications. Each genotype was represented by two seedlings per replicate. The inoculation room was maintained at 18°C.

Inoculation

Two inoculation methods were tested during this study. The first method was through injection where a 30 µl drop of *P. syringae* general inoculum suspension standardized to 108 cfu/ml was placed on each of the mature pair of leaves and a sharp sterile needle used to prick through each drop of the 30 µl of bacterial suspension inoculum. The second inoculation was conducted through a cut where mature leaf pair of the test genotypes was cut using a sharp sterilized blade and then 30 µl bacterial suspension was smeared on the cut edge.

Table 2. Means of disease score of the 13 coffee genotypes using both inoculation methods.

Injection method		Cut method		Both methods combined	
Genotype	**LS means**	**Genotype**	**LS means**	**Genotype**	**LS means**
Batian 1	2.167	Ruiru 11	1.500^{a}	Ruiru11	1.875^{a}
Sln 5A	2.167	Bourbon	1.583^{a}	Bourbon	2.042^{ab}
Ruiru 11	2.250	Batian 2	2.000^{b}	Batian2	2.250^{abc}
Bourbon	2.500	Rume Sudan	2.167^{bc}	Robusta	2.375^{abc}
Robusta	2.500	K7	2.250^{bcd}	SL28	2.375^{abc}
Batian 2	2.500	Robusta	2.250^{bcd}	Catimor134	2.458^{bcd}
SL28	2.500	SL28	2.250^{bcd}	RumeSudan	2.458^{bcd}
Catimor134	2.583	Catimor134	2.333^{cde}	K7	2.478^{bcd}
SL34	2.583	Batian 3	2.500^{de}	Sln5A	2.500^{bcd}
Rume Sudan	2.750	SL34	2.582^{ef}	SL34	2.583^{cd}
K7	2.792	Sln6	2.582^{ef}	Batian1	2.667^{cd}
Batian 3	2.917	Sln 5A	2.833^{f}	Batian3	2.708^{cd}
Sln6	3.250	Batian 1	3.167^{g}	Sln6	2.917^{d}
LSD (5%)	NS	LSD (5%)	0.292	LSD(5%)	0.427

Means marked with the same letter(s) are not significantly different at p=0.05; NS = not significant.

Data collection and analysis

Disease severity was recorded on a scale of 1 to 5, from the least to the most, based on the degree of necrosis reached after every 7 days, where: 1 = absence of the dark necrotic lesions, with yellow halo (bacterial blight); 2 = 1 to 15% diseased leaves; 3 = 16 to 30% leaves with bacterial blight; 4 = 31 to 45% leaves with bacterial blight; 5 = over 45% of leaves with dark necrotic lesions and dieback of some vegetative shoots (Ito et al., 2008). Recording of disease severity continued after every 7 days for 12 weeks. The peak data was subjected to analysis of variance (ANOVA) using XLSTAT version 2012 and effects declared significant at 5% level. Least significance difference ($LSD_{5\%}$) was used to separate the means. The seedlings that scored ≤2 were classified as resistant; those that scored >2 but ≤3 were classified as moderately susceptible, while the ones that scored >3 were classified as susceptible.

RESULTS AND DISCUSSION

The disease symptoms were observed from the first week after inoculation in all the genotypes except in the control. This was an indication of successful inoculation for both methods. In most cases, the symptoms (dark necrosis) started from the point of inoculation and spread to other parts though in some cases necrosis occurred away from the point of inoculation. Some genotypes reacted by shedding off the infected leaves as a way of managing the disease. The peak infection was achieved in the 7th and 9th week for injection and cut methods respectively. Using the injection method, a combined analysis of variance for all the three isolates conducted using the peak disease infection score that was reached at week 7 showed no significance difference ($p>0.05$) between the genotypes (Table 2). For the cut method, a combined analysis of variance for all the three isolates conducted using the peak disease infection score that was reached at week 9 showed highly significant ($p<0.001$) difference between the genotypes. Ruiru 11 recorded the lowest infection followed by Bourbon and Batian 2 with mean infection scores of 1.50, 1.58 and 2.0 respectively. Batian 1 was the most susceptible genotype with a mean score of 3.167 (Table 2). A combined analysis of variance for the two methods demonstrated highly significant differences between the genotypes ($p<0.001$), isolates ($p<0.0001$) and inoculation method ($p<0.01$).

For the injection method, there was no interaction between the genotypes and isolates indicating that the genotypes responded more or less the same to different isolates. However, for the cut method, there was high interaction between the genotypes and isolates indicating that the genotypes responded differently to different isolates. Following these contradicting results, it was not possible to clearly authenticate the reaction of the different genotypes to BBC. The injection method appeared to be more severe than the cut method representing a high disease score. The cut method however differentiated the varieties better in their level of resistance. Apparently the reaction of most of the genotypes tested ranged between moderately susceptible to susceptible except Ruiru 11 and Bourbon which portrayed appreciable tolerance to BBC for both methods of inoculation. Contrary to this finding, observations in the field especially in BBC prone areas in Kenya like in the Rift valley depict Ruiru 11 as being susceptible.

Although it was not possible to know the races in which the three isolates belonged, Ito et al. (2008) reported that SH_1 gene which confers resistance to some races of *Hemileia vastatrix* (causal agent of Coffee Leaf Rust) also confers resistance to some races of *P. s.* pv. *garcae*. The SH_1 gene is found in pure Arabicas of Ethiopian origin such as Dilla and Alghe. None of the 13 genotypes that were tested in this study is known to contain the SH_1

Table 3. Comparative effectiveness of the two methods.

Injection method		Cut method		Both methods combined	
Isolate	**LS means**	**Isolate**	**LS means**	**Isolate**	**LS means**
Control	1.385[a]	Control	1.231[a]	Control	1.308[a]
Kap-1/012	2.051[b]	Kap-1/012	1.385[a]	Kap-1/012	1.718[b]
Nak-1/012	3.128[c]	Kap-2/012	3.154[b]	Nak-1/012	3.295[c]
Kap-2/012	3.731[d]	Nak-1/012	3.462[c]	Kap-2/012	3.429[c]
LSD (5%)	0.446	LSD (5%)	0.162	LSD (5%)	0.233

Means marked with the same letter(s) are not significantly different at p=0.05.

gene. Previous studies conducted in Brazil also showed that apart from SH_1 gene, there are other resistance sources such as Catucaí (Petek et al., 2006), Hibrido de Timor (HDT) and Icatu (Sera et al., 1980). HDT is a derivative of *Coffea canephora* (Robusta coffee) while Catimor is a derivative of HDT. In addition, all improved Kenyan varieties, namely Ruiru 11, Batian 1, Batian 2 and Batian 3, also have HDT in their pedigree. Although these varieties recorded mixed reaction to BBC, it should be noted that none of them was selected for resistance to BBC and therefore the gene that confers resistance to BBC in HDT may have been lost during selection of the Kenyan varieties. Kenyan BBC isolates are also reportedly more virulent than Brazilian isolates (Kairu, 1997).

For both methods, the isolates depicted a highly significant difference (p>0.001) among themselves. The injection method portrayed isolate KAP-2/012 as the most virulent recording a score of 3.731, followed by isolate NAK-1/012 with a score of 3.128, and isolate KAP-1/012 with a score of 2.051. The control recorded the lowest infection score of 1.385 (Table 3). Unlike in the injection method, the cut method portrayed isolate NAK-1/012 as the most virulent recording a score of 3.46, followed by isolate BBC 18/012 with a score of 3.154, and isolate KAP-1/012 with a score of 1.385. The control recorded an infection score of 1.231 (Table 3). Possible contamination from the infected genotypes may have caused infection recorded in the control since all treatments were applied in the same environment. Differences in Kenyan isolates of *P. s.*pv*garcae* have also been reported by Kairu (1997) and Mugiira et al. (2011).

Conclusion

The two inoculation methods were found to be effective and can be recommended with slight improvements in the layout of the experiment e.g. use of many seedlings per replication. It was not possible to clearly authenticate the reaction of the different genotypes to BBC since the genotypes responded differently to different isolates and inoculation methods. Further study is therefore recommended using a wide range of genotypes and isolates under improved experimental set up. Field based evaluation studies can subsequently be conducted in BBC prone areas to determine whether there is any correlation between the laboratory results and those obtained under natural environment.

ACKNOWLEDGMENT

Thanks are due to the technical staff of CRF Breeding and Pathology sections who participated in this study. This work is published with the permission of the Director of Research, CRF, Kenya.

REFERENCES

Anthony F, Combes MC, Astorga C, Bertrand B, Graziosi G, Lashermes P (2002).The origin of cultivated *Coffeaarabica* L. varieties revealed by AFLP and SSR markers. Theor. Appl. Genet. 104:894-900.

Davis AP, Govaerts R, Bridson DM, Stoffelen P (2006). An annotated taxonomic conspectus of the genus *Coffea* (*Rubiaceae*). Bot. J. Linn. Soc. 152:465-512.

Gichimu BM, Omondi CO (2010). Early performance of five newly developed lines of Arabica Coffeeunder varying environment and spacing in Kenya. Agric. Biol. J. N. Am. 1(1):32-39.

Gichuru EK, Agwanda CO, Combes MC, Mutitu EW, Ngugi ECK, Bertrand B, Lashermes P (2008). Identification of molecular markers linked to a gene conferring resistance to Coffee Berry Disease (*Colletotrichum kahawae*) in *Coffea arabica.* Plant Pathol. 57:1117–1124

Ito DS, Sera T, Sera GH, Grossi LD, Kanayama FS (2008).Resistance to Bacterial Blight in Arabica coffee cultivars. Crop Breed Appl. Biotechnol. 8:99-103.

Jansen AE (2005). Plant protection in coffee: Recommendations for the common code for the coffee community-initiative. Deutsche Gesellschaft fur TechnischeZusammenarbeit, GmbH. Online Publication: http://www.evb.ch/cm_data/4C_Pesticide_Annex_final.pdf

Kairu GM (1997). Biochemical and pathogenic differences between Kenyan and Brazilian isolates of *Pseudomonas syringaepv. garcae.* Plant Pathol. 46:239-246.

Kairu GM, Nyangena CM, CrosseJE (1985).The effect of copper sprays on Bacterial Blight and Coffee Berry Disease in Kenya. Plant Pathol. 34:207-213

Lashermes P, Combes MC, Robert J, Trouslot P,Hont AD, Anthony F, CharrierA (1999). Molecular characterization and origin of the *Coffeaarabica* L. genome. Mol. Genet. 261:259-266

Mugiira RB, Arama PF, Macharia JM, Gichimu BM (2011). Antibacterial activity of foliar fertilizer formulations and their effect on ice nucleation activity of Pseudomonas syringarpvgarcae Van Hall; the causal agent of Bacterial Blight of Coffee. Int. J. Agric. Res. 6(7):550 – 561.

Pearl HM, Nagai C, Moore PH, Steiger DL, Osgood RV, Ming R (2004). Construction of a genetic map for Arabica coffee. Theor. Appl. Genet.

108:829-835.

Petek MR, Sera T, Sera GH, Fonseca IC de B, Ito DS (2006). Seleção de progênies de *Coffeaarabica*com resistênciasimultânea à manchaaureolada e à ferrugemalaranjada. Bragantia 65(1):65-73.

Ramos AH, Kamidi RE (1981) Seasonal periodicity and distribution of Bacterial Blight of coffee in Kenya. Plant Dis. 65:581–584.

Sera T, Cardoso RML, Mohan SK (1980) Resistênciaaocrestamentobacteriano (*Pseudomonas garcae*) emmateriaissegregantespararesistência à ferrugem (*Hemileiavastatrix*).In Proceedings of the 7th Congressobrasileirode pesquisascafeeiras, Araxá, Brazil, pp. 60-61. http://www.sbicafe.ufv.br/SBICafe/publicacao/frpublicacao.asp.

Silva MC, Várzea V, Guerra-Guimarães L, Azinheira HG, Fernandez D, Petitot AS, Bertrand B, LashermesP, Nicole M (2006). Coffee resistance to the main diseases: Leaf rust and Coffee Berry Disease. Braz. J. Plant Physiol. 18(1):119-147

Silveira SR, Ruas PM, Ruas CF, Sera T, Carvalho VP, Coelho ASG (2003). Assessment of genetic variability within and among coffee progenies and cultivars using RAPD markers. Genet. Mol. Biol. 26:329-336.

Van der Vossen HAM (1985). Coffee selection and breeding. In: M.N. Clifford and K.C. Willson (Eds.), Coffee botany, biochemistry and production of beans and beverage, pp. 48-96. Croom Helm, London.

4

Irradiation alone or combined with other alternative treatments to control postharvest diseases

Cemile Temur* and Osman Tiryaki

Department of Plant Protection, Faculty of Seyrani Agriculture, Erciyes University, Kayseri, 38039, Turkey.

The postharvest diseases are considered worldwide as the most significant issue for postharvest facilities. Although there are various methods to decrease postharvest losses, consumers are looking for agricultural product free of chemicals. It is therefore necessary to develop alternatives to synthetic chemical control to reduce environmental risks and raise consumer confidence. Several alternatives such as food irradiation show promise, but none alone is as effective as fungicides. A strategy must be developed that combines several of these alternatives to enhance their effectiveness. Therefore, there is a need for a method combining couple of methods together. A combination for this purpose can be irradiation with other treatment such as, heating, cooling and sodium carbonate/sodium bicarbonate treatment. The safety of irradiated food is declared by joint FAO/JAEA/WHO Expert Committee for food irradiation. In this study, advantages and disadvantages of irradiation, and combination with various other treatments were evaluated and recommendations were provided to minimize the postharvest losses.

Key words: Irradiation, postharvest disease, combined treatment.

INTRODUCTION

The Food and Agriculture Organization of the United Nations (FAO) estimated that 25% of all food products are wasted after harvest worldwide. The most economic losses of foods are due to infestation with insects, fungal contamination and premature germination (Harris, 1998; Braghini et al., 2009a). Postharvest diseases also limit the storage period and marketing life of fruit. Postharvest losses are 5 to 10% when postharvest fungicides are used; without fungicides, losses of 50% or higher have occurred in some years (Margosan et al., 1997). Postharvest losses are estimated to be 30 to 40% in Turkey and sometimes it may reach to 50% (Anonymous, 2011).

There are some registered fungicides such as fludioxonil and azoxystrobin in the USA for postharvest application to control decay in products. However, postharvest use of these fungicides in most European Union countries and Turkey are prohibited due to fungicide regulatory issues. In addition, public demands to reduce pesticide use, stimulated by greater awareness of environmental and health issues, as well as the development of resistance of some pathogens to fungicides limit the postharvest application of chemicals to agricultural products (Karabulut and Baykal, 2004). Many of the fungicides such as benzimidazole and dicarboximide, that are still available for use, are losing their effectiveness because of the development of resistance in postharvest pathogen of *Botrytis cinerea* (Lennox and Spotts, 2003).

It is necessary to find alternatives to control postharvest pathogens to reduce environmental risks and raise consumer confidence. Various methods have been investigated, and although they show promise, none alone has been found to be as effective as fungicides. Therefore it is necessary to develop a strategy which combines several of these alternatives that may equal the effectiveness of fungicides (Conway et al., 2005). One of the promised alternative methods is the use of gamma irradiation with the combination of other treatments such as antagonists, natural compounds, and physical treatments (Cia et al., 2007).

*Corresponding author. E-mail: cemiletemur@hotmail.com

Table 1. Permitted radiation dose for food irradiation in Turkish Food Codex (Anonymous, 1999).

Food type	Aim of irradiation	Maximum dose (kGy)
Bulb, root and tuber	İnhbition of shooting, germination, and budding during storage	0.2
Fresh vegetable and fruit	Delay maturation	1.0
	To prevent insects	1.0
	Extend the shelf life	2.5
	The quarantina control	1.0
Grain, milled grain products, nuts, oil seeds, legumes, dried fruit and vegetables	To prevent insects	1.0
	Reducing microorganisms	5.0
	Extend the shelf life	5.0
Diried vegetable, spices, dried herbs, condiments and herbal tea	Reducing some pathogenic microorganisms	10.0
	To prevent insects	1.0

The Joint FAO/ IAEA/ WHO Expert Committee for Food Irradiatin (JECFI) concluded that foods irradiated up to 10 kGy (1 Gy=100 rad) are safe and nontoxic (WHO, 1981). This limit is adapted to Codex Standard in 1983 (Anonymous, 1987). Later on, JECFI, for the evaluation of toxicological, nutritional, chemical and physical aspects of foods, declared that irradiated up to 10 kGy are safe and nutritionally adequate as long as they are produced according to good manufacturing practices (WHO, 1999). The limit of 10 kGy is also accepted for Turkish Food Codex in 1999 (Anonymous, 1999). Irradiation of fresh foods including fruit and vegetables is permitted to be irriadated at doses up to 1000 Gy (US FDA, 2004). "Food Irradiation Regulation" was published in Turkey in 1999. Agricultural products permitted for irradiation treatment were listed in this regulation (Table 1).

All foods are radioactive to some extent as a result of exposure to natural background radiation. Irradiation of food does not induce additional radioactivity, because the sources of radiation approved for use in food irradiation are limited to those producing energy too low to induce sub-atomic particles (Anonymous, 2000). Chain reactions cannot occur; therefore, no radioactivity is added. Neither the food nor the packaging materials become radioactive (Urbain, 1986). It is physically impossible for irradiated food to be radioactive just as your teeth are not radioactive after you have had a dental X-ray. Irradiation is radiant energy. It disappears when the energy source is removed (Brennand, 1995). It is concluded that gamma-rays with the energy of 5 MeV and accelerated electrons with the energy of 10 MeV, even if high doses, does not causes any radioactivity. Cobalt-60 (^{60}Co) and Cesium-137 (^{137}Cs) are generally used for irradiation purposes, with the energy of 1.33 and 0.66 MeV, respectively. Thus, radioactivity is not possible even if the high doses are used with these irradiation sources (Anonymous, 1999; CAC, 2003; Farkas, 2006). Radioactivity of food is only possible with the exposure to radioactive particle leak caused by nuclear accident and nuclear weapon tests (Çelebi, 2007). About 170 gamma facilities exist worldwide. Most facilities are used for medical sterilization, surgical or the preparation of packaging materials (Shea et al., 2000).

The ultimate goal of this review article is to devise a strategy that combines several of alternatives below mentioned that will equal the effectiveness of chemical control. The specific objective of this paper is to determine the effect of irradiation alone and in combination with other treatments such as sodium carbonate/sodium bicarbonate, heat treatment, chemicals, modified atmosphere packaging, cold storage and biocontrol agent.

IRRADIATION TREATMENT AGAINST THE POSTHARVEST DISEASES

Microbes in food fall into three categories. Some micro-rganisms, such as those that produce fermentation, create desirable changes in foods. Spoilage micro-organisms change the color, odor, and texture of food, rendering it unpalatable, but they do not cause human illness. Pathogens cause human disease and include invasive and toxigenic bacteria, toxigenic molds, viruses, and parasites. All food production techniques from the farm to the table are concerned with minimizing spoilage, eliminating pathogens, and prolonging shelf life. Gamma irradiation can contribute for the reduction of postharvest losses caused by fungi and reduce the use or doses of fungicides on disease control (Cia et al., 2007).

Irradiation has been used for the preservation and production of foods that are free of pathogenic micro-organisms and is therefore an important tool for the control of food contaminating microorganisms. In another words, irradiation of foods can reduce the risk of foodborne illness (Rustom, 1997; Braghini et al., 2009a).

This approach has also contributed to reduce economic losses resulting from food deterioration and to increase food safety, thus favoring the acceptance of products exported by developing countries (Loaharanu, 1994). Food irradiation is a process by which food is exposed to a controlled source of ionizing radiation to prolong shelf life and reduce food losses, improve microbiologic safety, and/or reduce the use of chemical fumigants and additives. It can be used to reduce insect infestation of grain, dried spices, and dried or fresh fruits and vegetables; inhibit sprouting in tubers and bulbs; retard postharvest ripening of fruits; inactivate parasites in meats and fish; eliminate spoilage microbes from fresh fruits and vegetables; extend shelf life in poultry, meats, fish, and shellfish; decontaminate poultry and beef; and sterilize foods and feeds (Brennand, 1995).

The dose of the ionizing radiation determines the effects of the process on foods. Radiation doses are measured in international units called Gray (Gy). Food is irradiated at levels from 50 Gy to 10 kGy, depending on the goals of the process. Low-dose irradiation (≤1 kGy) is used primarily to delay ripening of produce or kill or render sterile insects and other higher organisms that may infest fresh food. Medium-dose irradiation (1 to 10 kGy) pasteurizes food and prolongs shelf life. High-dose irradiation (>10 kGy) sterilizes food. The FDA has authorized the following 4 sources of ionizing radiation for food treatment: ^{60}Co, ^{137}Cs, machine-generated accelerated electrons not to exceed 10 MeV, and machine generated X-rays not to exceed 5 MeV. All petitioners for FDA approval of food irradiation must satisfy technical requirements that limit dose and specify conditions under which the food will be irradiated.the technical effect on the food, dosimetry, and environmental controls must be defined and in compliance with the Federal Food, Drug and Cosmetic Act (Shea et al., 2000).

Ionizing radiation has been widely recognized as a method of decontamination of foodstuffs. Many reviews have summarized the nutritional adequacy of irradiated foods. They clearly demonstrate that irradiation results in minimal, if at all noticeable, changes in the taste, provided that the optimal dose for each type of food is not exceeded. In general, irradiation to the recommended doses changes the chemical composition of foods very little. At doses below 1 kGy, nutritional losses are considered to be insignificant, and none of the chemical changes found in irradiated foods is harmful, dangerous or even lying outside of the limits normally observed (Braghini et al., 2009b). Doses of up to 10 kGy are highly effective in microbial decontamination and have no adverse effects on the nutritional quality of cereal grains (WHO, 1994; Aziz et al., 2006).

The use of gamma radiation to inactivate aflatoxins was investigated. The toxicity of a peanut meal contaminated with Aflatoxin B_1 (AFB_1) was reduced by 75 and 100% after irradiation with gamma-rays at a dose of 1 and 10 kGy, respectively. However, doses higher than 10 kGy inhibited the seed germination, and increased the peroxide value of the oil in gamma-irradiated peanuts. The presence of water has an important role in the destruction of aflatoxin by gamma energy, since radiolysis of water leads to the formation of highly reactive free radicals. These radicals can readily attack AFB_1, at the terminal furan ring, giving products of lower biological activity. The mutagenic activity of AFB_1 in an aqueous solution (5 pg ml^{-1} water) was reduced by 34, 44, 74 and 100% after exposure to gamma-rays at 2.5, 5, 10 and 20 kGy, respectively. Also, a dose of 10 kGy completely (100%) inactivated AFB_1, and destroyed 95% of AFG_1 in a dimethylsulphoxide-water (1:9, v/v) solution (Rustom, 1997).

Sclerotia of *Whetzelinia sclerotiorum* obtained from field grown peas and from laboratory cultures were exposed to gamma radiation from a ^{60}Co source. Over 2500 sclerotia irradiated at levels from 100 to 800 krads were observed. Sclerotia with moisture levels below 10% were highly resistant to radiation damage having an LD_{50} of up to 600 krads. An increase in moisture content resulted in a marked decrease in the LD_{50} (Blanchette and Tourneau, 1977).

Some studies in the use of ionizing radiation to control *Botrytis* rot in table grapes and strawberries were performed by Nelson et al. (1959). Cultures irradiated with 4 × 10^5 rep (Röntgen Equivalent Physique) made no growth after transfer to unirradiated media. The rate of spread of *Botrytis* rot among grape berries and strawberries was markedly reduced at doses of 1 × 10^5 and 2 × 10^5 rep

In a study showed that an irradiation dose of 200 000 rep inhibited brown rot for 10 days at 80 to 85°F. Unirradiated peaches were completely rotted within 5 days (Beraha et al., 1959).

Irradiation doses for the inhibition of fungal development are presented in Table 2 for several fungi. The important criteria for the evaluation of irradiation treatment against the fungi are D_{10} values. This term is radiaton dose that causes 90% decrease in population. The lower the D_{10} values, the higher the sensitivity of microorganism. D_{10} values for *Aspergillus flavus* and *A. parasiticus* which are aflatoxigenic were specified as 0.25 and 0.31 kGy, respectively (Table 3) (TAEK, 2001).

Advantages of irradiation treatment

The problems caused by diseases have been maximized by the development of pathogen resistance to fungicides and by the withdrawal of some products from the market. Moreover, consumers are looking for fruit free of chemical residues. Consequently, alternative control strategies, such as antagonists, natural compounds, and physical treatments have drawn attention. Gamma and UV-C (254 nm) irradiations are physical treatments that

Table 2. Radiation dose for fungal inhibition at room temperature (TAEK, 2001).

Fungus	Irradiation medium	Radiation source	Dose (kGy)
Aspergillus flavus	% 0.1 pepton	Electrons	1.6
Aspergillus niger	Malt extrakt agar	Gamma-ray	2.5
Aspergillus parasiticus	Water	Gamma-ray	1.6
Alternaria spp.	Malt ekstrakt agar	Gamma-ray	6.0
Botrytis cinerea	Malt ekstrakt agar	Gamma-ray	5.0

Table 3. Comparison of D_{10} values for fungi irradiated in aqueous suspension (TAEK, 2001).

Fungus	Irradiated with gamma-rays (kGy)
Aspergillus niger	0.245
A.flavus	0.250
Avicularia versicolor	0.282
A. parasitucus	0.310
Penicillium cyclopium	0.397
Alternaria alternata	2.409

can be used for the control of postharvest diseases Besides exhibiting fungicidal effects, these treatments can also induce resistance in fruit (Conway et al., 2005). Lu et al. (1993) reported that both UV-C and gamma-rays reduced storage rot and delayed ripening of peaches.

Gamma radiation is effective on all stages of the life cycle of a pest such as a fruit fly and it is ready to be used as an efficacious quarantine treatment (Cia et al., 2007). Green mold, caused by *Penicillium digitatum* and blue mold, caused by *P. italicum* are the most economically important postharvest diseases of citrus in Spain, California, and all citrus production areas characterized by low summer rainfall. Both diseases are primarily controlled worldwide by the application of synthetic fungicides such as imazalil, sodium ortho-phenyl phenate, or thiabendazole. However, alternative methods are needed because the widespread use of these agrochemicals in commercial packinghouses has led to proliferation of resistant strains of the pathogens (Kinay et al., 2007).

It is impossible to eradicate infections with fungicides without injuring the fruit excessively. The penetrating power of gamma-rays is more than fungicides. These rays reach decay organisms in aereas of fruits not accessible to chemicals (Tiryaki, 1990). The advantage of gamma radiation is the high penetrability and uniformity of the dose, which permits to treat products of different sizes and shapes (Jarrett, 1982).

Disadvantages of irradiation treatment

The process of irradiation essentially adds a small amount of energy to food. As such, many radiolytic products are generated, but in very small numbers. Heat processing forms the same general types of molecules, but in larger numbers, because the amount of energy added to foods is often greater than with irradiation. Induction of radiation-resistant microbial populations occurs when cultures are experimentally exposed to repeated cycles of radiation. Mutations in bacteria and other organisms develop with any form of food processing, including ionizing radiation, heat, drying, and ultraviolet light. Radiation does not produce mutations by unique mechanisms. Further, mutations from any cause can result in greater, less, or similar levels of virulence or pathogenicity from parent organisms. Although it remains a theoretical risk, several international reviews cite no reports of the induction of novel pathogens attributable to food irradiation. Similar concerns exist about mycotoxins. Experimental data are conflicting, but some studies show an increase in mycotoxin formation after irradiation. One theory is that the higher radio-resistance of molds and yeasts compared with bacteria results in a loss of competitive inhibition of mold and yeast growth. Any mold surviving under treatment with irradiation may be expected to grow more rapidly in the absence of competitors and eventually dominate the mycoflora. In the absence of temperature abuse in storage, the available evidence indicates that treating products with ionizing energy does not add to that hazard. More nutrients are made available for fungi by irradiation. This is an area in which additional study would be useful (Shea et al., 2000).

Palou et al. (2007) did not observe any resistance to green and blue molds on mandarins exposed to X-irradiation at doses from 195 to 875 Gy. Contrarily, green mold development was slightly favored in fruit treated at 875 Gy when *P. digitatum* was inoculated 6 days after irradiation. This might be related to a negative effect of X-rays at this dose on the physical and/or physiological condition of the fruit rind that would facilitate the fungal mycelial growth through the albedo and flavedo cells. The negative effect may include the induction of some incipient peel damage that was not readily visible.UV-C irradiation was not able to reduce the occurrence of *Colletotrichum gloeosporioides* lesions and caused browning in papaya fruit (Cia et al., 2007). Similar findings have been reported earlier indicating that smaller doses of UV-C reduced the development of *B. cinerea* in

Table 4. Lethal gamma radiation doses to young growing mycelium of some fungi (Beraha et all., 1960).

Organism	Source	Dose (*10^5 rad) on		
		Tochinai	Czapek	Host
Phytophthora infestans	Potato	0.25	0.25	0.25
Phomopsisi citri	Orange	0.44-0.96	0.44-0.96	0.91-1.45
Penicillium digitatum	Lemon	1.10-1.48	0.44-0.94	1.82-2.10
Penicillium italicum	Orange	1.43-1.47	1.19-1.43	1.57-1.82
Penicillium expansum	Apple	1.35-1.40	1.95-2.52	1.82-2.74
Botrytis cinerea	Grape, strawberry	0.95-1.86	0.95-1.86	2.74-4.56
Monilinia fructicola	Peach	1.38-1.85	0.90-1.38	1.37-1.82
Sclerotinia sclerotiorum	Bean	1.73-2.13	1.73-2.13	2.28-2.73
Rhizopus nigricans	Peach	2.74-3.52	3.52-4.43	1.82-2.28
Alternaria tenuis	Tomato	4.20-4.57	4.20-4.57	2.74-4.56

table grapes, but caused fruit browning (Camili et al., 2004).

Mechanism of action of irradiation on micro-organism

The content of the major phenolic compounds present in the peel of clementine mandarins significantly increased on fruit that had been previously irradiated with gamma-rays at 300 Gy. This increase was correlated with an enhancement of the activity of the enzyme phenylalanine ammonia-lyase (PAL). Ionizing radiation can stimulate the biosynthesis of constitutive and/or induced phenolic compounds that could extend storage life and in some cases induce fruit resistance against pathogens (Palou et al., 2007). In their study, if such bioactive compounds were actually synthesized, it was at levels not high enough to effectively induce disease resistance under our experimental conditions and the synthesis was not influenced by either X-ray dose, time between irradiation and pathogen inoculation, or incubation time after inoculation. They suggest, therefore, that the direct effects of irradiation on the fungal structures growing in the rind were more important for disease reduction than a possible indirect effect on the fruit mechanisms of defense. This assumption is further supported by the fact that X-irradiation considerably inhibited the sporulation of both *P. digitatum* and *P. italicum* on decayed mandarins.

A number of researchers indicated that gamma-rays inhibited fungal development and mycotoxin production during the food storage. The effect of irradiation depends on fungus type, application dose, moisture content and compostion of food, and storage conditions (Aziz et al., 2006; Kabak and Var, 2005). The effect is also depends on environmental factors such as, composition and moisture content of irradiated medium, temperature and presence of oxygen during the irradiation, being fresh or frozen (Smith and Pillai, 2004). In another explanation, surviving of microbial cells depends on the resisitence and recovery status of cell, irradiation dose, pH, atmospheric conditions and chemical composition of food (Monk et al., 1995).

Low irradiaton doses (for example 1 kGy) stimulated fungal development for both *invitro* and *in vivo* studies (Tiryaki, 1990). After 40 days irradiation, lesion diameters were 36.21 and 34.75 mm for 1 kGy and control treatment, respectively, in Ankara pears inoculated with *Penicillium expansum*. This supports stimulative effect of low gamma-rays (Tiryaki and Maden, 1991). Similarly 1 kGy of gamma irradiaton stimulated aflatoxin occurence. Whereas 3 to 4 kGy inhibited fungal and mycotoxin development (Kabak and Var, 2005).

Lethal gamma radiation doses required for pathogens in the host (*in vivo*) are higher than in the culture (*invitro*) media (Table 4) (Beraha et al., 1960). Irradiation dose rate is also important for inhibition of fungal development. Beraha (1964) worked on the effect of dose rate and demonstrated that high dose rate was more effective than low dose rate. *B. cinerea* infection was inhibited with the 125 to 150 krad of irradiation, at the dose rate of 25 krad/min; whereas, infection was not inhibited with the 200 krad at the 2.5 krad/min.

Irradiation kills microbes primarily by fragmenting DNA. The sensitivity of organisms increases with the complexity of the organism. Thus, viruses are most resistant to destruction by irradiation, and insects and parasites are most sensitive. Spores and cysts are quite resistant to the effects of irradiation, because they contain little DNA and are in highly stable resting states. Toxins and prions, which have few chemical bonds to disrupt, are resistant to irradiation. The conditions under which irradiation takes place (that is, temperature, humidity, and atmospheric content) can affect the dose required to achieve the food processing goal, but these are well-described and easily controlled (Shea et al., 2000).

Tiryaki et al. (1994) worked on pathogenicty of irradiated fungi. Effect of irradiation on pathogenicity cultural charasteristic and sporulation of fungi have been

Table 5. Effects of postharvest UV-C treatment on *B. cinerea* disease for freesia inflorescences (Darras et al., 2010).

Factor	Disease parameter		
	Disease severity (score 0-4)	Lesion number	Lesion diameter (mm)
(1) UV-C irradiation*			
Before inoculation	3.1[b]	83[b]	0.91[b]
After inoculation	1.3[a]	36[a]	0.79[a]
(2) UV-C doses (D) (kJ m^{-2})			
0.0	2.9[d]	75[c]	0.87[b]
0.5	1.8[b]	49[ab]	0.79[a]
1.0	1.5[a]	43[a]	0.80[a]
2.5	1.9[b]	52[b]	0.85[ab]
5.0	2.1[c]	55[b]	0.88[b]
(3) Interaction I×D			
	$P < 0.05$	$P < 0.05$	$P < 0.05$

*Within main factor means, numbers followed by the same letter are not significantly different at $P = 0.05$.

investigated by reisolation. It was found that there were no differences at these properties of fungi between irradiated and unirradiated samples. The gamma irradiation dose which inhibits decay was determined in apple, quince, and peach inoculated with *P. expansum*, *Monilinia fructigena* and *Rhizopus stolonifer*, respectively. Doses of 1, 2, 3 and 3.5 kGy did not inhibit decay on fruit, but infection was delayed for a certain period.

Time between inoculation and irradiation

Time interval between inoculation and irradiation affected the growth response of *Monilinia fructicola* infections and the irradiation dose needed for it control. When "firm-ripe"peaches were irradiated (200 krad) within 24 h after inoculation, only 10% of the inoculations formed lesions. Postponing irradiations to 36, 48 or 60 h after inoculation increased the incidence of lesion to 60, 80 and 90%, respectively (Kuhn et all., 1968).

A factor that could adversely influence the effectiveness of the treatments was the extended period of time between inoculation and irradiation (about 38 h at about 20°C). According to Spalding and Reeder (1985) the incidence of green mold was lower on grapefruit irradiated with gamma-rays 2 h after artificial inoculation with *P. digitatum* than on fruit irradiated 24 to 72 h after inoculation.

Moreover, it was observed that irradiation was more effective against citrus postharvest diseases when applied before extensive fungal development.

Nevertheless, satisfactory commercial control of citrus *Penicillium* decay usually requires the effective control of infections that were initiated in the field at least 24 h before the application of the antifungal treatment (Palou et al., 2007).

Irradiation before and after inoculation

Darras et al. (2010) searched germicidal and inducible host defense effects of UV-C irradiation on petal specking caused by *B. cinerea*. UV-C irradiation of freesia inflorescences after artificial inoculation with *B. cinerea* was more effective in reducing petal specking, compared to UV-C treatment before artificial inoculation. Cut freesia inflorescences exposed to 1 kJ m^{-2} UV-C after artificial inoculation with 10^4 conidia ml^{-1} displayed reduced disease severity scores, lesion numbers and lesion diameters by 74, 68 and 14%, respectively, compared to non-irradiated inflorescences. In contrast, UV-C irradiation with 1 kJ m^{-2} before artificial inoculation reduced lesion numbers and lesion diameters by 13 and 24%, compared to the non-irradiated controls. Higher UV-C doses of 2.5 or 5 kJ m^{-2} reduced disease severity scores, lesion numbers and lesion diameters when applied after artificial inoculation, but enhanced disease when applied before artificial inoculation.

Inflorescence irradiation following artificial inoculation with *B. cinerea* generally conferred significant ($P < 0.05$) disease reduction compared to irradiation prior to artificial inoculation (Table 5). Disease severity scores and lesion numbers on inflorescences irradiated after artificial inoculation was significantly ($P < 0.05$) lower than the ones irradiated prior to artificial inoculation at all UV-C doses tested. Irradiation of inflorescences with 0.5, 1, 2.5 or 5 kJ m^{-2} UV-C after inoculation reduced disease severity scores and lesion numbers by 44, 70, 74, and 59% and by 37, 62, 68 and 60%, respectively (Figure 1). UV-C irradiation suppressed petal specking caused by *B. cinerea* when applied after artificial inoculation (Darras et al., 2010). In a research carried out by Palou et al. (2007), irradiation performed before and after inoculation was evaluated with respect to fungal inhibition. There was

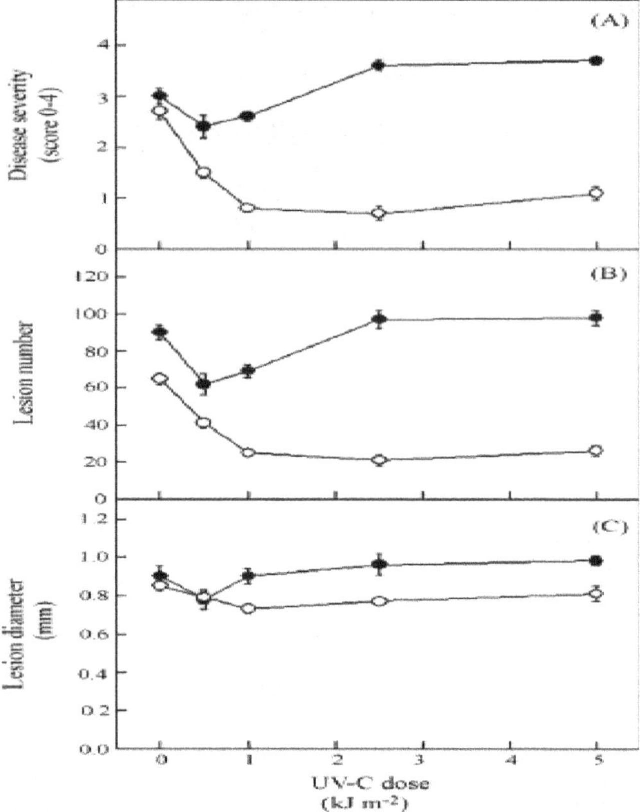

Figure 1. Disease severity scores (A), lesion numbers (B) lesion diameters (C) on freesia irradiated with UV-C (Darras, et al., 2010).

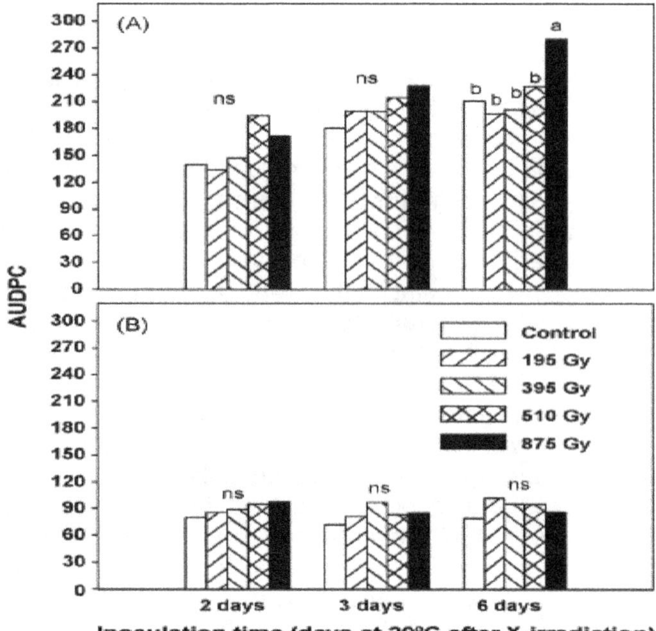

Figure 2. Area under the disease progress curve (AUDPC) of *P. digitatum* (A) and *P. italicum* (B) on clementine mandarins irradiated with X-rays (Palou et al., 2007).

not any difference between control fruit samples and treated fruit samples (that is, inoculated after irradiation) with respect to disease severity and disease incidence. It can be concluded that gamma and UV-C (254 nm) irradiations are physical treatments, entire effect on fungi occur during irradiation process.

Induction of fruit disease resistance by irradiation

Palou et al. (2007) carried out a detailed research about induction of fruit resistance to disease by X-ray irradiation. Non-inoculated mandarins that had been irradiated with X-rays at 0 (control), 195, 395, 510 and 875 Gy were kept at 20°C for 2, 3 or 6 days. After each of these time periods, irradiated fruit were inoculated with 10^5 spores' ml^{-1} of *P. digitatum* or *P. italicum* and incubated at 20°C for 7 days. Each pathogen was inoculated on different sets of fruit. For each pathogen, irradiation dose and inoculation time, four replicates of 10 fruits were used. Disease incidence and severity were evaluated after 3, 5 and 7 days of incubation. Irrespective of the X-irradiation dose and the time that inoculated fruit were incubated at 20°C (3, 5 or 7 days), neither the incidence (Figure 2) nor the severity (Figure 3) of both green and blue molds on artificially inoculated mandarins were significantly affected by exposure to X-rays. Significant differences among treatments were only observed for AUDPC on clementines inoculated with *P. digitatum* 6 days after irradiation; AUDPC was significantly higher on fruit treated at 875 Gy than on control fruit or fruit irradiated with other doses. Therefore, under experimental conditions, fruit resistance to disease not only was not increased but also was reduced by X-irradiation. Therefore, X-ray treatment did not induce disease resistance in the rind of irradiated fruit.

To check the possibility of resistance induction by irradiation, papayas were also inoculated after the treatments. It seems that papaya inoculation 24 h after treatment did not induce resistance since lesion diameter was not reduced. It can also be seen that UV-C did not reduce pathogen sporulation on fruit lesions. Thus, it is possible that fruit inoculation 24 h after the treatments did not stimulate defense responses in the fruit (Cia et al., 2007).

COMBINATION OF IRRADIATION WITH OTHER TREATMENTS AGAINST THE POSTHARVEST DISEASES

Recently, combined treatment is recommended to control the postharvest diseases. The main purpose of combination is to increase the effectiveness, to decrease the negative effect of application by exposure to lower doses compared to single application. The effect of irradiation is more promising when applied in combination

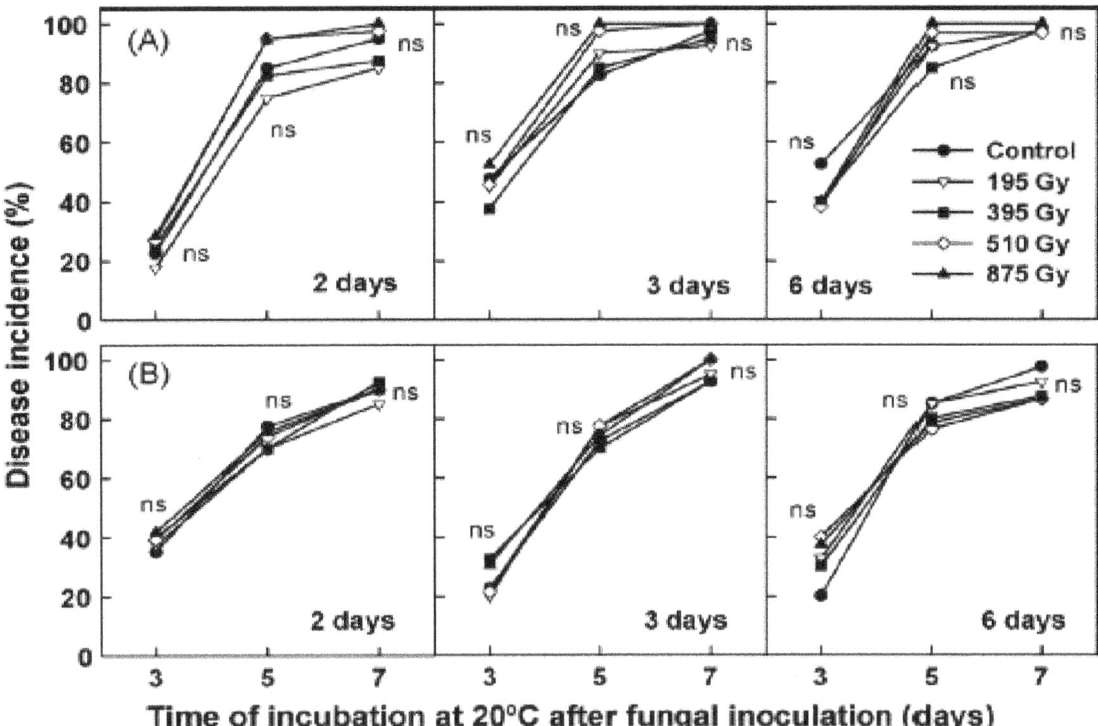

Figure 3. Incidence of *P. digitatum* mold (A) and *Penicillium italicum* mold (B) on clementine mandarins (Palou et al., 2007).

with hot water treatment, chemicals, such as SO_2 fumigation and cold storage treatment (Beraha et al., 1960; Tiryaki et al., 1994). Sodium carbonate (SC) treatment is also another alternative combination against the postharvest diseases.

Consumers are demanding less chemical residue on produce, and many fungi are developing resistance to commonly used fungicides. Since the use of fungicides is becoming more restricted due to health concerns, many alternatives to chemical control have been investigated, but none was able to provide the level of control of synthetic fungicides. While heat treatment virtually eliminates decay if fruit are inoculated prior to heating, it has little effect when infection occurs after heating, therefore, having no protective effect. Likewise, sodium bi-carbonate (SBC) does not provide persistent protection of the fruit from re-infection after treatment. The major limitations with biocontrol are the lack of eradicative activity, and a narrower spectrum of activity than is found with synthetic fungicides. The effect of environmental factors on biological control is also generally greater than the effect of fungicides. A combination of three methods described above may complement one another to overcome the shortcomings of each. Combination of several of these alternatives increases their effectiveness (Conway et al., 2004). But some workers revealed that both UV-C and gamma-rays reduced storage rot, but the combination of UV and gamma showed no advantage over the use of UV or gamma alone (Lu et al., 1993).

Combination with sodium carbonate

Palou et al. (2007) investigated the effects of X-ray irradiation and SC treatments on postharvest *Penicillium* decay in mandarins. As shown in Figure 4, by storage at 20°C, SC treatment with 875 Gy is more effective with respect to disease severity, disease incidence and sporulation of *P. digitatum* and *P. italicum.* Green mold severity as lesion diameter was only significantly reduced by irradiation at the highest dose of 875 Gy. Similarly, blue mold severity was not significantly reduced by SC treatment alone, but it was 25 to 35 mm by treatment at both X-ray doses (Figure 4). Comparable results were obtained with mandarins cold-stored at 5°C for 21 days. For both molds, SC treatment alone did not significantly affect either disease incidence and severity or pathogen sporulation (Figure 5). Incidence of both diseases on previously SC-treated fruit was markedly reduced by X-irradiation at both doses after 7 days of cold storage.

Palou et al. (2007) observed remarkably lower efficiencies of sodium carbonate on citrus postharvest green and blue molds than Smilanick et al. (1999). Factors that could account for this lack of effectiveness in reducing either disease incidence and severity or pathogen sporulation may include: The high initial susceptibility to

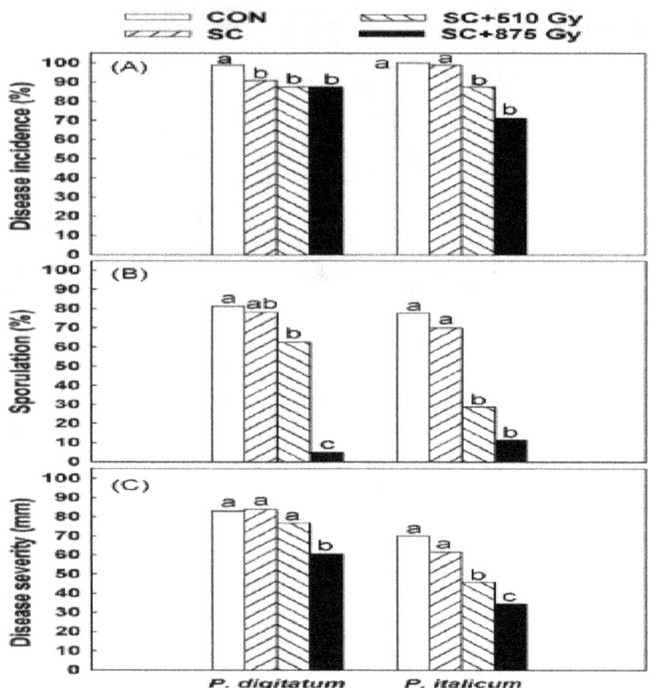

Figure 4. Percentage of infected (A) and sporulated (B) fruit and lesion size (C) on clementine mandarins incubated at 20°C for 7 days (Palou et al., 2007).

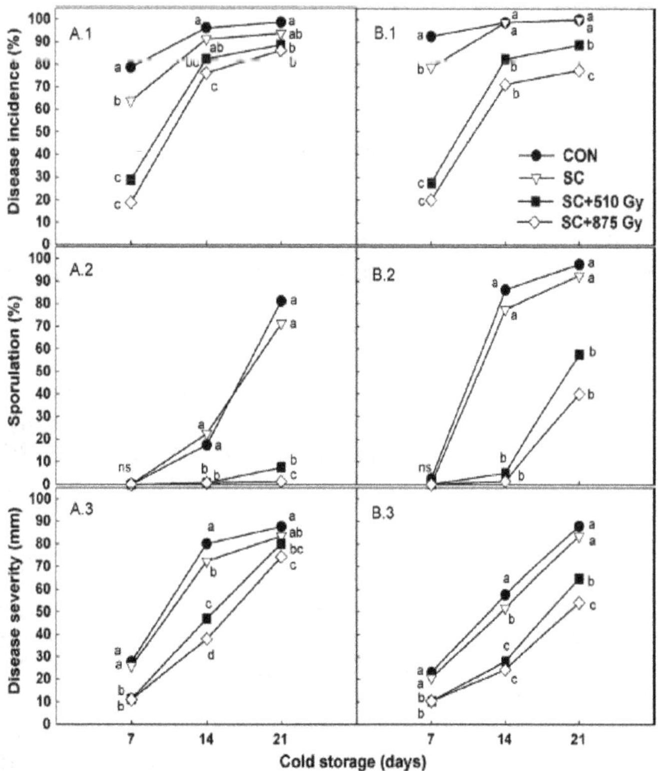

Figure 5. Percentage of infected (1), sporulated (2) fruit and lesion size (3) on clementine mandarins artificially inoculated with *P. digitatum* (A) or *P. italicum* (B) and incubated at 5°C for 21 days (Palou et al., 2007).

decay of the fruit used in the experiments, the use of non-heated SC solutions, and the rinse of treated fruit with tap water. It has been shown that the effects of SC, SBC, and other low-toxicity food additives on *Penicillium*-infected citrus fruit are fungistatic, not very persistent, and highly dependent on the host species and its physical and physiological condition.In contrast to fungal growth, pathogen sporulation was clearly inhibited on inoculated mandarins by the combined treatments (irradiation and SC), especially that of *P. italicum* on fruit incubated at 20°C and that of *P. digitatum* on fruit cold-stored at 5°C. Since SC does not exert anti-sporulant activity, this effect should be attributed to irradiation. The reduction of spore production has commercial value because stored citrus fruit are usually treated with fungicides; if resistance develops among these pathogens, the treatment would reduce the proliferation of resistant spores and presumably would prolong the useful life of postharvest fungicides. Further, *Penicillium* spores that are produced from stored fruit are a significant source of contamination for healthy adjacent fruit, and for packages, walls, and floors of rooms. Thus, irradiation treatment could greatly reduce the load of airborne pathogenic spores (Palou et al., 2007). Results from this work suggest that some technological aspects of the integration of SC and X-ray treatments should be improved for satisfactory control of established infections of *P. digitatum* and *P. italicum* on clementine mandarins. Heating SC solutions, non-rinsing SC-treated fruit, reducing the time between inoculation and irradiation, or even applying first the irradiation then the SC treatments could presumably enhance the effectiveness of the combined treatments.

Carbonic acid salts, such as sodium carbonate (SC, Na_2CO_3, soda ash) and sodium bicarbonate (Palou et al., 2001), are common food additives allowed with no restrictions for many applications. SC has been re-examined during recent years as a potential alternative to synthetic fungicides to manage citrus postharvest diseases because it is inexpensive, readily available, and can be used with a minimal risk of injury to the fruit. In general, carbonic acid salts are considered to be good candidates to be used in combination with other chemical, physical, or biological methods for the integrated control of postharvest diseases (Palou et al., 2002).

Rinsing the fruit at low pressure has been an effective method to avoid potential negative effects (loss of weight and firmness during cold storage) of SC treatments on the quality of clementine mandarins. However, in contrast to what was observed on oranges (Smilanick et al., 1999), rinsing of SC-treated clementines resulted in a significant loss of SC effectiveness against green mold. In order to preserve fruit quality and avoid potential interactions of X-rays with SC residues present on the surface of the mandarins, it was decided to rinse the fruit after SC treatment. This fact could help explain why SC effectiveness was less in this work than in previous experiments with unrinsed clementines (Palou et al., 2007).

Spores of both *P. digitatum* and *P. italicum* were found in early research to be more sensitive to irradiation than spores of other major citrus postharvest pathogens such as *B. cinerea, Diplodia natalensis, R stolonifer*, or *Alternaria citri* and were killed at very high rates (>95%) with a dose of 1000 Gy. However, effective control of established infections of *Penicillium* on oranges or lemons required irradiation doses higher than 1000 Gy and, in general, such doses induced apparent rind injury.

Therefore, the effectiveness of irradiation treatments alone or the potential synergy between SC and X-ray treatments against *Penicillium* established infections were not evaluated. Instead, the objective was to assess directly the disease control ability of the combined treatments. The integration of SC and X-irradiation, especially at the highest dose of 875 Gy, significantly reduced disease incidence and severity of both green and blue molds on mandarins stored at either 20 or 5°C (Palou et al., 2007).

Combination with heat treatment

A detailed study on radiation-heat synergism for inactivation of market disease fungi of stone fruits was carried out by Sommer et al. (1967). Synergistic effects of combined gamma radiation and heat treatments were compared with the single treatments for inactivation of spores of postharvest pathogens of *Prunus* spp. Interaction between treatments sometimes caused a 5- to 10-fold increase in inactivation.

The amount of synergism and the preferred sequence of application for maximum fungicidal effect depended upon the pathogen. With *R. stolonifer,* the maximum effect occurred if irradiation was first. In all other species studied, the reverse sequence resulted in greatest inactivation. Studies with fruit inoculated with *M. fructicola* demonstrated the advantage of heat sensitization before irradiation in brown rot control.

Hot water reduces decay substantially, but the risk of injury, weight loss, and the lack of antifungal residues has made this treatment a less attractive option than the relative ease of application, efficacy, and persistent protection offered by fungicides (Margosan et al., 1997).

Palou et al. (2007) did not use heated solutions in their experiments because according to previous reports, the combination of hot water with irradiation resulted in detrimental effects on the quality of treated clementines. Furthermore, the integration of hot water and irradiation for the control of green mold on oranges and grapefruit yielded contradictory results in previous research. While in some studies such combination resulted in synergistic effects, in other studies no benefits were observed.

Various non-chemical approaches have been investigated or proposed in recent years. Several studies have shown that hot water treatments have the potential to control postharvest diseases of peaches. In addition, biological control of postharvest diseases of stone fruits has been pursued actively by using bacteria and yeast antagonists (Karabulut and Baykal, 2004).

Heat treatment (38°C for 4 days) was effective in eradicating *P. expansum* Link on apples initially but exhibited no residual activity. The mode of action of the heat treatment seems to be both through direct interactions with the fungus itself, and *via* physiological responses of the fruit tissue. *In vitro* studies showed that both germination and growth declined when fungi were exposed to extended periods at higher temperatures. Heat treatment may also alter the susceptibility of the host to pathogens by the formation of an inhibitory substance in the peel (Conway et al., 2004).

The effect of irradiation is more promising when applied in combination with hot water treatment and chemicals, such as SO_2 fumigation, and cold storage treatment. It is possible to obtain more hopeful result with decreasing the time between irradiation and harvest. Less decay is observed when fruits were irradiated soon after picking than after a storage period (Tiryaki et al., 1994).

Combination with chemicals

Previous study has demonstrated that radiation, heat and chemical is the most efficient combination for the inhibition of postharvest diseases. Benomyl at 1000 ppm was the best single treatment for control of *P. expansum* in apples at about 25°C although it was not completely effective. Double combinations of hot water followed by irradiation and one triple combination of 50°C/10 min with irradiation and aureofungin could provide 80 to 100% control depending on quantity of inoculum. The double combination of 250 ppm benomyl preceded by hot water (50°C/5 min) at an interval of ½ h or the triple combinations of 50°C/10 min -150 krad –irradiation-250 ppm benomyl and 56°C/4 min 150 krad of radiation -1000 ppm aureofungin in these sequences completely controlled the 2-day–old infections of apples during the 3-week holding period at 25 1°C and extended the storage life (Roy, 1975).

Georgiev (1983) worked on combined treatment with irradiation and chemicals against the *B. cinerea* infection in "Bolgar" grapes at 80% relative humudity and 4 to 10°C storage. Captan +1kGy treatment was more effective against the *B. cinerea* infection than 2 and 3 kGy treatment. Similar findings were reported by Shirzad and Langarek (1984). They investigated irradiation and SO_2 treatment to increase shelf-life of grapes.

Combination with modified amosphere packaging

Several studies demonstrated the inhibitory effect of modified atmosphere packing (MAP) on postharvest

Table 6. Lesion diamater (mm) of Ankara pears inoculated with *B. cinerea* (Tiryaki and Maden, 1991).

Dose (kGy)	Day after irradiation			
	10	20	30	35
0	25.63^{A*a**}	35.04^{Aa}	40.58^{Bb}	46.29^{Bb}
1	22.83^{Bb}	34.54^{Aa}	49.46^{Aa}	53.33^{Aa}
2	15.79^{Cc}	25.88^{Bb}	43.58^{Bb}	50.63^{ABab}
3	13.21^{Dd}	20.88^{Cc}	32.13^{Cc}	37.54^{Cc}

* Figures followed by different capital letters differ significantly at p< 0.05; *** figures followed by different lower-case letters differ significantly at p<0.01.

Table 7. Lesion diamater (mm) of Ankara pears inoculated with *P. expansum* (Tiryaki and Maden, 1991).

Dose (kGy)	Day after irradiation			
	10	20	30	40
0	12.13^{A*a**}	19.20^{Aa}	26.29^{Aa}	34.75^{Aa}
1	11.29^{Aa}	17.92^{Aa}	25.63^{Aa}	36.21^{Aa}
2	10.13^{Bb}	14.71^{Bb}	19.21^{Bb}	28.46^{Bb}
3	9.67^{Bb}	12.08^{Bb}	15.71^{Bb}	22.88^{Cc}

* Figures followed by different capital letters differ significantly at p< 0.05; *** figures followed by different lower-case letters differ significantly at p<0.01.

Table 8. Development of rot (lesion diameter, mm) on quince fruits wound-inoculated with *Monilinia fructigena* (Tiryaki et al., 1994).

Dose (kGy)	Day after irradiation				
	10	14	24	31	39
0	34.87^{a*}	44.95^{a}	59.95^{a}	71.15^{a}	81.55^{a}
1	29.35^{a}	40.70^{a}	55.40^{a}	65.70^{ab}	76.35^{ab}
2	20.00^{b}	25.90^{b}	37.90^{b}	49.15^{c}	59.50^{c}
3	28.10^{ab}	36.00^{ab}	45.90^{ab}	55.30^{bc}	62.00^{bc}

* Figures followed by different letters differ significantly at p< 0.05.

pathogens (Karabulut and Baykal, 2004). The effect of combining low-dose irradiation (1.75 kGy) with MAP on the microbiological and sensory quality of pork chops stored at refrigeration temperatures was studied by Grant and Patterson (1991). The microflora of irradiated MAP pork was almost exclusively composed of lactic acid bacteria, predominantly *Lactobacillus* spp. Modified atmospheres containing either 25 or 50% CO_2, balance N_2, resulted in the best microbial control in irradiated pork held at 4°C, compared to an unirradiated MAP control, and these atmospheres were subsequently used in sensory studies. The atmosphere containing 25% CO_2, 75% N_2 maintained the uncooked color and odour of irradiated pork chops more effectively than 50% CO_2, 50% N_2. Therefore packaging in a modified atmosphere containing 25% CO_2, balance N_2, followed by irradiation to a dose of 1.75 kGy is recommended to improve the microbiological and sensory quality of pork chops.

Combination with cold storage

A few workers reported that it is not possible to inhibit postharvest fungi by irradiation. It delays only fungal development at different degrees (fungustatic effect). Combination of irradiation with cold storage is more promising (Beraha et al., 1960; Tiryaki and Maden, 1991). In a study carried out by Tiryaki (1990), the degree of sensitivity of these storage pathogens on PDA to gamma-rays at 3 to 4°C was found (from resistant to sensitive): *B. cinerea*>*Alternaria tenuissima*>*P. expansum*>*R. stolonifer.*

'Golden Delicious' apple fruit inoculated with *Colletotrichum acutatum* did not decay during storage at 0°C for four months, confirming earlier observations with this pathogen. Therefore, fruit were stored at 20°C for an additional two weeks to allow decay to develop so that the effectiveness of the various treatments could be determined. *P. expansum*, however, caused extensive decay, even under cold storage conditions, indicating that it is a much more aggressive pathogen than *C. acutatum* (Conway et al., 2004).

In a study performed by Tiryaki and Maden (1991), required gamma-irradiation dose was determined for the inhibition of infection in Ankara pears inoculated with *P. expansum, B. cinerea* and *Rhizopus nigricans* at the 0°C and 85 to 90% relative humidity. After 10 days treatment, diameter of lesion was 25.63 mm in unirradiated Ankara pears inoculated with *B. cinerea.* Whereas irradiated with 3 kGy dose lesion diameter was 13.21 mm. After 20 days irradiation these values were 20.88 and 35.04 mm for 3 kGy and control samples, respectively (Table 6). As to come, *P. expansum*, after 10 days treatment, lesion diameters were 12.1, 11.29, 10.13 and 9.67 mm in 0, 1, 2 and 3 kGy treated pears, respectively. Ankara pears inoculated with *B. cinerea*. After 40 days irradiation, lesion diameter was 36.21 and 34.75 mm for 1 kGy and control treatment in Ankara pears inoculated with *P. expansum* (Table 7). This supports stimulative effect of low gamma rays.

The gamma irradiation doses inhibiting decay were determined in apple and quince inoculated with *M. fructigena* (Tiryaki et al., 1994). The diameters of rot on quince which were wound-inocolated with *M. fructigena* at each dose level are shown (Table 8). Although the differences between 2 and 3 kGy was not statisticaly significant at P< 0.05, the most inhibitory irradation dose for quince rot after 39 days was 2 kGy. In general, doses of 1, 2, 3 and 3.5 kGy did not inhibit decay on fruit, but infection was delayed for a certain period (Tiryaki et al., 1994).

Combination with biocontrol agent

Biological control is another alternative to chemical control that shows effectiveness in controlling postharvest diseases. The reduction of decay by biological control is generally more variable than for fungicides since biocontrol is affected more by environmental factors. As fruit mature, higher concentrations of the biocontrol antagonist must be used to achieve the same level of control as on immature fruit (Conway et al., 2005).

D'hallewin et al. (2005) worked on combination of ultraviolet-C irradiation and biocontrol treatments to control decay caused by *P. digitatum* in orange fruit. The combination of the yeast *Candida oleophila* strain '13L' with UV-C irradiation evidenced a synergistic effect in reducing *P. digitatum* mould and only 11% of the artificially inoculated wounds were infected. Adversely, when the bacteria *Bacillus subtilis* strain 'B160' was combined with UV-C irradiation no synergistic effect was achieved. By using only yeast, bacteria or UV-C treatments the decay percentage was reduced by 79.6, 55 and 75%, respectively. The phytoalexin scoparone accumulation was high in all treatments where UV-C was applied but the highest values were found when combined with the yeast. Population growth of bacteria *in vivo* was halved when fruit was irradiated, whereas direct irradiation of bacteria did not affect their growth *in vitro*. An inhibitory effect of the phytoalexin toward the bacteria is suggested as the reason for the growth inhibition *in vivo* when the bacterial treatment was combined with UV-C irradiation.

QUALITY ASSESMENT OF IRRADIATED FRUIT

Beraha et al. (1959) found that textural and skin-color abnormalities (softening and skin browning) were noted following irradiation at 400 000 rep or higher but not at 300 000 rep.

The effect of irradiation doses of 150, 200 and 250 krad on the shelf-life and eating quality of Veteran peaches was studied. These levels did not affect flavor, texture or color as evaluated by a taste panel, but were effective in controlling rot for four weeks (Larmond and Hamilton, 1968).

A study of the effect of gamma irradiation on table grapes has shown that a total dose of 200 krad will increase shelf life by 2 to 3 weeks. This conclusion results from organoleptic and biochemical analyses, and from observation on the colour, flavourand consistency of the grapes, as well as their resistance to attack microorganisms (Donini and Pansolli, 1970).

Previous works demonstrated that it is not possible to kill the post-harvet fungi by irradiation. It delays only fungal development at different degrees (fungustatic effect). Radiation dose required to kill fungi has negative effect in skin colour and texture of stored fruit and vegetables. Therefore, it is important to balance fungal inhibition and organoleptic properties in food irradiation (Beraha et al., 1960; Tiryaki and Maden, 1991).

As stated earlier, rinsing the fruits after SC treatment was an effective method to avoid potential negative effects. But rinsing of SC-treated clementines may result in a significant loss of SC effectiveness against green mold. In order to preserve fruit quality, it was recommended to rinse the fruit after SC treatment (Palou et al., 2007; Smilanick et al., 1999).

In general, the effects of ionizing radiation on the quality of fresh horticultural perishables are affected by factors related to the type of radiation and energy level, the produce itself and the postharvest handling. Although X-irradiation at doses up to 875 Gy followed by either 14 days at 20°C or 60 days at 5°C caused very slight rind pitting, minor decreases in fruit firmness, and modest increases in juice acetaldehyde and ethanol contents, these changes had no practical impact on fruit quality. Rind color, titratable acidity, soluble solids concentration, maturity index and juice yield were not influenced by irradiation. 'Clemenules' can be considered as a clementine cultivar highly tolerant to X-irradiation (Palou et al., 2007).

Physicochemical and orgonoleptic alterations of apple varieties (Golden Delicious, Royal Delicious, Red Delicious and Rich-A-Red) irradiated with 0.1, 0.2, 0.4 and 0.6 kGy and stored at 2 to 4°C for 6 months were investigated and the best results with regard to preservation of organoleptic properties, minimal alteration in texture, amount of total soluble solids, acidity and vitamin-C content were observed in variety Rich-A-Red with 0.1 kGy radiation. It was stated that radiation could be used as an alternative preservation technique for apple varieties (Korel and Orman, 2005).

CONCLUSION

Postharvest losses are very significant issue for storage facilities as indicated above. The lack of an effective postharvest treatment against postharvest decay of fruit highlights the need for developing new control methods (Karabulut and Baykal, 2004). It is therefore necessary to develop alternatives to synthetic chemical control to reduce environmental risks and raise consumer confidence (Conway et al., 2004). Furthermore, concerns about human health risks and protection of the environment associated with fungicide residues, have increased the need for alternatives to fungicide usage (Palou et al., 2007). Although there are various methods to control the postharvest disease, a combination may be needed for better postharvest preservation. In this way, effectiveness of single treatment can be increased; negative impact of each treatment can be minimized by applying low doses of each treatment (Cia et al., 2007).

Food irradiation utilizes a source of ionizing energy that passes through food to destroy harmful bacteria and other organisms. It is capable of improving the safety of many foods, and extending their shelf life.

It is not possible to kill the post-harvet fungi by irradiation. It delays only fungal development at different degrees (fungistatic effect). Radiation dose required to kill fungi has negative effect in skin colour and texture of stored fruit and vegetables. Therefore, it is important to balance fungal inhibition and organoleptic properties in food irradiation studies (Beraha et al., 1960; Tiryaki and Maden, 1991).

In general, the effects of ionizing radiation on the quality of fresh horticultural perishables are affected by factors related to the type of radiation and energy level, the produce itself and the postharvest handling (Palou et al., 2007).

REFERENCES

Anonymous (1987). International Atomic Energy Agency, Food Irradiation Newsletter 11(2):6.

Anonymous (1999). Gıda ışınlama yönetmeliği. Resmi gazete 06.11.1999, 23868.

Anonymous (2000). International atomic energy agency facts about food irradiation. available at: http://www.iaea.org/worldatom/inforesource/other/ food/q&a.html. Accessed May 12.

Anonymous (2011). Hasat sonrası kayıplar. http://www.bahcesel.com/forumsel/taze-meyve-ve-sebzelerin-muhafazasi/20385-taze-meyve-ve-sebzelerde-hasat-sonrasi/ Accessed May 15.

Aziz NH, Souzan RM, Azza AS (2006). Effect of gamma irradiation on the occurrence of pathogenic microorganism and nutritive value off our principal cereal grains. Appl. Radiat. Isot. 64:1555-1562.

Beraha L, Ramsey GB, Smith MA, Wright WR (1959). Effects of gamma radiation on brown rot and *Rhizopus* rot of peaches and the causal organisms. Phytopathology 49:354-356.

Beraha L, Ramsey GB, Smith MA, Wright WR (1960). Gamma radiation dose response some decay pathogens. Phytopathology 50:474-476.

Beraha L (1964). Influence of gamma radiation dose rate on decay of citrus, pears, peaches on Penicillium italicum and botrytis cinerea in-vitro. Phytopathology 54(7):755-759.

Blanchette B, Tourneau DL (1977). The effects of gamma radiation on sclerotia of *Whetzelinia sclerotiorum*. Environ. Exp. Bot. 17:49-54.

Braghini R, Sucupira M, Rocha LO, Reis TA, Aquino S, Correˆa B (2009a). Effects of gamma radiation on the growth of *A.alternata* and on the production of alternariol and alternariolmonomethyl ether in sunflower seeds. Food Microb. 26:927-931.

Braghini R, Pozzi CR, Aquino S, Rocha LO, Correˆa B (2009b). Effects of gamma radiation on the fungus *Alternaria alternata* in artificially inoculated cereal samples. Appl. Radiat. Isotopes 67:1622-1628.

Brennand CP (1995). Food Irradiation The Radiation Information Network. Idaho State University, http://www.physics.isu.edu/radinf/food.htmAccesse May 15, 2011.

CAC (Codex Alimententarius Commission) (2003). Codex general standard for irradiated foods, Codex Stan.106-1983, Rev. 1-2003.

Camili EC, Benato A, Pascholati SF, Cia P (2004). Avaliac¸ ˜ao de irradiac¸ ˜ao UV-C aplicada em p´os-colheita na protec¸ ˜ao de uva 'It´alia' contra *Botrytis cinerea*. Summa. Phytopathologica 30:306-313 (in Portuguese with English abstract).

Cia P, Pascholati SF, Benato EA, Camili EC, Santos CA (2007). Effects of gamma and UV-C irradiation on the postharvest control of papaya anthracnose. Postharvest Biol. Technol. 43:366-373.

Conway WS, Leverentz B, Janisiewicz WF, Blodgett AB, Saftner RA, Camp MJ (2004). Integrating heat treatment, biocontrol and sodium bicarbonate to reduce postharvest decay of apple caused by *Colletotrichum acutatum* and *P. expansum*. Postharvest Biol. Technol. 34:11-20.

Conway WS, Leverentz B, Janisiewicz WF, Saftner RA, Camp MJ (2005). Improving biocontrol using antagonist mixtures with heat and/or sodium bicarbonate to control postharvest decay of apple fruit. Postharvest Biol. Technol. 36:235-244.

Çelebi (2007). İyonlaştırıcı radyasyonun etkiselliği ve ışınlamanın bir gıda işleme ve güvenliği tekniği olarak kullanımı. Vegapaks, www.vegapaks.com.

Darras AI, Joyceb DC, Terrya LA (2010). Postharvest UV-C irradiation on cut Freesia hybrida L. inflorescences suppresses petal specking caused by *Botrytis cinerea*. Postharvest. Biol. Technol. 55:186-188.

D'hallewin GD, Arras G, Venditti T, Rodov V, Ben-Yehoshua S (2005). Combination of ultraviolet-C irradiation and biocontrol treatments to control decay caused by *Penicillium digitatum* in ′Washıngton Navel′ orange fruit. IshsActaHorticulturae 682:V International Postharvest Symposium: 2007-2012.

Donini MLB, Pansolli P (1970). Gamma irradiation of table grapes of the "Hoanes" variety. Food Irrad. 10(4):15-21.

Farkas J (2006). Irradiation for better foods. Trends Food Sci. Technol. 17:148-152.

Georgiev I (1983). The effect of gamma irradiation in conjuction with the Fungicide, Captan on suppressing Gray rot during storage of cv "Bolgar" grapes. Hort. ViticulT. Sci. 20:6 Sofia.

Grant IR, Patterson MF (1991). Effect of irradiation and modified atmosphere packaging on the microbiological and sensory quality of pork stored at refrigeration temperatures. Int. J. Food Sci. Technol. 26(5):507-519.

Harris B (1998). The battle to minimise losses due to mycotoxins. World Poultry Magazine on Production, Processing and Marketing 14(4):52-54.

Jarrett RD (1982). Isotope radiation sources. In: Josephson, E.S., Peterson, M.S.(Eds) Preservation of Food by Ionizing Radiation. CRC Press, Boca rato′ n. pp. 137-163.

Kabak B, Var I (2005). Işınlamanın Küf Gelişimi ve Mikotoksin Kontrolü Üzerine Etkisi. Gıda 30(1):197-201.

Karabulut OA, Baykal N (2004). Integrated control of postharvest diseases of peaches with a yeast antagonist, hot water and modified atmosphere packaging. Crop Protect. 23:431-435.

Kinay P, Mansour MF, Mlikota Gabler F, Margosan DA, Smilanick JL (2007). Characterization of fungicide-resistant isolates of *Penicillium digitatum* collected in California. Crop Prot. 26:647-656.

Korel F, Orman S (2005). Gıda Işınlaması, Uygulamaları ve Tüketicinin Işınlanmış Gıdaya Bakış Açısı. HR.U.Z.F.Dergisi 9(2):19-27.

Kuhn GD, Merkley MS, Dennison RA. (1968). Irradiation inactivation of brown rot infections on peaches. Food Technol. 22(7):91-92.

Larmond E, Hamilton HA (1968). The effect of low level gamma irradiation on peaches. Food Irradiation 8(4): 2-9.

Lennox CL, Spotts RA (2003). Sensitivity of populations of *Botrytis cinerea* from pear-related sources to benzimidazole and dicarboximide fungicides. Plant Dis. 87:645-649.

Loaharanu P (1994). Food irradiation in developing countries a pratical alternative. IAEA Bull. 36:30-35.

Lu JY, Lukombo SM, Stevens C, Khan VA, Wilson CL, Pusey PL, Chaultz E (1993). Low dose UV and gamma radiation on storage rot and physicochemical changes in peaches. J. Food Qual. 16:301-309.

Margosan DA, Smilanick JL, Henson DJ (1997). Combination of hot water and ethanol to control postharvest decay of peaches and nectarines. Plant Dis. 81:1405-1409.

Monk JD, Beuchat LR, Doyle MP (1995). Irradition inactivation of foodborne microorganisms. J. Food Prot. 58(2):197-208.

Nelson K E, Maxie E C, Eukel W (1959). Some studies in the use of ionizing radiation to control *Botrytis* rot in table grapes and strawberries. Phytopathology 49:475-480

Palou L, Smilanick JL, Usall J, Vinas I (2001). Control of postharvest blue and green molds of orange by hot water, sodium carbonate, and sodium bicarbonate. Plant Dis. 85(4):371-376.

Palou L, Usall J, Munoz J, Smilanick JL, Vinas I (2002). Hot water, sodium carbonate, and sodium bicarbonate for the control of postharvest green and blue molds of clementine mandarins. Postharvest Biol. Technol. 24:93-96.

Palou L, Marcilla A, Rojas-Argudo C, Alonso M, Jacas JA, Angel del Rio M (2007). Effects of X-ray irradiation and sodium carbonate treatments on postharvest Penicilliumdecay and quality attributes of clementine mandarins. Postharvest Biol. Technol. 46, 252-261.

Roy MK. (1975). Radiation heat and chemicals combines in the extensi on of shelf life of apples infected with blue mold rot (*Penicilium expansum*), Plant Dis. Rep. 59(1):61-64.

Rustom IYS (1997). Aflatoxin in food and feed: occurrence, legislation and inactivation by physical methods. Food Chem. 59 (1):57-67.

Shea KM, MD, MPH, and the Committee on Environmental Health (2000). Technical Report: Irradiation of Food. American Academy of Pediatrics, Pediatrics. 106(6):1505-1510.

Shirzad BM, Langarek DI (1984). Gamma radiation techonological feasibility of increasing shelf-life of table garpes. ActaAlimentaria. 13:47-64.

Smilanick JL, Margosan DA, Mlikota F, Usall J, Michael IF (1999). Control of citrus green mold by carbonate and bicarbonate salts and the influence of commercial postharvest practices on their efficacy. Plant Dis. 83:139-145.

Smith JS, Pillai S (2004). Irradiation and food safety. Food Technol. 58(11):48-55.

Sommer NF, Fortlage RT, Buckeley MP, Maxie EC (1967). Radiation heat synergism for inactivation of market disease fungi of stone fruits. Phytopathology 57:428-433.

Spalding DH, Reeder WF (1985). Effect of hot water and gamma radiation on postharvest decay of grapefruit. Proc. Fla. State Hortic. Soc. 98:207-208.

TAEK (Türkiye Atom EnerjisiKurumu) (2001). Türkiye Atom Enerjisi Kurumu Ankara Nükleer Tarım ve Hayvancılık Araştırma Merkezi Gıda Işınlama Kurs Notları. 26 Şubat-1 Mart, Ankara.

Tiryaki O (1990). Inhibition of *Penicilllium expansum, Botrytis cinerea*, *Rhizopus stolonifer*, and *Alternaria tenuissima*, which were isolated from Ankara pears, by gamma irradiation J. Turkish. Phytopathol. 19(3):133-140.

Tiryaki O, Maden S (1991). *Penicillium expansum*, *Botrytis cinerea* ve *Rhizopus nigricans* ile Enfekteli Ankara Armutlarında Gamma Radyasyonunu ile Standart Depolama Koşullarında Çürümenin Engellenmesi. VI. Türkiye Fitopatoloji Kongresi. 7-11 Ekim, İzmir.

Tiryaki O, Aydın G, Gürer, M (1994). Post harvest disease control of apple, quince, onion and peach, with radiation treatment. J. Turk. Phytopath. 23 (3):143-152.

Urbain WM (1986). Food Irradiation. Orlando, FL: Academic Press Inc. p. 16.

US FDA (US Food and Drug Administration) (2004). Irradiation in the production, processing and handling of food final rule. Fed. Reg. 69:76844-76847.

World Health Organization (WHO) (1981). Wholesomeness of irradiated food, WHO Technical ReportsSeries, 659 (pp. 34) Geneva: World Health Organisation.

World Health Organization (WHO) (1994). Safety and Nutritional Adequacy of Irradiated Food. Geneva, Switzerland: World Health Organization.

World Health Organization (WHO) (1999). High-dose irradiation: wholesomeness of food irradiated with doses above 10 kGy. Report of a Joint FAO/IAEA/WHO Study Group, Technical Report Series N_890. Geneva, Switzerland:WHO.

5

Drug bioavailability and traditional medicaments of commercially available papaya: A review

P. L. Saran and Ravish Choudhary

Indian Agricultural Research Institute, Regional Station, Pusa, Samastipur (Bihar)-848 125, India.

Papaya (*Carica papaya* L.) is regarded as an excellent source of ascorbic acid, a good source of carotene, riboflavin and a fair source of iron, calcium, thiamin, niacin, pantothenic acid, vitamin B-6 and vitamin K. Each and every part of papaya plant from root to shoot is used for medicament purposes. Ripe papaya fruit is used in jam, jelly, marmalade, puree, wine, nectar, juice, frozen slices, mixed beverages, ice-cream, powder, baby food, cooked in pie, pickled, sweet meat, concentrated and candied items. Young leaves are cooked and eaten like spinach, animal feed, tenderize meat, stomach trouble, purgative effects and abortion may result from consumption of the dried papaya leaves. The flowers are sometimes candied and used for making sprays. Papaya seeds are rich source of amino acids; scented oil was extracted, used in treatment of sickle cell disease and poisoning related disorders. Papain is used in food processing to tenderize meat, clarify beer and juice, produce chewing gum, coagulate milk, prepare cereals, and produce pet food, also to treat wool and silk before dying, de-hair hides before tanning, adjunct in rubber manufacturing and proteolytic enzymes (papain and chymopapain). In folk medicine, latex is used on boils, warts, freckles, abortion, expel roundworms, salt making, relieve asthma stomach troubles, purgative for horses, treatment for genito-urinary ailments, tumer destroying, making herbal tea, digestive and aid in chronic indigestion, weight loss, obesity, arteriosclerosis, high blood pressure, blood purifier and weakening of heart etc. It has also several antibiotic, allergic, anti-nutritional and toxic properties.

Key words: Pawpaw, drug bioavailability, traditional medicaments, industrial uses, papain, nutritive value.

INTRODUCTION

Papaya (*Carica papaya* L.) belongs to the family Caricaceae, one of the most important fruits cultivated throughout the tropical and subtropical regions of the world (Anonymous, 2000). Papaya known by different names in world *viz.* Arabic (fafay, babaya); Burmese (thimbaw); Creole (papayer,papaye); English (bisexual pawpaw, pawpaw tree, melon tree, papaya); Filipino (papaya, lapaya, kapaya); French (papailler, papaye, papayer); German (papaya, melonenbraum); Indonesian (gedang, papaya); Javanese (kates); Khmer (lhong, doeum lahong); Lao (Sino-Tibetan) (houng); Luganda (papaali); Malaya (papaya, betek, ketalah, kepaya); Sinhala (pepol); Spanish (figuera del monte, fruta bomba, papaya, papaita, lechosa); Swahili (mpapai); Thai (ma kuai thet, malakor, loko); Tigrigna (papayo); Vietnamese (du du). In Australia, red and pink-fleshed cultivars are often known as 'papaya' to distinguish them from the yellow-fleshed fruits, known as 'pawpaw'. In India, locally known as pappaiya (Bengali), papeeta (Hindi), papaya (English) and pappali or pappayi (Tamil). It is native of tropical America and introduced from Philippines through Malaysia to India during 16th Century. It is cultivated in

Figure 1. Mature pawpaw tree cv. Pusa Dwarf.

Figure 2. Pawpaw tree showing ripe and unripe fruit of cv. Pusa Majesty.

the world in an area of 3.83 lakh ha with a production of 8.05 million tones. Brazil is one of the largest producer worldwide that continues to show rapid growth. In India, it is cultivated in 73,000 ha with a production of 23.17 lakh tones (Singh et al., 2010). Pawpaw has gained more importance owing to its high palatability, fruitability throughout the year, early fruiting and highest productivity per unit area and multifarious uses like food, medicine and industrial input. Being a highly remunerative and short duration fruit crop, it has tremendous impact on economic and nutritional propitiations. It needs plentiful rainfall or irrigation but must have good drainage. Papayas grow and produce well on a wide variety of soils. The tree (Figure 1) often develops a strong taproot shortly after planting. The well-drained or sandy loam soil with adequate organic matter is most important for the papaya cultivation (Anonymous, 2009).

Pawpaw fruit is consumed at unripe and ripe stages (Figure 2). Unripe fruits are cooked and utilized as vegetables, processed products and as a source of papain (Mendoza, 2007). Ripe papaya is consumed as a fresh fruit and is also used for processing. At unripe stage, the fruit is consumed as a cooked vegetable where papaya is widely grown (Mano et al., 2009). In Thailand, unripe fruits are used as ingredients in papaya salad and cooked dishes (Sone et al., 1998). In Puerto Rico, unripe fruits are canned in sugar syrup and sold either in local markets or exported (Morton, 1987). The preserved unripe papaya fruit, which contains high sugar content, is used as an additive in ice cream. Green papaya fruit must be cooked (often boiled) prior to consumption to denature the papain in the latex (Odu et al., 2006). Ripe papaya fruits and papaya products are consumed by humans for their flavour and nutritional value (Saran, 2010). Unripe papaya fruits are consumed both as a cooked vegetable and processed products (Morton, 1987).

NUTRITIVE VALUE

The main constituent of *Carica papaya* fruit is water like other fruits. The dry matter content increases during fruit development from unripe to ripe stages (Chavasit et al., 2002). Agro-climatic conditions or cultivation practices, climate, seasons, site and cultivars all these factors can influence the nutrient content of papaya (Hardisson et al., 2001; Wall, 2006; Marelli de Souza et al., 2008; Charoensiri et al., 2009). Stages of maturity also affect the nutrient content of fruits like the vitamin C content of pawpaw increases with ripening (Bari et al., 2006). Consequently, when comparing the nutrient content of papaya fruits, it is important to compare fruits harvested and stored under similar conditions (Hernandez et al., 2006).

The major components of papaya dry matter are carbohydrates (USDA, 2009). There are two main types of carbohydrates in papaya fruits, the cell wall polysaccharides and soluble sugars. During an early stage of fruit development, glucose is the main sugar. The sucrose content increases during the ripening process and can reach levels up to 80% of total sugars (Paull, 1993). Among the major soluble sugars in ripe fruits (glucose, fructose and sucrose), sucrose is most prevalent. During fruit ripening, the sucrose content was shown to increase from 13.9 ± 5.0 mg/g fresh weight in green fruit to 29.8 ± 4.0 mg/g fresh weight in ripe fruits (Gomez et al., 2002). The total dietary fiber content of ripe fruit varies from 11.9 to 21.5 g/100 g dry matter (Puwastien et al., 2000; Wills et al, 1986; Saxholt et al, 2008). The crude protein content ranges from 3.74 to 8.26 g/100 g dry matter and aspartic acid is the most abundant amino acid in ripe fruits followed by glutamic acid. chemists of Italy and Somalia collaboratively identified 18 amino acids in seeds as given in descending order *viz.*, glutamic acid, arginine, proline, and aspartic acid in the endosperm; and proline, tyrosine, lysine, aspartic acid, and glutamic acid in the sarcotesta. The

Table 1. Food value per 100 g of edible portion.

Ingredient	Fruit	Leaves
Calories	23.10-25.80	-
Moisture	85.90-92.60 g	83.30%
Protein	0.081-.34 g	5.60%
Fat	0.05-0.96 g	0.4%
Carbohydrates	6.17-6.75 g	8.3%
Crude fiber	0.50-1.30 g	1.0%
Ash	0.31-.66 g	1.4%
Calcium	12.90-40.80 mg	0.406% (CO)
Phosphorus	5.3-22.0 mg	-
Iron	0.25-0.78 mg	0.0064%
Carotene	0.005-0.676 mg	28,900 I.U.
Thiamine	0.021-0.036 mg	-
Riboflavin	0.024-058 mg	-
Niacin	0.23-555 mg	-
Ascorbic acid	35.5-71.3 mg	38.6%
Tryptophan	4-5 mg	-
Methionine	1 mg	-
Lysine	15-16 mg	-
Magnesium	-	0.035%
Phosphoric acid	-	0.225%

major organic acids found in ripe papaya are citric acid (335 mg/100 g FW) followed by L-malic acid (209 mg/100 g FW), qiunic acid (52 mg/100 g FW), succinic acid (52 mg/100 g FW), tartaric acid (13 mg/100 g FW), oxalic acid (10 mg/100 g FW), and fumaric acid (1.1 mg/100 g FW) (Hernandez et al., 2009).

Pawpaw is regarded as an excellent source of vitamin C (ascorbic acid); a good source of carotene, riboflavin and a fair source of iron, calcium, thiamin, niacin, pantothenic acid, vitamin B-6 and vitamin K (Bari et al., 2006; Adetuyi et al., 2008; Saxholt et al., 2008). Carotenoid content (13.80 mg/100 g dry pulp) of papaya is low compared to mango (50 to 260 mg/100 g dry pulp), carrot and tomato (Saran, 2010). The major carotenoid is cryptoxanthin. Carotenoids are responsible for the flesh colour of papaya fruit mesocarp. Red-fleshed papaya fruits contain five carotenoids, *viz.* beta-carotene, beta-cryptoxanthin, beta-carotene-5-6-epoxide, lycopene and zeta-carotene. Yellow-fleshed papaya contains only three carotenoids, *viz.* beta-carotene, beta-cryptoxanthin and zeta-carotene (Chandrika et al., 2003). Pawpaw is a source of vitamin C with amounts varying between the maturation stages (Bari et al., 2006; Hernandez et al., 2006). The total lipid content in ripe papaya fruit varies between 0.92 and 2.2 g/ 100 g dry matter. Papaya contains a low level of fatty acids. Palmitic acid and linolenic acid are two major fatty acids in papaya. Fatty acid composition change during fruit ripening and no significant difference are observed in lipid composition with maturity of papaya fruits. The minimum and maximum levels of constituents in ripe papaya fruit and current mature leaves were given in Table 1. The edible portion of fruit contains macro-minerals include sodium, potassium, calcium, magnesium and phosphorus. The micro-minerals include iron, copper, zinc, manganese and selenium (USDA, 2009).

HOME CONSUMPTION

Fruit

Ripe papaya fruit is most commonly consumed like a melon. It can be peeled, the seeds removed, cut into pieces and served as a fresh fruit. It can also be cut into wedges and then served with lime or lemon. Ripe pawpaw is also used in jam, jelly, marmalade and other products containing added sugar. Other processed products include puree or wine, nectar (Matsuura et al., 2004;), juice, frozen slices or chunks, mixed beverages, papaya powder, baby food, concentrated and candied items (OECD, 2005; OGTR, 2008). Papaya puree is prepared from fully ripe peeled fruit with the seeds removed. Papaya flesh is pulped, passed through a sieve and thermally treated. Papaya puree is an important intermediate product in the manufacture of several products such as beverages, ice cream, jam and jelly (Brekke et al., 1972). Papaya nectar is prepared from papaya puree and consumed either alone or with other fruit juices such as passion fruit juice and pineapple juice (Brekke et al., 1972). It should be stored at or below 24°C to maintain acceptable quality (Brekke et al., 1976). Ripe papayas are most commonly eaten fresh, merely peeled, seeded, cut in wedges and served with a half or quarter of lime or lemon. Sometimes a few seeds are left attached for those who enjoy their peppery flavor but not many should be eaten because the seed extract of papaya causing sterility in mammals also. The flesh is often cubed and served in fruit salad or fruit cup. Firm-ripe papaya may be seasoned and baked for consumption as a vegetable. Ripe flesh is commonly made into sauce for short cake or ice cream sundaes, or is added to ice cream just before freezing; or is cooked in pie, pickled, or preserved as marmalade or jam. Papaya and pineapple cubes, covered with sugar syrup, may be quick-frozen for later serving as dessert. Half-ripe fruits are sliced and crystallized as a sweet meat for consumption. Unripe papaya is never eaten raw because of its latex content. Raw green papaya is frequently used in Thai and Vietnamese cooking. Even for use in salads, it must first be peeled, seeded, and boiled until tender, then chilled. Green papaya is frequently boiled and served as a vegetable. Cubed green papaya is cooked in mixed vegetable soup. Green papaya is commonly canned in sugar syrup in Puerto Rico for local consumption and for export. Green papayas for canning in Queensland must be checked for nitrate levels. High nitrate content causes damage of ordinary cans, and all papayas with over 30 ppm nitrate must be packed in cans lacquered on the inside. Australian growers are hopeful that the papaya can be bred for low nitrate uptake. A lye process for batch peeling of green papayas has proven feasible in Puerto Rico. The fruits may be immersed in boiling 10% lye solution for 6 min, in a 15% solution for 4 min, or a 20% solution for 3 min. They are then rapidly cooled by a cold water bath and then sprayed with water to remove all softened tissue. Best proportions are 0.45 kg of fruit for every 3.8 L of solution.

Drying and freeze drying are used to reduce the moisture content of papaya chunks and slices. Powdered or dried papaya can be used as a flavoring agent, meat tenderizer or as an ingredient in soup mixes (Sing field, 1998). Papaya seeds are a good source of 18 amino acids and edible oil. Seeds are sometimes also used to adulterate with whole black pepper (Morton, 1987).

Leaves, pomace and fruit skin

Young leaves are cooked and eaten like spinach in the East Indies. Mature leaves are bitter and must be boiled with a change of water to eliminate much of the bitterness. Crushed leaves may be used to tenderize meat; however, stomach trouble, purgative effects and abortion may result from consumption of the dried papaya leaves (Morton, 1987). Pawpaw leaves contain the bitter alkaloids; carpaine and pseudocarpaine, which act on the heart and respiration like digitalis, but are destroyed by heat. In addition, two previously undiscovered major piperideine alkaloids, dehydrocarpaine I and II, more important than carpaine, were reported. Papaya pomace, skins, leaves, and other by-products of papaya processing may find use in animal feed applications (Babu et al., 2003; Fouzder et al., 1999; Munguti et al., 2006; Reyes and Fermin, 2003; Alobo, 2003; Ulloa et al., 2004).

Flower and stem

Sprays of male flowers are sold in Asian and Indonesian markets and in New Guinea for boiling with several changes of water to remove bitterness and then eating as a vegetable. In Indonesia, the flowers are sometimes candied. Young stems are cooked and served in Africa. Older stems, after peeling, are grated, the bitter juice squeezed out, and the mash mixed with sugar and salt.

INDUSTRIAL USES

Many biologically active phytochemical from different parts of papaya tree (latex, seed, leaf, root, stem, bark and fruit) have been isolated from papaya and studied for their potency (Table 2). Some of these parts are known to be analgesic, amoebicidic, antibacterial, cardiotonic, cholagogue, digestive, emenagogue, febrifuge, hypotensive, laxative, pectoral, stomachic and vermifugic (Boshra and Tajul, 2013).

Seed

Papaya seeds are rich source of amino acids especially in the sarcotesta. A yellow to brown, faintly scented oil was extracted from the sundried, powdered seeds of unripe papayas at the Central Food Technological Research Institute, Mysore, India. White seeds yielded 16.1% and black seeds 26.8% and it were suggested that the oil might have edible and industrial uses. The seeds are used in treatment of sickle cell disease. Air dried papaya seeds with honey showed significant effect on human intestinal parasites without significant side effect. Consumption of papaya seed is cheap, natural, harmless, readily available, mono-therapeutic, and prevent against intestinal parasitosis especially in tropical communities.

In India, the fruit is widely classified as harmful in pregnancy, hence pregnant women are strictly forbidden from eating it for fear of its teratogenic and abortifacient effects (Adebiyi et al., 2003). Chinoy et al. (2006) proved the anti-fertility, anti-implantation and abortifacient properties of extracts from papaya seeds. It has been established in males that the seeds of *C. papaya* are potential anti-fertility drugs (Lohiya et al., 2005). Pawpaw seeds are used to produce an indigenous Nigerian food condiment called 'daddawa', the Hausa word for a fermented food condiment (Dakare, 2004). Fermented seeds have no effects on litters of rats (Abdulazeez et al., 2009), whereas, those effects were apparent when the unfermented extract was administered (Abdulazeez, 2008). Normal consumption of ripe papaya during pregnancy may not be dangerous, however unripe or semi-ripe papaya (which contains high amount of latex that produces marked uterine contraction) could be unsafe for consumption during pregnancy (Krishna et al., 2008). Antihelmintic activity of papaya seed has been predominantly attributed to carpaine (an alkaloid) and carpasemine (later identified as benzyl thiourea). Carpaine has an intensively bitter taste and a strong depressant action on health. It is present not only in papaya fruit and seed but also in its leaves. Benzylisothiocyanate (BITC), the main bioactive compound in *C. papaya* seeds (Kermanshai et al., 2001) has been shown to be responsible for the anti-fertility effect (Adebiyi et al., 2003). BITC is capable of damaging the endometrium, making the uterus non-receptive and, thus, affecting adversely the implantation (Adebiyi et al., 2003).

Green fruit

Papaya latex is obtained by cutting the green fruit surface

Table 2. Constituents and medicinal uses of different parts of the papaya tree (Krishna et al., 2008; Boshra and Tajul, 2013).

Part	Constituents	Medicinal uses
Fruits		
Ripe fruits	Protein, fat, fiber, carbohydrates, mineral: calcium, phosphorous, iron, vitamin-C, thiamine, riboflavin, niacin and carotene, amino acids.	Stomachic, digestive, carminative diuretic, dysentery and chronic diarrhoea, expectorant, sedative and tonic, relieves obesity, bleeding piles, wound of urinary tract, ringworm and skin disease psoriasis
Unripe fruits	Citric and malic acids.	Laxative, diuretic, dried fruit reduces enlarged spleen and liver, use snakebit to remove poison, abortifaciant, anti- implantation activity and antibacterial activity.
Juice	N-butyric acids, n-hexanoic and n-octanoic acids, lipids, myristic, planets, stars, linolec, linolenic, *cis*-vaccenic and oleic acid	Flower: Jaundice, emmenagogue, febrifuge and pectoral properties
Seed	Fatty acids, crude protein, crude fiber, papaya oil, carpaine, benzylisothiocynate, benzylglucosinolate, glucotropacolin, bemzylthiourea, hentriacontane, ß-sitostrol, caressing and enzyme myrosin.	Seed: Carminative, emmenagogue, vermifuge, abortifaciant, counter irritant, as paste in the treatment of ringworm, pasoriasis and anti-fertility agent.Seed juice: Bleeding piles and enlarged liver and pectoral properties
Root	Carposide and enzyme myrosin	Abortifacient, diuretic, checking irregular bleeding from the uterus, piles, antifungal activity.
Leaves	Alkalodis carpain, pseudocarpain and dehyrocarpaine, choline, carposide, vitamin C and E.	Young leaves as vegetable, Jaundice (fine paste), urinary complaints, gonorrhoea (infusion) dressing wound fresh leave and antibacterial.
Bark	ß-sitosterol, glucose, fructose, sucrose and xylitol.	Jaundice, anti-haemolytic activity, STD, store teeth (inner bark) and anti-fungal activity.
Latex	Proteolytic enzymes, papain, chemopapain, glutamine, cyclortransferase, chymopapains A, B, C, peptidase A, B and lysozymes.	Anathematic, relieves dyspepsia, cures diarrhoea, pain of burn, topical use, bleeding haemorrhoids, stomachic and whooping cough.

with containers over a couple of days. The latex is then sun dried or oven dried, and ground into powder. A proteolytic enzyme, papain, is purified from papaya latex and used in the food and feed industries, as well as the pharmaceutical and cosmetic industries (OGTR, 2008). Papain is used in food processing to tenderize meat, clarify beer and juice, produce chewing gum, coagulate milk, prepare cereals, and produce pet food (Morton, 1987). The latex of the papaya plant and its green fruits contains two proteolytic enzymes, papain and chymopapain. The latter is most abundant but papain is twice as important. The latex is obtained by making incisions on the surface of the green fruits early in the morning and repeating every 4 or 5 days until the latex ceases to flow. The tool is of bone, glass, sharp-edged bamboo or stainless steel (knife or razor blade). Ordinary steel stains the latex. Tappers hold a coconut shell, clay cup, or glass, porcelain or enamel pan beneath the fruit to catch the latex, or a container like an "inverted umbrella" is clamped around the stem. The latex coagulates quickly and, for best results, is spread on fabric and oven-dried at a low temperature, then ground to powder and packed in tins. Sun-drying tends to discolor the product. One must tap 1,500 average-size fruits to gain 680 g of papain. The lanced fruits may be allowed to ripen and can be eaten locally, or they can be employed for making dried papaya "leather" or powdered papaya, or may be utilized as a source of pectin. Because of its papain content, a piece of green papaya can be rubbed on a portion of tough meat to tenderize it. Sometimes a chunk of green papaya is cooked with meat for the same purpose. One of the best known uses of papain is in commercial products marketed as meat tenderizers, especially for home use. A modern development is the injection of papain into beef cattle a half-hour before slaughtering to tenderize more of the meat than would normally be tender. Papain treated meat should never be eaten but should be cooked sufficiently to inactivate the enzyme. The tongue, liver and kidneys of injected animals must be consumed quickly after cooking or utilized immediately in food or feed products, as they are highly perishable.

Papain has many other practical applications. It is used to clarify beer, also to treat wool and silk before dying, to de-hair hides before tanning, and it serves as an adjunct in rubber manufacturing. It is applied on tuna liver before extraction of the oil which is thereby made richer in vitamins A and D, It enters into toothpastes, cosmetics and detergents, as well as pharmaceutical preparations to aid digestion. Carpaine, an alkaloid found in papaya leaves, has also been used for medicinal purpose (Sankat and Maharaj, 2001).

Papaya can be used as a diuretic (the roots and leaves), anthelmintic (the leave and seed) and to treat bilious conditions (the fruit). Parts of the plant are also used to combat dyspepsia and other digestive disorders and a liquid portion has been used to reduce enlarged tonsils. In addition, the juice is used for warts, cancers, tumors, corns and skin defects while the root is said to help tumors of the uterus. Root infusion is also used for syphilis, and the leaf is smoked to relieve asthma attacks. The papaya eating prevents rheumatism and the latex is used for psoriasis, ringworm and the removal of cancerous growth (Nwofia et al., 2012).

Papain has been employed to treat ulcers, dissolve membranes in diphtheria, and reduce swelling, fever and adhesions after surgery. With considerable risk, it has been applied on meat impacted in the gullet. Chemopapain is sometimes injected in cases of slipped spinal discs or pinched nerves. Precautions should be taken because some individuals are allergic to papain in any form and even to meat tenderized with papain.

ANTIBIOTIC ACTIVITY

The extracts of ripe and unripe papaya fruits and of the seeds are active against gram-positive bacteria. Strong doses are effective against gram-negative bacteria. The substance has protein-like properties.

The fresh crushed seeds yield the aglycone of glucotropaeolin benzyl isothiocyanate (BITC) which is bacteriostatic, bactericidal and fungicidal. A single effective does is 5 g seeds (25-30 mg BITC) for the same. Papaya also found effective in curing post-operative infection in a kidney-transplant patient by strip laid on the wound and left for 48 h.

Allergens

The flowers pollen has induced severe respiratory reactions in sensitive individuals (Blanco et al., 1998). Papaya pollen in papaya cultivating areas can contribute to aeropollen and aeroallergen loads as reported by Chakraborty et al. (2005). Papaya contains four cysteine endopeptidases including papain, chymopapain, glycyl endopeptidase and caricain. Papain is commonly found in papaya latex (Azarkan et al., 2003). The recorded level of papain in papaya latex is 51,000 to 135,000 mg/kg (OGTR, 2008). Papain can also induce IGE-mediated allergic reactions through oral, respiratory or contact routes of exposure. The typical symptoms include bronchial asthma, rhinitis or both (Van Kampen et al., 2005). One case of a life threatening anaphylaxis due to occupational exposure to papain was also reported (Freye, 1988). Thereafter, such people react to contact with any part of the plant and to eating ripe papaya or any food containing papaya, or meat tenderized with papain. Skin irritation in papaya harvesters because of the action of fresh papaya latex, and of the possible hazard of consuming undercooked meat tenderized with papain.

Anti-nutrients

Peel and pulp of ripe papaya fruits contains low amounts of anti-nutrients like tannin (10.16 mg/100 g of dry matter), phytate (3.29 mg/100 g of dry matter) and oxalate (1.89 mg/100 g of dry matter) creating incopatability problems as reported by Onibon et al. (2007).

Toxicants

Carpaine is a major alkaloid found in various parts of papaya, but is primarily found in leaves (Morton, 1987; Duke, 1992; Krishna et al., 2008). The major natural toxicants found in papaya are benzylglucosinolate (BG), benzyl isothiocyanate (BITC) and alkaloids. Fruit and seed extracts have pronounced bactericidal activity. The seeds of unripe fruits are rich in benzyl isothiocyanate, a sulphur containing chemical that has been reported to be an effective germicide and insecticide. These substances are important for plant natural defense mechanisms (El Moussaoui et al., 2001). Although both BG and BITC are found in papaya peel, pulp and seed, the highest levels of BG and BITC are found in seeds, 1269.3 and 461.4 µmol/100 g fresh weights, respectively. The levels of BG and BITC in papaya pulp were <3.0 µmol/100 g fresh weight (Nakamura et al., 2007). The concentration of BITC decreases in pulp and increases in seeds during fruit ripening (Tang, 1971).

FOLK USES

In folk medicine, the fresh latex is smeared on boils, warts and freckles and given as a vermifuge in African countries. In India, it is applied on the uterus as an irritant to cause abortion. The unripe fruit is sometimes hazardously ingested to achieve abortion. Seeds, too, may bring on abortion. The root is ground to a paste with salt, diluted with water and given as an enema to induce abortion. A root decoction is claimed to expel roundworms. Roots are also used to make salt.

Crushed leaves smeared around tough meat will tenderize it in overnight. The leaf also functions as a vermifuge and a primitive soap substitute in laundering. Dried leaves have been smoked to relieve asthma or as a tobacco substitute. It also infusion is taken for stomach troubles in Ghana and they say it is purgative and may cause abortion. Packages of dried, pulverized leaves are sold by "health food" stores for making tea, despite the fact that the leaf decoction is administered as a purgative for horses in the Ivory Coast. It is also used as treatment for genito-urinary ailments. The leaf tea or extract has reputation as tumor destroying agent (Walter, 2008). The fresh green tea is act as antiseptic and dried leaves are best as a tonic and blood purifier (Nwofia et al., 2012). The tea also promote digestive system and aid in chronic indigestion, weight loss, obesity, arteriosclerosis, high blood pressure and weakening of heart (Mantok, 2005).

REFERENCES

Abdulazeez AM, Ameh DA, Ibrahim S, Ayo J, Ambali SF (2009). Effect of fermented and unfermented seed extracts of *Carica papaya* on pre-implantation embryo development in female Wistar rats (*Rattus norvegicus*). Scientific Res. Essay 4(10):1080-1084.

Abdulazeez MA (2008). Effect of fermented and unfermented seed extract of *Carica papaya* on implantation in Wistar rats (*Rattus norvegicus*). Thesis submitted to Department of Biochemistry, A.B.U Zaria

Adebiyi A, Adaikan PG, Prasad RNV (2003). Tocolytic and toxic activity of papaya seed extract on isolated rat uterus. Life Sci. 74:581-592.

Adetuyi FO, Akinadewo LT, Omosuli SV, Lola A (2008). Antinutrient and antioxidant quality of waxed and unwaxed pawpaw *carica papaya* fruit stored at different temperatures. Afr. J. Biotech. 7:2920-2924.

Alobo AP (2003). Proximate composition and selected functional properties of defatted papaya (*Carica papaya* L.) kernel four. Plant Food Hum. Nutr. 58:1-7.

Anonymous (2000). Organic farming in the tropics and subtropics (exemplary description of 20 crops). Naturlande.V– 1st edition.

Anonymous (2009). Cultivating papayas. Department of Agriculture, Forestry and Fisheries, Republic of South Africa.

Azarkan M, Moussaoui A, Van Wuytswinkel D, Dehon G, Looze Y (2003). Fractionation and purification of the enzymes stored in the latex of *Carica papaya.* J. Chromat. 790:229-238.

Babu AR, Rao DS, Parthasarathy M (2003). *In sacco* dry matter and protein degradability of papaya (*Carica papaya*) pomace in buffaloes. Buffalo Bull. 22:12-15.

Bari L, Hassen P, Absar N, Haque ME, Khuda MIIE, Pervin MM, Khatun S, Hossain MI (2006). Nutritional analysis of two local varieties of papaya (*Carica papaya*) at different maturation stages. Pak. J. Biol. Sci. 9:137-140.

Boshra V, Tajul AY (2013). Papaya - an innovative raw material for food and pharmaceutical processing industry. Health Environ. J. 4(1):68-75.

Brekke JE, Cavaletto CG, Nakayama TOM, Suehina R (1976). Effects of storage temperature and container lining on some quality attributes of papaya nectar. J. Agric. Food Chem. 24:341-343.

Brekke JE, Chan Jr HT, Cavaletto CG (1972). Papaya puree: a tropical flavor ingredient. Food Prod. Devel. 6:36-37.

Chakraborty P, Ghosh D, Chowdhury I, Roy I, Chatterjee S, Chanda S, Gupta-Bhattacharya S (2005). Aerobiological and immunochemical studies on *Carica papaya* L. pollen: an aeroallergen from India. Clin. Allergy 60:920-926.

Chandrika UG, Jansz ER, Wickramasinghe SN, Warnasuriya ND (2003). Carotenoids in yellow- and red-fleshed papaya (*Carica papaya* L). J. Sci. Food Agri. 83:1279-1282.

Charoensiri R, Kongkachuichai R, Suknicom S, Sungpuag P (2009). Beta-carotene, lycopene, and alpha-tocopherol contents of selected thai fruits. Food Chem. 113:202-207.

Chavasit V, Pisaphab R, Sungpung P, Jittinandana S, Wasantwisut E (2002). Changes in β-carotene and vitamin a contents of vitamin a-rich foods in thailand during preservation and storage. J. Food Sci. 67:375-379.

Chinoy NJ, Dilip T, Harsha J (2006). Effect of *Carica papaya* seed extract on female rat ovaries and uteri. Phytother. Res. 9(3): 169-165.

Dakare M (2004). Fermentation of *Carica papaya* seeds to be used as "daddawa". An MSc thesis; Department of Biochemistry, A.B.U Zaria.

Duke JA (1992). Handbook of phytochemical constituents of GRAS herbs and other economic plants. CRC Press, Ann Arbor, MI, pp. 136-137.

El Moussaoui A, Nijs M, Paul C, Wintjens R, Vincentelli J, Azarkan M, Looze Y (2001). Revisiting the enzymes stored in the laticifers of *Carica papaya* in the context of their possible participation in the plant defence mechanism. Cell. Mol. Life Sci. 58:556-570.

Fouzder SK, Chowdhury SD, Howlider MAR, Podder CK (1999). Use of dried papaya skin in the diet of growing pullets. Brit. Poult. Sci. 40:88-90.

Freye HB (1988). Papain anaphylaxis: a case report. Allergy Proc. 9:571-574.

Gomez M, Lajolo F, Cordenunsi B (2002). Evolution of soluble sugars during ripening of papaya fruit and its relation to sweet taste. J. Food Sci. 67:442-447.

Hardisson A, Rubio C, Baez A, Martin MM, Alvarez R (2001). Mineral composition of the papaya (*carica papaya* variety sunrise) from tenerife island. Eur. Food Res. Tech. 212:175-181.

Hernandez Y, Lobo MG, Gonzalez M (2006). Determination of vitamin c in tropical fruits: a comparative evaluation of methods. Food Chem. 96:654-664.

Hernandez Y, Lobo MG, Gonzalez M (2009). Factors affecting sample extraction in the liquid chromatographic determination of organic acids in papaya and pineapple. Food Chem. 114:734-741.

Kermanshai R, McCarry BE, Rosenfeld J, Summers PS, Weretilnyk EA, Sorger GJ (2001). Benzylisothiocyanate is the chief or sole anthelmintic in papaya seed extracts. Phytochemistry 57(3):427-435.

Krishna KL, Paridhavi M, Patel JA (2008). Review on nutritional and pharmacological properties of papaya (*Carica papaya* Linn.). Nat. Prod. Radian. 7:364-373.

Lohiya NK, Pathak N, Mistra PK, Maniovannan B, Bhande SS, Panneerdoss S, Sriram S (2005). Efficacy trial on the purified compounds of the seeds of *Carica papaya* for male contraception in albino rats. Reprod. Toxicol. 20(1):135-148.

Mano R, Ishida A, Ohya Y, Todoriki H, Takishita S (2009). Dietary intervention with okinawan vegetables increased circulating endothelial progenitor cells in healthy young woman. Atherosclerosis 204:544-548.

Mantok C (2005). Multiple usage of green papaya in healing a Tao garden. Tao garden health spa and resort, Thailand. Retrieved from: www.tao.garden.com.

Marelli de Souza L, Ferreira KS, Chaves JBP, Teixeira (2008). L-Ascorbic acid, β-carotene and lycopene content in papaya fruits (*Carica papaya*) with or without physiological skin freckles. Sci. Agricola *(Piracicaba, Braz.)* 65:246-250.

Matsuura FCAU, Folegatti MIDS, Cardoso RL, Ferreira DC (2004). Sensory acceptance of mixed nectar of papaya, passion fruit and acerola. Sci. Agricola (Piracicaba, Braz.) 61:604-608.

Mendoza EMT (2007). Development of functional foods in the Philippines. Food Sci. Tech. Res. 13:179-186.

Morton J (1987). Papaya. In: Fruits of warm climates. Morton JF, Miami FL (Eds.). Available on the website of the New Crop Resource Online Programme, Purdue University,http://www.hort.purdue.edu/newcrop/morton/papaya_ars.html, pp. 336-346.

Munguti JM, Liti DM, Waidbacher H, Straif M, Zollitsch W (2006). Proximate composition of selected potential feedstuffs for Nile tilapia (*Oreochromis niloticus* L.) production in Kenya. Die Bodenkultur

57:131-141.

Nakamura Y, Yoshimoto M, Murata Y, Shimoishi Y, Asai Y, Park EY, Sato K, Nakamura Y (2007). Papaya seed represents a rich source of biologically active isothiocyanate. J. Agric. Food Chem. 55:4407-4413.

Nwofia GE, Ogimelukwe P, Eji C (2012). Chemical composition of leaves, fruit pulp and seed in some morphotypes of *C. papaya* L. morphotypes. Int. J. Med. Arom. Plant 2:200-206.

Odu EA, Adedeji O, Adebowale A (2006). Occurrence of hermaphroditic plants of *Carica papaya* L. (Caricaceae) in Southwestern Nigeria. J. Plant Sci. 1:254-263.

OECD (2008). Consensus document on compositional considerations for new varieties of tomato: key food and feed nutrients, toxicants and allergens. Series on the Safety of Novel Foods and Feeds, OECD Environment Directorate, Paris 17.

OECD (Organisation for the Economic Cooperation and Development) (2005). Consensus document on the biology of papaya (*Carica papaya*). Series on Harmonisation of Regulatory Oversight in Biotechnology, OECD Environment Directorate, Paris, P. 33.

OGTR (Office of the Gene Technology Regulator, Australia) (2008). The biology of *Carica papaya* L. (papaya, papaw, paw paw), Version 2: February 2008, Australian Government, Dpt. of Health and Ageing, OGTR, website http://www.ogtr.gov.au/internet/ogtr/publishing.nsf/Content/papaya/$FILE/biologypapaya08.pdf.

Onibon VO, Abulude FO, Lawal LO (2007). Nutritional and antinutritional composition of some nigerian fruits. J. Food Tech. 5:120-122.

Paull RE (1993). Pineapple and papaya in biochemistry of fruit ripening. Chapman and Hall, Boundary Row, London.

Puwastien P, Burlingame B, Raroengwichit M, Sungpuag P (2000). ASEAN Food Composition Tables of Nutrition. Mahidol University, Thailand.

Reyes OS, Fermin AC (2003). Terrestrial leaf meals or freshwater aquatic fern as potential feed ingredients for farmed abalone *Haliotis asinina* (Linnaeus 1758). Aquacul. Res. 34:593-599.

Sankat CK, Maharaj R (2001). Papaya. In: Postharvest physiology and storage of tropical and subtropical fruits, Mitra S (Ed.). Faculty of Horticulture, CAB International, West Bengal, India, pp. 167-185.

Saran PL (2010). Screening of papaya cultivars under Doon Valley conditions. Pantnagar J. Res. 8(2):246-47.

Saxholt E, Christensen AT, MØller A, Hartkopp HB, Hess Ygil K, Hels OH (2008). Danish food composition databank. revision 7, Department of Nutrition, National Food Institute, Technical University of Denmark, website: http://www.foodcomp.dk/.

Sing field P (1998). Papaya and Belize. Belize Development Trust, website http://www.belize1.com/BzLibrary/trust19.html, 19.

Singh K, Ram M, Kumar A (2010). Forty years of papaya research at pusa, bihar, india. Acta Hort. 851:81-88.

Sone T, Sakamoto N, Suga K, Imai K, Nakachi K, Sonlkin P, Sonklin O, Lipigorngoson S, Limtrakul P, Suttajit M (1998). Comparison of diets among elderly female residents in two suburban districts in Chiang Mai Provence, Thailand, in dry season –survey on high- and low-risk districts of lung cancer incidence. Appl. Human Sci. 17:49-56.

Ulloa JB, van Weerd JH, Huisman EA, Verreth JAJ (2004). Tropical agricultural residues and their potential uses in fish feeds: the costa rican situation. Waste Manage. 24:87-97.

USDA (United States Department of Agriculture) (2009). Agricultural Research Service, National Nutrient Database for Standard Reference, Release 22, Nutrient Data Laboratory Home Page, http://www.ars.usda.gov/ba/bhnrc/ndl.

Van Kampen V, Merget R, Bruning T (2005). Occupational allergies to papain. Pneumologie 59:405-410.

Wall MM (2006). Ascorbic acid, vitamin a and mineral composition of banana (*Musa* sp.) and papaya (*Carica papaya*) cultivars grown in Hawaii. J. Food Comp. Anal. 19:434-445.

Walter L (2008). Cancer remedies" retrieved from: health-science-sprite.com/cancer6-remedies.

Wills RBH, Lim JSK, Greenfield H (1986). Composition of Australian foods -31. Tropical and Sub-tropical Fruit. Food Tech. Australia 38:118-123.

6

Purification and characterisation of a plant peroxidase from rocket (*Eruca vesicaria* sbsp. Sativa) (Mill.) (syn. *E. sativa*) and effects of some chemicals on peroxidase activity *in vitro*

Hayrunnisa Nadaroglu[1], Neslihan Celebi[1], Nazan Demir[2] and Yasar Demir[2]

[1]Department of Food Technology, Erzurum Vocational Training School, Ataturk University, 25240 Erzurum, Turkey.
[2]Department of Chemistry, Science Faculty, Mugla University, 48000 Mugla, Turkey.

Rocket (*Eruca vesicaria* sbsp. Sativa) (Mill.) (syn. *E. sativa*) was grown and used widely in Turkey as a garnish in salads. A peroxidase (POD) from leaves of rocket (*Eruca vesicaria* sbsp. *Sativa*) was purified using sequential $(NH_4)_2SO_4$ precipitation, CM-Sephadex and Sephacryl S-200 chromatographies. A peroxidase (POD) was purified 220.3-fold from the Rocket (*E. vesicaria* sbsp. *Sativa*) with an overall yield of 80.79%. The purified enzyme has an optimum pH, 6.0 and its optimum temperature was 40°C. The V_{max} and K_M values were determined by Lineweaver-Burk graphics using different substrates. The purification degree and the molecular mass of the enzyme (34 kDa) were determined by SDS-PAGE and gel filtration chromatography. POD enzyme activity was strongly inhibited by Ca^{2+}, Mn^{2+}, Hg_2^{2+}, Zn^{2+} and Fe^{2+} as metal ions and SDS, EDTA, ascorbic acid, dithioeritritol as chemicals. But, Ni^{2+}, Co^{2+}, Cu^{2+} slightly activated the enzyme. They inhibited in the different range of peroxidase activity. Changes of POD enzyme's kinetic parameters were most important during chemicals and metal ions metabolism, because they were risk for environmental pollution.

Key words: Rocket (*Eruca vesicaria* sbsp. Sativa), peroxidase (POD), purification, metal ions.

INTRODUCTION

Peroxidase (POD) (EC 1.11.1.7), an oxidoreductase enzyme catalyzes reactions between compounds which hydrogen atoms tend to give and H_2O_2 as the receiver of atoms. POD catalyzes the oxidation of the organic and inorganic substrates by using hydrogen peroxide. In addition, POD also catalyzes dehydrogenation reaction of a large number of aromatic compounds such as phenols, hydroquinone and hidrokinonid amines (Pütter and Becker, 1987; Van Huytstee, 1987). Peroxidases examined in different groups according to the arrangement of amino acids and structure as animal peroxidases, plant, fungal and bacterial peroxidases. These species of peroxidase is due to the differences of amino acid sequences (Welinder, 1979). In plants, peroxidase enzyme has vital functions such as hormonal regulation, defense mechanisms, lignin biosynthesis and adjustment of the amount of indole acetic acid during the catch up the fruits and vegetables (Welinder, 1979; Wakamatsu and Takahama, 1993; Agostini et al., 1997; Adams, 1978; Duarte-Vazquez et al., 2001). In the contemporary world, environment pollution is one of the most important problems of the living that as a result, it constitute the extreme penetration growth, rapid urbanization and advanced technology, and it is threaten to the natural resources. Heavy metal species are some of the most common pollutants that are found in industrial waste-waters. Because of their toxicity, these species can have a serious impact if released into the environment as a result of bioaccumulation and they may be extremely toxic even in trace quantities.

High concentration of heavy metals in the environment can be detrimental to a variety of living species (Mahvi and Bazrafshan, 2007; Nadaroglu and Kalkan, 2012; Nadaroglu et al., 2010). Rocket (*Eruca vesicaria* sbsp. Sativa) (Mill.) (syn. *E. sativa*) is in the family Brassicaceae and it is widely grown in Turkey. In Turkey, rocket is existing as wild in nature as well as are being presently cultivated widely for agricultural purposes. Rocket was often used as garnish Turkey and rocket salad, but it was cooked as food in some countries. This plant is also used as medicine in the stomach disease among the people (Alqasoumi et al., 2009; Lamy et al., 2008). The present investigation reports the isolation, purification and biochemical characterization of peroxidase from rocket leaves. In addition, in the second phase of the research, it was investigated the effects of metal ions, detergent residues and some chemicals on the activity of purified enzyme, because the plant had a very high risk of exposure to these chemicals in the environmental conditions.

MATERIALS AND METHODS

Chemical

Guaiacol, 2,2'-azino-bis(3-ethylbenzothiazoline-6-sulphonic acid (ABTS), pyrogallol, 4-methylcatechol, hydrochinon (1,4-dihydroxybenzol), ethanol, 2-propanol, sodium acetate (CH_3COONa), bovine serum albumin (BSA), CM-sephadex, and sephacryl S-200, ethylenediaminetetraacetic acid (EDTA), dithiothreitol, *β*-mercaptoethanol and agents for SDS-PAGE were purchased from Sigma (USA). Ammonium sulfate [$(NH_4)_2SO_4$], trichloroacetic acid (TCA, 99%), sodium chloride (NaCl) and potassium hydrogen phosphate were purchased from Merck (Darmstadt, Germany). All other chemicals were of analytical grade.

Plant material and storage conditions

Rocket (*Eruca vesicaria* sbsp. *Sativa*) was collected from the Erzurum region of Turkey and was stored at -20°C till further use.

Purification of peroxidase enzyme

All procedure was carried out at 0 to 4°C unless otherwise stated, and the working buffer was 50 mM phosphate buffer (pH 7).

Preparation of crude extract

In the study, leaves of rocket (*Eruca vesicaria* sbsp. *Sativa*) (20 g) were ground in liquid N_2 and then homogenized in a blender with 50 ml of 50 mM KH_2PO_4 (pH: 7) buffer including 0.5% PVP by shaking and centrifuged at 5,000x*g* for 30 min. The homogenates were centrifuged and precipitates were removed. For the purification of the peroxidase enzyme, the following procedure was implemented (Havir and Mchale, 1987; Nadaroglu, 2009).

Ammonium sulfate fractionation

The collapse of $(NH_4)_2SO_4$ was done from 0 to 90% in supernatant with the internals of 0 to 10, 10 to 20, 20 to 30, 30 to 40, 40 to 50, 50 to 60, 60 to 70, 70 to 80 and 80 to 90. Significant activity was not observed below at a range of 0 to 40% $(NH_4)_2SO_4$. The majority of activity was found in the 40 to 60% precipitate. Solid $(NH_4)_2SO_4$ was added to the supernatant to increase the concentration of $(NH_4)_2SO_4$ from 40% of the fraction to 60%. After mixing it in an ice-bath for 1 h with magnetic stirring, it was centrifuged (10,000x*g*, 30 min and 4°C). The supernatant was discarded and the precipitate was dissolved in 0.1 M KH_2PO_4 (pH: 7) buffer and dialyzed against the same buffer (Havir and Mchale, 1987; Nadaroglu, 2009).

Cation-exchange chromatography

The dialyzed suspension after ammonium sulfate precipitation from the aforementioned step was subjected to cation-exchange chromatography on CM-Sephadex fast flow column preequilibrated with 100 mM phosphate buffer, pH 7.0. The column was washed thoroughly with the same buffer until no protein was detected in the eluate. The bound proteins were eluted with the same buffer using a linear gradient of NaCl from 0 to 1 M. Fractions of 3 ml volume were collected at a flow rate of 3 ml/min. Protein elution was monitored spectrophotometrically by measuring the absorbance at 280 nm. Activity was measured by using guaiacol as the assay substrate. The active fractions from each peak were pooled separately from the other peaks and stored at 4°C.

Gel filtration

Active and homogenous fractions from the cation exchange were pooled, desalted and concentrated using Sephadex G25. The resulting enzyme preparation was subjected to gel filtration on a Sephacryl S-200 column (120 × 1 cm) pre-equilibrated with 25 mM phosphate buffer at pH 7 containing 0.5 M NaCl, and the column was eluted isocratically. All the fractions were analyzed as described earlier. The active and homogenous fractions were pooled, concentrated and stored at 4°C for further use (Whitaker, 1963).

Protein concentration

Protein concentration was determined spectrophotometrically (absorbance at 280 nm) as well as by Bradford's method (Bradford, 1976), using bovine serum albumin (BSA) as the standard.

Determination of peroxidase enzyme activity

Peroxidase (POD) activity was carried out spectrophotometrically using guaiacol/H_2O_2 as substrate (Lobarzewski et al., 1990). The increase in the absorption as a result of the formation of the oxidized product (tetraguaiacol) was measured at 470 nm. Reaction mixture contained 100 mM phosphate buffer (pH 6.0), 5 mM guaiacol, and 0.5 mM H_2O_2 at 25°C. The changes in absorbance were read for 3 min using a UV-vis spectrophotometer (T80 UV-VIS Spectrophotometre). Substrate specificity and classification of rocket peroxidase enzyme was determined using different substrates with similar reaction mixture and assay conditions. All the substrates and H_2O_2 were at a fixed concentration of 0.5 mM.

The rate of oxidation of guaiacol, was followed at 470 nm, (ε_{470} = 26.6 mM^{-1} cm^{-1}), ABTS at 734 nm (ε_{734} = 1.5 × 10^4 $M^{-1}cm^{-1}$), pyrogallol at 430 nm (ε_{430} = 2.47 mM^{-1} cm^{-1}), 4-methylcatechol at 412 nm, (ε470 = 1010 M^{-1} cm^{-1}) and hydroquinone (1.03 mM^{-1} cm^{-1}). One unit of enzymatic activity was defined as the amount of enzyme that oxidizes 1 µmol/min of hydrogen donors under assay conditions.

SDS polyacrylamide gel electrophoresis

SDS polyacrylamide gel electrophoresis was performed after the purification of the enzyme. It was carried out in 3 and 10% acrylamide concentrations for the stacking and running gels, respectively, each of them containing 0.1% SDS (Laemmli, 1970). The sample (20 µg) was applied to the electrophoresis medium. Brome tymol blue was used as tracking dye. Gels were stained in 0.1% Coomassie Brilliant Blue R-250 in 50% methanol, 10% acetic acid and 40% distilled water for 1.5 h. It was destained by washing with 50% methanol, 10% acetic acid and 40% distilled water several times (Laemmli, 1970). The electrophoretic pattern was photographed (Figure 3).

Molecular weight determination by gel filtration

A column (3 × 70 cm) of Sephadex G100 was prepared. The column was equilibrated with the buffer (0.05 M Na_2HPO_4, 1 mM dithioerythritol, pH: 7) until the absorbance was zero at 280 nm. The standard protein solution (bovine serum albumin, 66 kDa; egg ovalbumin, 45 kDa; pepsin, 34 kDa; trypsinogen, 24 kDa; *β*-lactoglobulin and lysozyme, 14 kDa) was added to the column. The purified peroxidase enzyme was added into the column separately and then eluted under the same conditions. The flow rate through the column was 20 ml/h. The elution volume was compared with standard proteins (Whitaker, 1963).

Dependence of enzyme activity on pH and temperature

The effect of pH on the enzymatic activity of the purified enzyme was determined within the range of pH 2.0 to 11.0. The buffers used were glycine-HCl (pH 2.0 to 3.0), sodium acetate (pH 4.0 to 5.5), sodium phosphate (pH 6.0 to 7.5), Tris (pH 8.0 to 10.0) and glycine (pH 10.5 to 12.0). Activity measurements were separatelly conducted as described earlier by using guaiacol, ABTS, pyrogallol, 4-methylcatechol and hydrochinon as a substrate. Similarly, an analysis of the effect of temperature on the enzyme activity was conducted to determine the optimum temperature. Enzyme samples were incubated at different temperatures in the range of 20 to 90°C for 15 min, and an aliquot was used for activity measurement at the same temperature for substrates of guaiacol, ABTS, pyrogallol, 4-methylcatechol and hydrochinon, separately.

Effect of pH and temperature on the peroxidase stability

The ability of the peroxidase to retain its activity under conditions of varying pH (2 to 11) and temperature (40 to 80°C) was investigated. The enzyme was incubated under specified conditions of pH for 24 h, and the residual activity was determined. In the case of temperature stability measurements, the sample was incubated at the desired temperature for 15 min and the residual activity was measured as described earlier. Kinetic parameters V_{max} and K_M values were separately determined with ABTS/H_2O_2, guaiacol/H_2O_2, ABTS/H_2O_2, pyrogallol/H_2O_2, 4-methylcatechol/H_2O_2 and hydrochinon/H_2O_2 as a substrate. Kinetic parameters were determined for each substrates using the Lineweaver–Burk double reciprocal plot.

Effect of various inhibitors on the peroxidase activity

Effect of some metal ions on the peroxidase activity

The effect of various metal ions (Ca^{2+}, Ba^{2+}, Mn^{2+}, Ni^{2+}, Co^{2+}, Cu^{2+}, Zn^{2+}, Fe^{2+} and K^{+}) on purified peroxidase were investigated. Each inhibitor solution was prepared at six different concentrations (0.084 to 1.67 mM), and each solution was added in a cuvette containing 0.5 m enzyme. Its total volume was adjusted to 3 ml with buffer solution. A control assay of the enzyme activity was done without inhibitors and resulting activity was taken as 100%. The effect of each agent was determined by measuring the enzyme activity using the guaiacol as a substrate.

Effect of some compounds on the peroxidase activity

The effect of various compounds on the activity of purified peroxidase enzyme was determined using thiol specific inhibitors, activators, non-specific compounds and detergent. The used compounds were sodium dodecyl sulfate (SDS), ethylene-diaminetetraacetic acid (EDTA), ascorbic acid, dithioerythritol, 5 mM. The effects of these compounds on the activity of purified peroxidase enzyme were performed as described earlier.

Statistical analysis

All of the tests were conducted in triplicate for determination of the peroxidase activities of samples. Data were expressed as means ± standard errors. Statistical analyses were performed using SPSS version 10.0 software (SPSS Inc., Chicago, IL, USA), and the significant differences were determined with a 95% confidence interval (p < 0.001 and 0.05) using Tukey's test.

RESULTS AND DISCUSSION

A new plant peroxidase from the leaves of rocket *(Eruca vesicaria* sbsp. Sativa*)* was purified and characterized by precipitating in $(NH_4)_2SO_4$ followed by cation-exchange and gel filtration chromatograph. Guaiacol was used as a substrate in the determination of activity in the protein eluted from the CM-Sephadex column and Sephacryl S200 column. The results pertaining to purification of peroxidase using all purification techniques are summarized in Table 1. Ammonium sulfate $(NH_4)_2SO_4$ fractionation, an extensively used technique in enzyme purification was performed as a first purification step. The enzyme obtained from the crude extract of the rocket (*Eruca vesicaria* sbsp. Sativa), using 40 to 60% $(NH_4)_2SO_4$ saturation with 27.96-fold purification and 33.45% recovery, was subjected to ion-exchange chromatography, which gave one peak with peroxidase activity. Passage from CM-sephadex column further purified the enzyme to 100.5-fold with a recovery of 62.2% and speciWc activity of 6.85 EU/mg. Then, only one peak with peroxidase activity was obtained when the partially purified enzyme was applied to Sephacryl S 200

Table 1. Purification scheme for rocket peroxidase.

Fractions	Volume (ml)	Activity (EU/ml)	Total activity		Protein (mg/ml)	Spesific activity (EU/mg protein)	Purification
			EU	%			Fold
Homogenate	50	18.13	5906.5±0.12	-	8.53±0.11	2.13	-
$(NH_4)_2SO_4$ precipitetion (40-60%)	20	8.55	1651.0±1.21	27.96	0.12±0.65	71.25	33.45
CM-Sephadex ıon exchange chromatography	15	6.85	1026.75±0.21	62.19	0.032±0.45	214.06	100.5
Sephacryl S 200 gel filtration chromatography	15	5.63	829.5±0.78	80.79	0.012±1.65	469.17	220.27

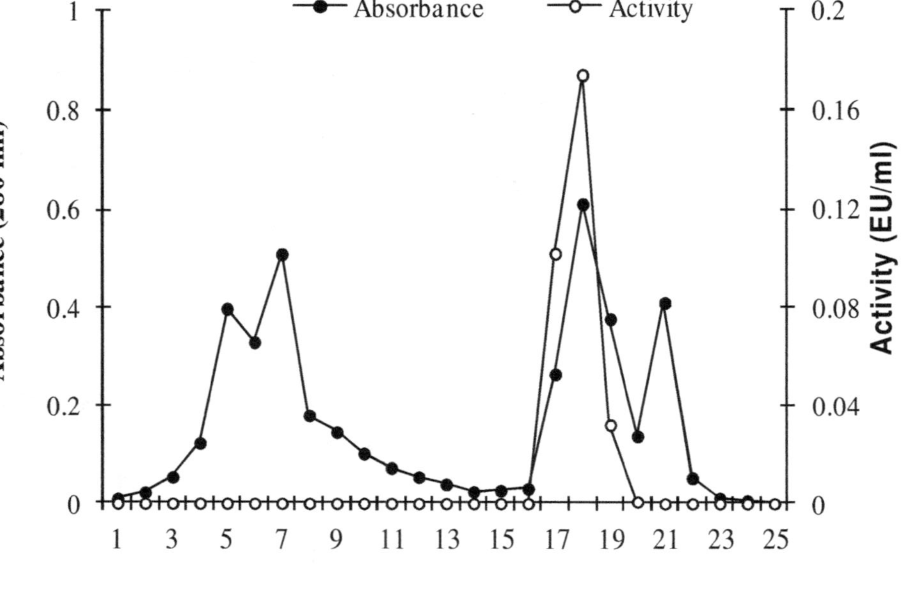

Figure 1. Typical elution profile for the chromatography of rocket peroxidase on CM-Sephadex column.

column.

An overall 220.3-fold purity was achieved with a yield of 80.8% and specific activity of 5.63 EU/mg. The final purification of 220.3-fold suggested that the peroxidase is highly abundant in the rocket (*Eruca vesicaria* sbsp. Sativa). The elution profiles consisting of a peroxidase, which was purified from the rocket (*E. vesicaria* sbsp. Sativa) using cation-exchange chromatography and gel filtration chromatography, was shown in Figures 1 and 2.

The purified peroxidase was examined by SDS electrophoresis (Figure 3). As shown in Figure 3,

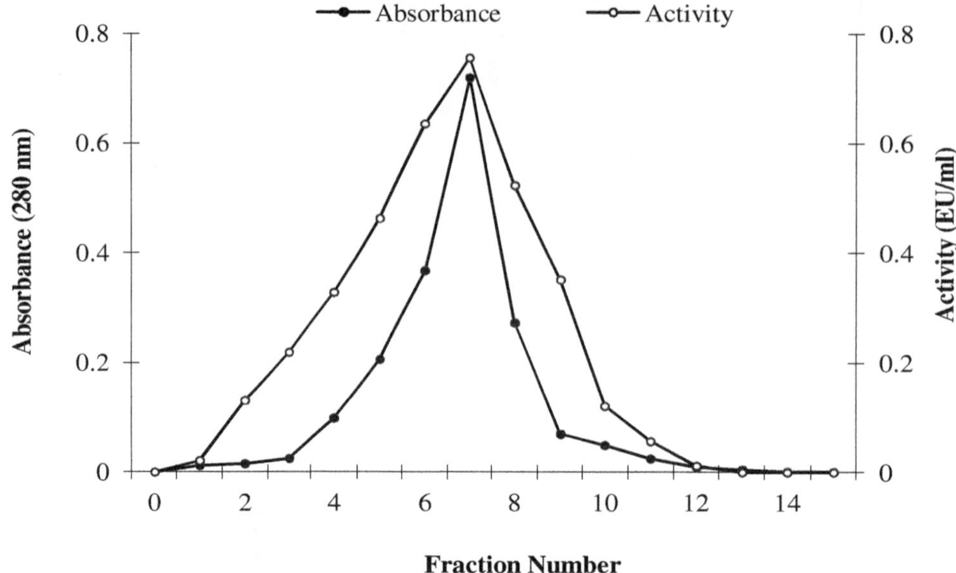

Figure 2. Gel filtration of rocket peroxidase on Sephacryl S-200 column.

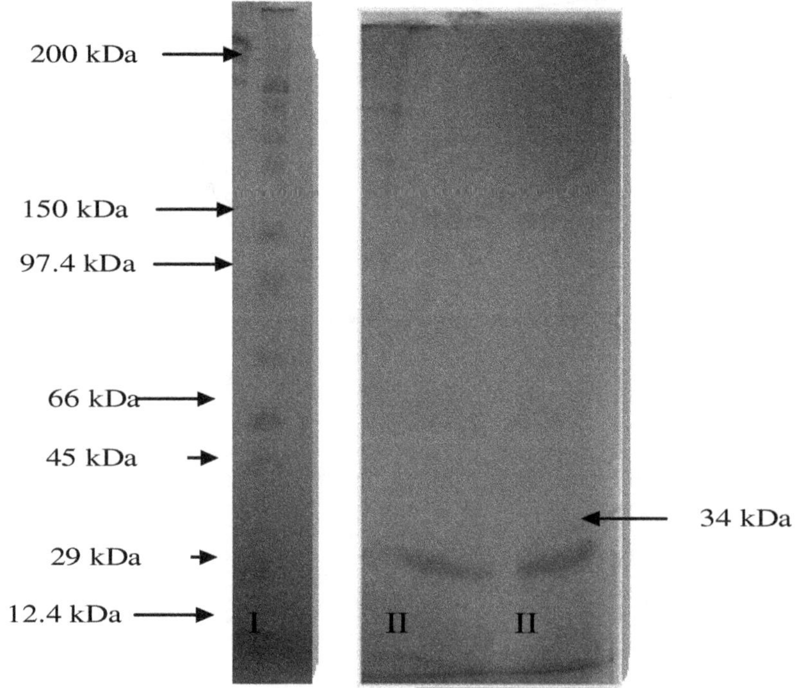

Figure 3. SDS-PAGE electrophoretic pattern of peroxidase [standart protein (β-Amylase, sweet potato, 200 kDa; alcohol dehydrogenase, yeast, 150 kDa bovine serum albumin, 97.4 kDa; rabbit muscle phosphorylase A, 66 kDa; egg ovalbumin, 45 kDa; pepsin, 29 kDa; carbonic anhydrase); cytochrome c, horse heart 12.4 kDa (I); purified laccase enzyme from rocket peroxidase (II)].

SDS-PAGE revealed a single protein band. The molecular weight of the enzyme was determined as 34 kDa by using the gel filtration chromatograph and compared with known standard proteins. The molecular weight of the purified peroxidase was determined by using the following protein standards: bovine albumin (66

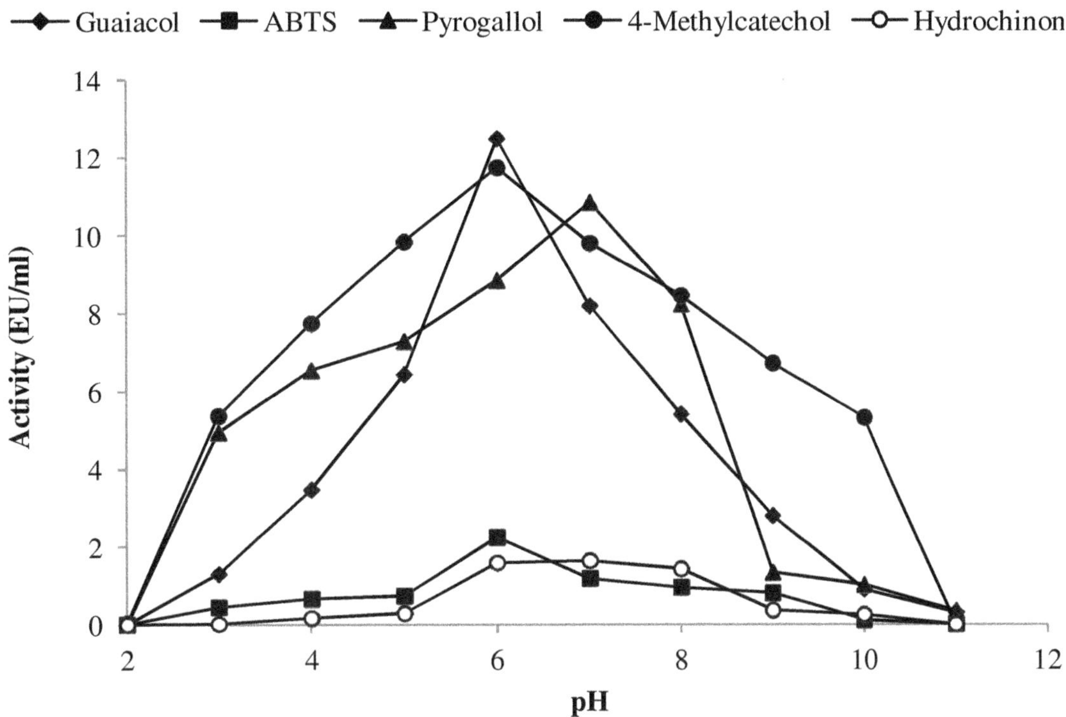

Figure 4. Effect of pH on the activity of the purified peroxidase. Enzymes and substrate were dissolved either in 10 mM buffers of various pH. Other conditions were as given for the standard assay method.

kDa), egg albumin (45 kDa), glyceraldehyde-3-phosphate dehydrogenase (36 kDa), carbonic anhydrase (29 kDa), trypsinogen (24 kDa), soybean trypsin inhibitor (20.1 kDa) and α-lactalbumin (14.2 kDa).

The molecular weight for most plant peroxidases has within the range 40 to 50 kDa. It is indicated that they are a slightly wider range of 20 to 54 kDa for peroxidases from oranges (22 to 44 kDa, peanuts (40 to 42 kDa), horseradish (44 kDa), *Raphanus sativus* (44 kDa), sweet potato (37 kDa) and olive fruit (18 to 20 kDa) (Clemente, 1998; Hu et al., 1989; Kim and Lee, 2005; Leon et al., 2002; Welinder, 1992; Saraiva et al., 2007; Robinson, 1991). These plant peroxidases are similar with the present study.

Effect of pH on rocket peroxidase activity

Optimum enzyme activity was observed at pH 6.0 for substrates of guaiacol and ABTS, at 7 for 3-methylcathecol and 7.5 for hydrochinon as shown in Figure 4. This broad range of acidic pH dependence for activity made this enzyme interesting for industrial applications especially in the food industry and it also suggest that this enzyme has better function under acid conditions (Tipawan and Barrett, 2005). This pH optimum of the rocket peroxidase closer to palm (*Roystonea regia*) (Sakharow et al., 2002), clover peroxidase (Criquet et al., 2001) and *Prangos ferulacea* (Apiaceae) (Nadaroglu, 2009) by using pyrogallol, guaiacol, ABTS as a substrate.

Effect of temperature of the purified peroxidase

By using guaiacol/H_2O_2, ABTS/H_2O_2, pyrogallol/H_2O_2, 4-methylcatechol/H_2O_2 and hydrochinon/H_2O_2 as subtrate, the optimal temperature for enzyme activity was found to be 40°C, and about 35.0 and 19.2% of the optimum activity was still detectable at 70 and 80°C, as shown in Figure 5. The peroxidase was stable at 40°C and lost 18 and 19% of its activity after 50 and 60 min at 40°C, respectively (Figure 6). This purified peroxidase has high thermal stability. So it indicates that it is an excellent enzyme for the pharmaceutical, food and detergent industries. Most plant peroxidases show optimum activity in the temperature range of 30 to 45°C (Criquet et al., 2001).

Determination of V_{max} and K_M values

V_{max} and K_m values were determined by using different substrates (guaiacol/H_2O_2, ABTS/H_2O_2, pyrogallol/H_2O_2, 4-methylcatechol/H_2O_2 and hydrochinon/H_2O_2) at optimum pH: 6 and 40°C by means of the Linewaver-Burk graph. K_m and V_{max} values for five different substrates are shown in Table 2. The peroxidase exhibited the greatest activity with ABTS (2631 U/mg).

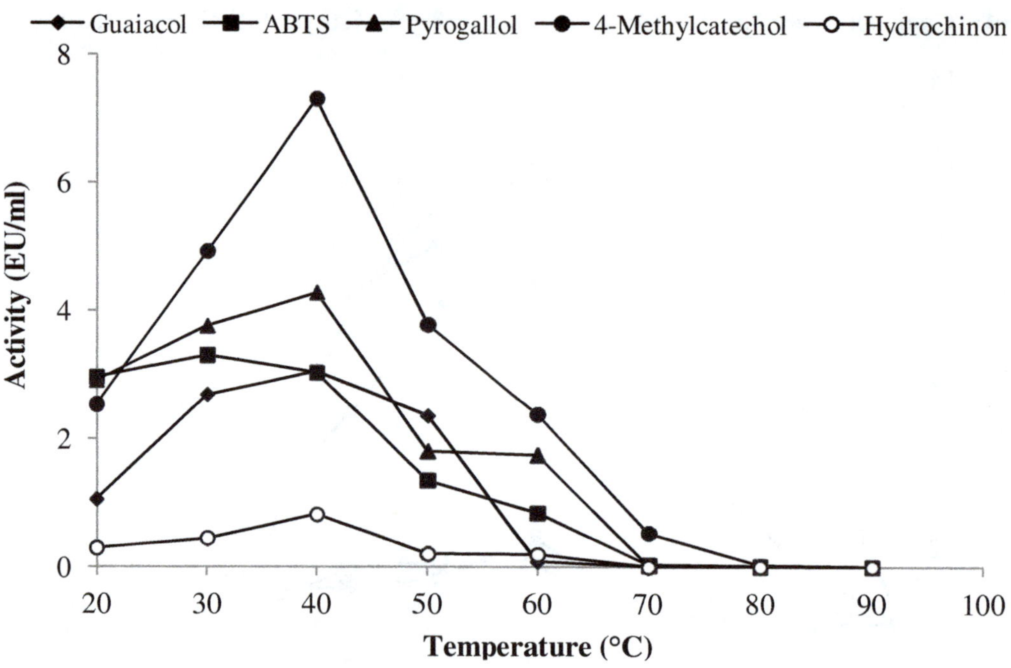

Figure 5. Effect of temperatures on the activity of purified peroxidase. Activity was determined at different temperatures and at pH 6.0 over 10 min using the standard assay method.

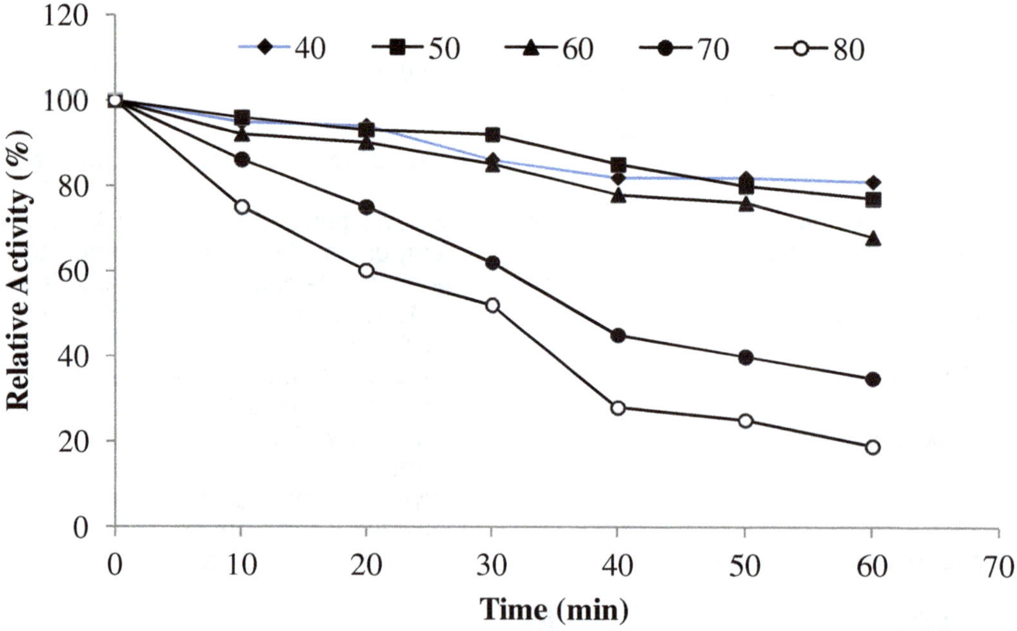

Figure 6. Effect of the temperature on the stability of the purified peroxidase. Enzymes were incubated at pH:6 for 1 h at different different temperature, and the residual activity was measured the standard assay method.

Effect of inhibitors on activity of rocket peroxidase

Although, several metal cations had partial inhibitory effects toward rocket peroxidase, Ni^{2+}, Co^{2+}, Cu^{2+} and Ba^{2+} activated the activity 193, 205, 135 and 106% at 1.67 mM, respectively (Table 3). Mn^{2+}, Hg_2^{2+}, Zn^{2+} and K^+ had inhibited the rocket peroxidase activity at about 50% at 1.67 mM. The enzyme retained less than 40% of its

Table 2. Determination of K_M and V_{max} for different substrates.

Substrat	K_M (mM)	V_{max} (µmol/L.dak)
Guaikol	375.74	0.314
ABTS	2631.56	5.376
Pyrogallol	510.14	0.89
4-Methyl cathecol	1535.65	3.65
Hidrokinon	65.19	0.303

Table 3. Effect of metal cations on rocket peroxidase.

Concentration (mM)	$CaCl_2$	$BaCl_2$	$MnCl_2$	$NiCl_2$	$CoCl_2$
	Relative activity (%)				
0	100	100	100	100	100
0.084	62.53	53.18	92.15	143.80	152.66
0.167	68.85	54.18	74.18	150.63	172.15
0.5	58.48	67.85	73.92	162.0	179.24
0.84	47.1	60.76	58.73	173.16	186.08
1.167	37.2	83.80	56.23	184.8	192.91
1.67	36.2	105.57	55.699	193.42	205.06

Concentration (mM)	Hg_2Cl_2	$ZnCl_2$	$FeCl_2$	KCl	$CuCl_2$
	Relative activity (%)				
0	100	100	100	100	100
0.084	80.5	76.1	73.58	99.16	156.95
0.167	76.1	74.2	46.16	93.44	155.44
0.5	73.5	69.06	42.58	88.69	136.95
0.84	71.7	68.18	36.98	86.31	136.68
1.167	71.38	63.08	35.22	83.21	135.95
1.67	69.81	60.12	32.33	80.0	135.19

Table 4. Effect of some chemicals on rocket peroxidase.

Concentration (mM)	SDS	EDTA	β-mercaptoethanol	Ditihierithritol
	Relative activity (%)			
0	100	100	100	100
0.084	56.35	109.64	32.15	9.37
0.167	43.52	95.53	23.22	2.77
0.5	11.45	12.15	10.05	0.66
0.84	6.42	8.9	2.12	0.28
1.167	4.09	7.68	0	0
1.67	2.58	3.2	0	0

activity in the presence of 1.67 mM Ca^{2+} and Fe^{2+}. In addition, SDS, EDTA, β-mercaptoethanol and dithioeritritol caused strongly inhibitory effects toward rocket peroxidase activity (Table 4). Inhibition by EDTA was 3.2%; this indicated that the enzyme has metal ion as cofactor. The effect of metal ions on peroxidase activity has been largely described previously. Similar results have been observed with peroxidase of plants (Wong, 1995; Ajila and Rao, 2009; Marquez et al., 2008; Onsa et al., 2004).

Conclusion

This study describes the purification and enzymatic properties of an acidic peroxidase from the rocket (*E.vesicaria* sbsp. Sativa). The results indicate that purified peroxidase wide substrate specificity and stability over a wide range of pH, temperature, metal ions and some chemicals. Therefore, rocket leaves are a potential source of peroxidase for bioanalytical or biotechnological applications, such as enzymatic reagents for clinical diagnosis, food analysis, biotransformation and degredation of various chemicals.

ACKNOWLEDGEMENTS

This research was performed under the project numbered 2011/331 and supported by the Research Development Center of Ataturk University. The authors acknowledge the support of Ataturk University, Turkey for this work.

ABBREVİATİONS

POD, Peroxidase; **PAGE,** polyacrylamide gel electrophoresis; **SDS,** sodium dodecyl sulfate; **EDTA,** ethylenediaminetetraacetic acid; **ABTS,** 2,2'-azino-bis(3-ethylbenzothiazoline-6-sulphonic acid); **$(NH_4)_2SO_4$,** ammonium sulfate; **TCA,** trichloroacetic acid.

REFERENCES

Adams JB (1978). The inact ivationandre generation of peroxidasein relation to the high temperature-shorttime processing of vegetables. J. Food Technol. 13:281-297.

Agostini E, Medina MI, Milrad de Forchetti SR, Tigier H (1997). Properties of two anionic peroxidase isoenzymes from turnip (*Brassica napus L.*) roots. J. Agric. Food Chem. 45:596-598.

Ajila CM, Rao PUJS (2009). Purification and characterization of black gram (*Vigna mungo*) husk peroxidase. J. Mol. Catal. B. Enzym. 60:36-44.

Alqasoumi S, Al-Sohaibani M, Al-Howiriny T, Al-Yahya M, Rafatullah S (2009). Rocket "*Eruca sativa*": A salad herb with potential gastric anti-ulcer activity. World J. Gastroenterol. 15(16):1958-1965.

Bradford MM (1976). Rapid and sensitive method for the quantitation of microgram quantities of protein utilizing the principle of protein-dye binding. Anal. Biochem. 72:248-254.

Clemente E (1998). Purification and thermostability of isoperoxidase from oranges. Phytochemitry 49:29-36.

Criquet S, Joner EJ, Leyval C (2001). 2,7 diaminofluorene is a sensitive substrate for detection and characterization of plant root peroxidase activities. Plant Sci. 161:1063-1066.

Duarte-Vazquez MA, Garcia-Almendarez BE, Regalado C, Whitaker JR (2001). Purification and properties of a neutral peroxidase ısozyme from turnip (*Brassica napus* L. V*ar. purpletop* white globe) roots. J. Agric. Food Chem. 49:4450-4456.

Havir EA, Mchale NA (1987). Biochemical aad developmental characterization of multiple forms of catalases-peroxidases. FEBS Lett. 492:177-182.

Hu C, Smith R, Van Huystee R (1989). Biosynthesis and localiza-tion of peanut peroxidases. A comparison of the cationic and anionic isozymes. Plant Physiol. 135:391-397.

Kim SS, Lee DJ (2005). Purification and characterization of a cationic peroxidase Cs in *Raphanus sativus*. J. Plant Physiol. 162:609-617.

Laemmli UK (1970). Cleavage of structural proteins during the assembly of the head of bacteriophage. Nature 227:680-685.

Lamy E, Schröder J, Paulus S, Brenk P, Stahl T, Mersch-Sundermann V (2008). Antigenotoxic properties of *Erucasativa* (Rocket Plant), erucin and erysolin in human hepatoma (Hepg2) cells towards benzo(A) pyrene and their mode of action. Food Chem. Toxicol. 46:2415-2421.

Lobarzewski J, Brzyska M, Wojcik A (1990). The influence of metal ions on the soluble and immobilized cytoplasmic cabbage peroxidase activity and its kinetics. J. Mol. Catal. 59:373-383.

Mahvi AH, Bazrafshan E (2007). Removal of cadmium from industrial effluents by electrocoagulation process using aluminum electrodes. World Appl. Sci. J. 2(1):34-39.

Marquez O, Waliszewski KN, Oliarta RM, Pardio VT. Marquez O, Waliszewski KN, Oliarta RM, Pardio VT (2008). Purification and characterization of cell wall-bound peroxidase from Vanilla bean. LWT 41:1372-1379.

Nadaroglu H (2009). Purification and properties of peroxidase from prangos ferulacea (apiaceae) and ınvestigation by some chemicals. Asian J. Chem. 21(7):5768-5776.

Nadaroglu H, Kalkan E (2012). Alternative absorbent industrial red mud waste material for cobalt removal from aqueous solution. Int. J. Phys. Sci. 7(9):1386-1394.

Nadaroglu H, Kalkan E, Demir N (2010). Removal of copper from aqueous solution using red mud. Desalination 153:90-95.

Onsa GH, Saari Nbin, Selamat J, Bakar J (2004). Purification and characterization of membrane-bound peroxidases from *Metroxylon sagu*. Food Chem. 85:365-376.

Pütter J, Becker R (1987). Methods of Enzymathic Analysis:Peroxide Bergmeyer, Third Edition, VCH, Newyork. P. 286.

Robinson DS (1991). Peroxidases and Catalases in Foods In Oxidative Enzymes in Food, 1st ed, Robinson D S, Eskin N A M, Eds, Elsevier Applıed Scıence: London, UK. pp. 1-37.

Sakharow IY, Vesga Blanco MK, Galaev IY, Sakharova IV, Pletjushakina OY (2002). Peroxidase from leaves of royal palm Tree *Roystonea regia*: Purification and some properties. Plant Sci. 161:853-860.

Saraiva JA, Nunes CS, Coimbra MA (2007). Purification and characterization of olive (*Olea europaea* L.) peroxidase-evidence for the occurrence of a pectin binding peroxidase. Food Chem. 101(4):1571-1579.

Tıpawan T, Barrett M (2005). Heat inactivation and reactivation of broccoli peroxidase. J. Agric. Food Chem. 53:3215-3222.

Van Huytstee RB (1987). Some molecular aspects of plant peroxidase biosynthetic studies. Annu. Rev. Plant Physiol. 38:205-219.

Wakamatsu K, Takahama U (1993). Changes in peroxidase activity and peroxidase isoenzymes in *Carrot callus*. Physiol. Plantarum. 88:167-171.

Welinder KG (1979). Amino acid sequence studies of horseradish peroxidase. Europ. J. Biochem. 96(3):483-502.

Welinder KG (1992). Super family of plant, fungal and bacterial peroxidases. Curr. Opin. In Struc. Biol. 2(3):388-393.

Whitaker JR (1963). Determination of molecular weight of proteins by gel filtration on sephadex. Anal. Chem. 35:1950-1953.

Wong DW (1995). Horseradish Peroxidase WS Dominic (Ed), Food Enzymes Structure and Mechanism. Champman and Hall., New York. pp. 321-345.

7

Isolation and cloning of *ech*36 gene from *Trichoderma harzanium*

Radheshyam Sharma and Sumangala Bhat

Institute of Agri-Biotechnology (IABT), College of Agriculture, University of Agricultural Sciences, AC, Dharwad-580 005 Karnataka, India.

A lab experiment was conducted to screen for the presence of *ech*36 gene in 80 isolates of *Trichoderma.* Further, using gene specific primers *ech*36 gene were cloned into pTZ57R/T from *Trichoderma harzanium* IABT1042. The clone was confirmed through PCR amplification and restriction analysis. The clone were sequenced and analyzed for homology at nucleotide and protein level to find out conserved domains of the putative gene and protein. The gene encoding endochitinase from the species have 99 and 100% homology with reported sequences both at nucleotide and protein level. The cloned *ech*36 has a size of 1212 bp, of which 12 bp corresponds to the 3' untranslated region, with a 1030 bp open reading frame. The amino acid sequence of gene has signal peptide sequence ranges from1 to 25. The nucleotide sequence analysis using GENETOOL software revealed presence of three exon and introns and has unique restriction sites for *HindIII, BglII*, and *HaeIII* at 908, 387 and 672 positions respectively

Key words: *Trichoderma harzanium, ech*36, protein.

INTRODUCTION

The present day world is facing various problems regarding food security. The traditional agriculture is affected by various problems such as drought, pest and diseases, reduced availability of the land, increase in population. Among them, pests and diseases cause major losses. Iindiscriminate use of pesticide and fungicides can have drastic effects on the environment and the consumers. Chemical methods with repeated use are not economical in the long run because they pollute the atmosphere, damage the environment, leave harmful residues, and lead to the development of resistant strains among the target organisms (Naseby et al., 2000). A reduction or elimination of synthetic pesticide applications in agriculture is highly desirable. Therefore, efforts have been made to breed resistant cultivars and to develop biocontrol agents (BCA) for the control of various fungal plant pathogens. *Trichoderma* spp. is a well known BCA against plant diseases. One of the most promising means to achieve this goal is by the use of new tools based on BCA for disease control alone, or to integrate with reduced doses of chemicals in the control of plant pathogens resulting in minimal impact of the chemicals on the environment (Chet and Inbar, 1994; Harman and Kubicek, 1998). *Trichoderma* spp. is among the most frequently isolated soil fungi present in plant root ecosystems (Harman et al., 2004). Species of the genus *Trichoderma* are widely known for their biotechnological interest; however their use as bio-control agents requires a comprehensive analysis of the biological principles of their action. Their antagonistic abilities are described as a combination of several mechanisms, including nutrient competition and direct mycoparasitism, which involves the production of antifungal metabolites and cell wall-degrading enzymes (Vieira et al., 2013). These fungi are opportunistic, avirulent plant symbionts, and function as parasites and antagonists of many phytopathogenic

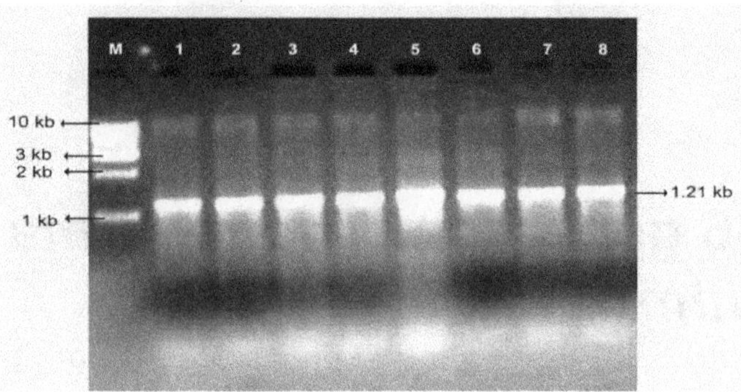

Figure 1. PCR amplification of *ech*36 gene. M, 1 kb ladder; 1, *T. virens* (IABT 1010); 2, *T. viride* (IABT 1012); 3, *T. viride* (IABT 1013) 4 – *T. viride* (IABT 1014); 5, *T. virens* (IABT 1017); 6, *T. harzianum* (IABT 1042); 7, *T. viride* (IABT 1021); 8, *T. viride* (IABT 1022).

fungi, thus protecting plants from diseases. So far, Trichoderma spp. are the most studied fungal BCA and commercially marketed as biopesticides, biofertilizers and soil amendments (Harman et al., 2004; Lorito et al., 2004). Depending upon the strain, *Trichoderma* spp. (notably *H. lixii/T. harzianum, H. virens/T. virens, T. atroviridis/T. atroviride* and *T. asperellum)* are used as biocontrol agents against various diseases of crops, vegetables, and fruits (Harman et al., 2004). They have evolved numerous mechanisms that are involved in attacking other fungi. These mechanisms include competition for space and nutrients (Elad et al., 1999), mycoparasitism (Haran et al., 1996; Lorito et al., 1996), production of inhibitory compounds (Sivasithamparam and Ghisalberti, 1998), inactivation of the pathogen's enzymes (Roco and Perez, 2001) and induced resistance (Kapulnik and Chet, 2000).

The antifungal mechanism of *Trichoderma*, an extensively studied and widely used biocontrol fungus, mainly relies on cell wall degrading enzymes such as chitinases and glucanacses (Lorito, 1998) and is being exploited to control a variety of plant pathogens. Several chitinases and glucanacses are isolated from *Trichoderma* and transferred to plants to impart resistance to several fungal plant pathogens. Chitinase encoding genes are being used to improve plant defense against fungal pathogens. These enzymes are capable of degrading the linear homopolymer of ß-1, 4-N-acetyl- D-glucosamine, the main cell wall component of most phytopathogenic fungi, showing strong inhibitory activity *in vitro* on germination and hyphal growth (Lorito et al., 1996). Plants do have chitinases, but are not as effective as microbial chitinases. Therefore, cloning and characterization of genes from biocontrol microbes such as *Trichoderma* is very important. There are many evidences to show that fungal chitinases alone has increased the resistance of transformed plants against pathogenic fungus. The rice plant transformed with an endochitinase gene (*ech*33) from the biocontrol fungus *T. atroviride* increased the resistance to sheath blight caused by *Rhizoctonia solani* and rice blast caused by *Magnaporthe grisea* (Liu et al., 2004). The identification of the Trichoderma genes involved in these mechanisms and analysis of their expression profiles can provide researchers with biotechnological tools that exhibit anti-fungal activity and that could potentially be used as transgenes capable of inducing resistance to pathogens in economically valuable plants (Vieira et al., 2013). The present study is aim to produce transgenic plants expressing either plant or microbial chitinase. In recent years, considerable progress has been made in producing disease-resistant and high-yielding transgenic plants. It may be necessary to integrate different resistance genes together in order to extend the host defenses.

MATERIALS AND METHODS

Isolation of genomic DNA from fungus

For isolation and cloning of gene 80 *Trichoderma* isolates were inoculated in100 ml potato dextrose broths at 30°C in incubator. The complete growth occured in 2 to 5 days depending on the species. About 100 mg of fungus mycelium was taken in 1.5 ml micro centrifuge tube and 500 µl of lysis buffer (400 mM Tris-HCl [pH 8.0], 60 mM EDTA [pH 8.0], 150 mM NaCl, 1% sodium dodecyl sulfate), was added. Mycelium was finely macerated using micro-pestle and vortexed for 5 min. The suspension was extracted with equal volume of phenol: chloroform: indole acetic acid (25:24:1) and centrifuged at 10,000 rpm for 10 min. The supernatant was taken into a fresh tube and RNase at the rate of 100 µg per ml was added and this solution was incubated for 20 min at 55°C on water bath and then equal volume of isopropanol was added at room temperature, mixed by gentle inversion and kept for 10 min at room temperature. The DNA was recovered by centrifugation at 10,000 rpm for 10 min at 4°C. The DNA pellet was washed with 70% ethanol, air dried and resuspended in 50 µl of TE (10 mM Tris-Cl and 1 mM EDTA, pH 7.5). Concentration of DNA was estimated using ethidium bromide spotting method as described by Sambrook and Russel (2001).

PCR amplification

PCR was carried out from *Trichoderma genomic* DNA. The *ech*36 gene sequence was downloaded from NCBI gene bank and primers were designed using online software primer-3. PCR amplification using two gene specific primer were used- 5'ATGACACGCCTCCTCGAC3'and RP 3'TACCCATATTCATTAAAGGCTATCAA5'. Reaction mixture for PCR (20 µl) contained 10 mM Tris-HCl (pH 8.3), 50 mM KCl, 1.5 mM $MgCl_2$, 10 mM each of dATP, dCTP, dGTP and dTTP, 5 pM primer, 1 µl genomic DNA (100 ng) and one unit of Taq polymerase. Amplification was performed in 0.2 ml tube using thermocycler (Eppendorf 2231, Hamburg, Germany). Initial denaturation was carried out at 94°C for 5 min. Thirty five cycles of the following programme were used for amplification; denaturation at 94°C for 2 min, annealing at 41°C for 2 min and extension at 72°C for 10 min. The amplified products were separated by electrophoresis on 1.2% gel stained in ethidium bromide. The gel was observed and photographed using UV transilluminator. The amplification showed ~1.12 kb amplicon (Figure 1).

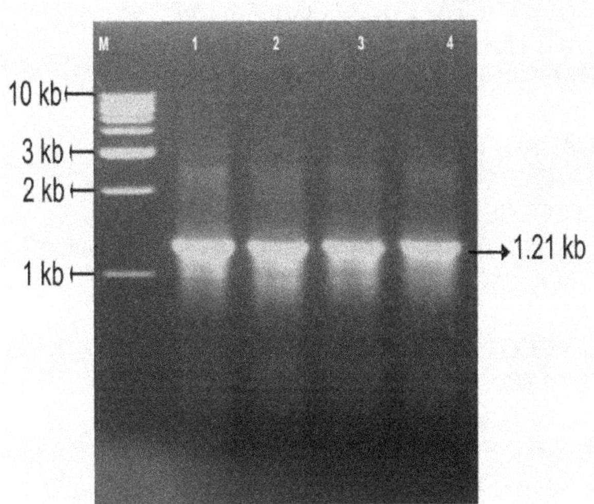

Figure 2. PCR confirmation of *ech*36 cloned gene. M = 1KB DNA Ladder; 1. *T. Harzianum* 2 *T. Harzianum;* 3. *T. Harzianum.*

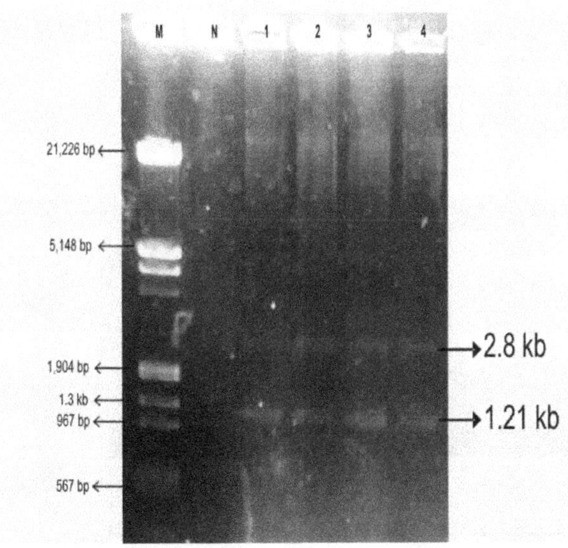

Figure 3. Restriction confirmation of *ech*36 gene. M = λ*EcorI*/*Hind*III double digest Ladder; 1. *T. Harzinum;* 2. *T. Harzinum;* 3. *T. Harzinum*; 4. *T. Harzinum*

Cloning of endochitinase gene

The specific size of band was eluted using elution kit (Bangore geni India). The bands (~1.12 kb) corresponding to ech36 from *T. harzanium* IABT1042, were ligated to pTZ57R/T vector (2886 bp) as described in InsT/A clone™ PCR product cloning kit (#k1214) from MBI Fermentas USA. The ligation products were used to transform *E. coli* DH5α.

Transformation of *E. coli* DH5α with recombinant construct

The competent cells of *E. coli* DH5α were prepared following the protocol mentioned by Sambrook and Russell (2001). About 100 μl of freshly prepared competent cells were taken in a chilled centrifuge tube and 10 μl of ligated mixture was added into the tube and was mixed gently. The mixture was chilled in ice for 45 min. Later, heat shock was given by shifting the chilled mixture to preheated 42°C water bath for exactly 2 min. immediately it was transferred onto ice to chill for 5 min. To this, 800 μl of Luria broth was added and incubated at 37°C at 200 rpm for 45 min, to allow bacteria to recover and express the antibiotic marker encoded by the plasmid. The culture was centrifuged at 13,000 rpm for 1 min and about 700 μl of supernatant was discarded and the pellet was dissolved in remaining supernatant and spread on the plates having Luria agar with Amp_{100}, X-gal IPTG and incubated overnight at 37°C.

The recombinant clones were identified by blue/white assay (Sambrook and Russel, 2001). After incubation only white colonies, were picked up and streaked on plates having Luria agar with Amp_{100}, X-gal, IPTG and incubated at 37°C overnight and checked further for the presence of construct through PCR and restriction confirmation.

Sequencing and in silico analysis of the clones

The recombinant plasmids were sequenced using M13 universal forward and reverse primers at Bangalore Genei Private Ltd., Bangalore. The sequence was subjected for analysis after removing vector sequence, through vecscreen service available in NCBI website. The available sequence information from cloned fragments was subjected to analysis using BLAST algorithm available at http://www.ncbi.nlm.nih.gov. In silico translation was done using GENETOOL software. Dual and multiple alignments for homology search were performed using the Clustal W algorithm in BioEdit software (Hall, 1999). The general features of the protein (amino acid composition) were assessed using the GENETOOL and the presence of a putative signal sequences was predicted using Signal P 3.0 Verson (Bendtsen et al. 2004; http://www.cbs.dtu.dk/services/SignalP/). All other bioinformatics like searching domain, catalytic active sites were performed using tools that are accessible via different links on the proteomics service of the Swiss Institute of Bioinformatics (Zdobnov and Apweiler2001; http://www.ebi.ac.uk/InterProScan/).

RESULTS

Among the 80 *Trichoderma* isolates only 72 isolates gave amplification of *ech*36 gene. Based on previous pathogensity bioassay in our laboratory on *Rhizoctani solani* the efficient strain *T. harzanium* (IABT1042) was selected for cloning. The cloned *ech*36 gene from *T. harzanium* (IABT1042) 12 colonies were observed on selection medium of which 4 were white. Further, these colonies were screened for the presence of *ech36* and only three clones showed the presence of ~1.12 kb insert when checked through PCR with specific primers and restriction analysis (Figures 2 and 3).

One of the clones corresponding to *ech*36 was named as pBRS-21. The clone was sequenced using M13 forward and reverse primers at Bangalore Genie Pvt. Ltd. The complete sequence of nucleotides (Figure 4) was found after removing vector sequence through vecscreen service of the NCBI website. The available sequence information from cloned gene was subjected to analysis

ATGACACGCCTCCTCGACGCCAGCTTTCTGCTGCTGCCTGCTATCGCATCGACGCTAT
TTGGCACCGCCTCTGCACAGAATGCGATATGCGCACTTAAGGGAAAGCCGGCTGGCA
AAGTCTTGATGGGATATTGGGAAAATTGGGATGGAGCAGCCAACGGTGTTCACCCTG
GATTTGGTTGGACACCAATCGAAAACCCCATCATTAAACAGAATGGCTACAATGTGA
TCAACGCCGCCTTCCCCGTTATTCTGTCAGATGGCACAGCGTTATGGGAAAACGACA
TGGCTCCTGACACACAAGTCGCAACTCCAGCTGAAATGTGTGAGGCTAAAGCAGCTG
GAGCCACAATTCTGCTGTCAATTGGAGGTGCTACTGCTGGCATAGATCTCAGCTCCA
GTGCAGTCGCTGACAAGTTCATCGCCACCATTGTACCAATCTTGAAGCAGTACAATTT
TGACGGCATTGATATAGACATTGAGACGGGGTTGACCAACAGCGGTAATATCAACAC
ACTCTCCACATCCCAGACCAACTTGATTCGCATCATTGATGGTGTTCTTGCTCAGATG
CCTTCCAACTTCGGCTTGACTATGGCACCTGAGACAGCGTACGTTACAGGCGGTAGC
ATCACGTATGGCTCTATTTGGGGAGCGTACCTACCTATCATCCAGAAATATGTTCAAA
ACGGCCGGCTGTGGTGGCTAAACATGCAATATTACAACGGCGACATGTACGGTTGCT
CTGGCGACTCTTACGCAGCTGGCACTGTCCAAGGATTCATCGCTCAGACTGACTGCCT
AAATGCAGGACTTACCGTCCAAGGCACCACAATCAAGGTTCCGTACGACATGCAAGT
ACCAGGTCTACCTGCGCAATCAGGAGCTGGCGGTGGTTATATGAATCCAAGCTTGGT
TGGACAAGCATGGGATCACTACAACGGTGCTCTGAAAGGCTTGATGACGTGGTCAAT
CAACTGGGATGGAGCGGGTAACTGGACATTTGGCGACAATTTGCTCACTCGCATTGG
CTAGAAGAAGATGACGGGAGGAAGATTAGGCGTTGAAGTCAATATATTTCTTGATTT
CATACCAGCGATATGCCATGAATTAGTCAATGATATATATACCTTCCAGCAGTATATA
GTTGGACTGACTATTTAGTCTATGCACATAAAAAACTAGCTTGATAGCCTTTAATGAA
TATGGGTA

Figure 4. Complete nucleotide sequence of cloned endochitinase gene (*ech36*) from *T. harzanium* (IABT1042). 5'UTR- 0; 3'UTR- 12; Exon-----1-1035; Introns----1036-1112.

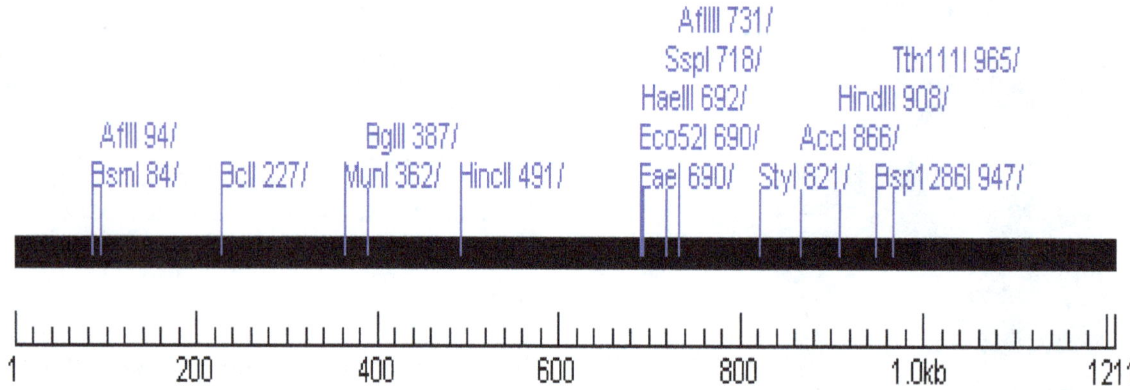

Figure 5. Restriction map of cloned endochitinase gene (*ech36*) sequences from *T. harzanium* (IABT1042) with common enzymes.

using BLAST algorithm available at http://www.ncbi.nlm.nih.gov. It showed homology with conserved domain of CHI-18, chitinase like superfamily (Figure 7). The nucleotide sequence of *ech*36 showed 99% homology with the published *Trichoderma harzianum* endochitinase Chit36Y (chit36Y) gene, complete cds (AF406791.1), and 96% with *Trichoderma asperellum* chitinase mRNA, complete cds (DQ663089.1). The cloned *ech*36has a size of 1212 bp, of which 12 bp corresponds to the 3' untranslated region, with open reading frame present in the DNA (Figure 10). The nucleotide sequence was translated to amino acid using GENETOOL software and code for 258 amino acids. The amino acid sequence of gene is shown in Figure 8. It has signal peptide sequence ranges from 1 to 25. The nucleotide sequence analysis using GENETOOL software revealed presence of three exon and introns (Figure 9). Nucleotid alignment of *ech*36 referance with cloned endochitinase gene (*ech*36) sequences from *T. harzanium* (IABT1042) was done and shown in Figure 11. Similarly the amino acid alignment was also done and sequence is shown in Figure 12.

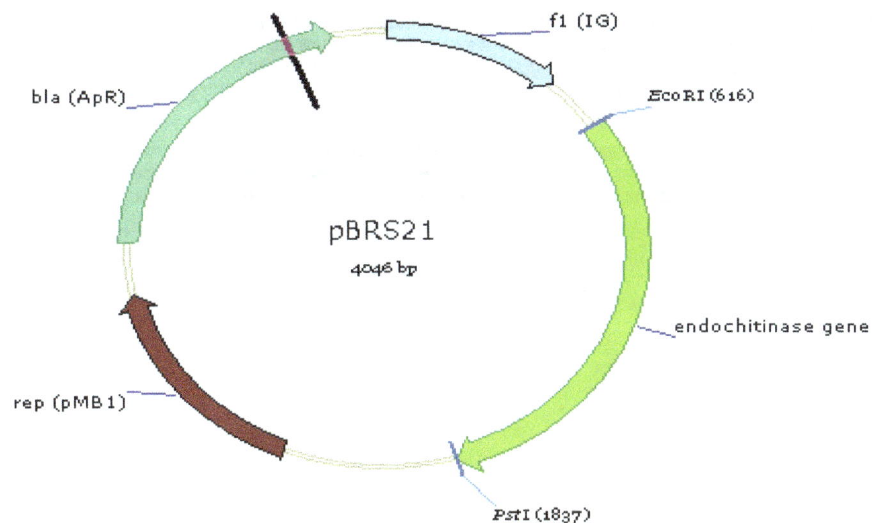

Figure 6. Construct map of pBRS-21 clone.

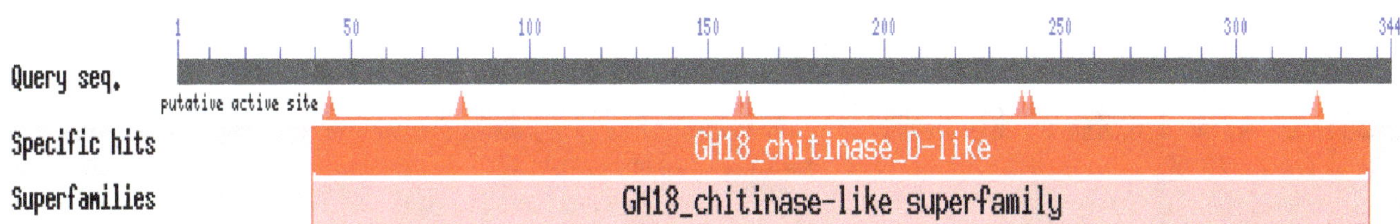

Figure 7. rps BLAST results of cloned endochitinase gene (*ech36*) sequences from *T. harzanium* (IABT1042) showing conserved domain. Cd02871, GH18_chitinase_D-like, GH18 domain of Chitinase D (ChiD). ChiD, a chitinase found in *Bacillus circulans* hydrolyzes the 1, 4-beta-linkages of N-acetylglucosamine.

MTRLLDASFLLLPAIASTLFGTASAQNAICALKGKPAGKVLMGYWENWDGAANGVHPGFGWTPIE
NPIIKQNGYNVINAAFPVILSDGTALWENDMAPDTQVATPAEMCEAKAAGA**TILLSIGG**ATAGID
LSSSAVADKFIATIVPILKQYN**FDGIDIDIE**TGLTNSGNINTLSTSQTNLIRIIDGVLAQMPSNF
GLTMAPETAYVTGGSITYGSIWGAYLPIIQKYVQNGRLWWLNMQYYNGDMYGCSGDSYAAGTVQG
FIAQTDCLNAGLTVQGTTIKVPYDMQVPGLPAQSGAGGGYMNPSLVGQAWDHYNGALKGLMTWSI
NWDGAGNWTFGDNLLTRIGKKMTGGRLGVEVNIFLDFIPAICHELVNDIYTFQQYIVGLTISMHI
KNLDSLIWV

Figure 8. Deduced amino acid sequences of cloned endochitinase gene (*ech36*) from *T. harzanium* (IABT1042). Signal peptide.........1-25; Chitinase family active site-79-87; Chitin binding domain.......97-106

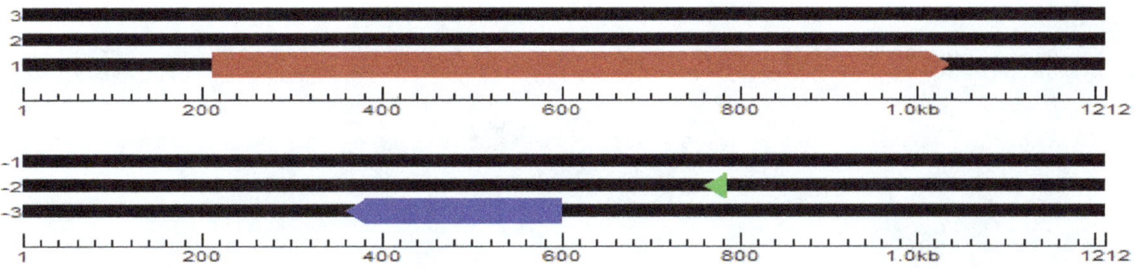

Figure 9. Exon map of cloned endochitinase gene (*ech36*) from *T. harzanium* (IABT1042).

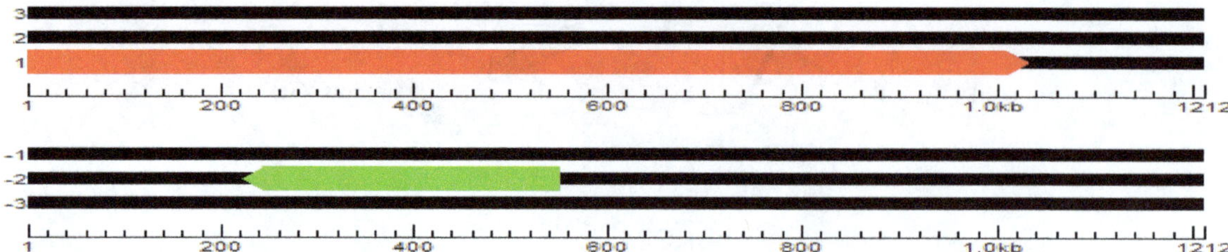

Figure 10. ORF of cloned endochitinase gene (*ech36*) from *T. harzanium* (IABT1042).

Ref ech36	1	ATGACACGCCTCCTCGACGCCAGCTTTCTGCTGCTGCCTGCTATCGCATCGACGCTATTCGGCACCGCCTCTGCACAGAA
ech36 cloned	1	ATGACACGCCTCCTCGACGCCAGCTTTCTGCTGCTGCCTGCTATCGCATCGACGCTATTTGGCACCGCCTCTGCACAGAA
Ref ech36	81	TGCGACATGCGCACTTAAGGGAAAGCCGGCAGGCAAAGTCTTGATGGGATATTGGGAAAATTGGGATGGAGCAGCCAACG
ech36 cloned	81	TGCGATATGCGCACTTAAGGGAAAGCCGGCTGGCAAAGTCTTGATGGGATATTGGGAAAATTGGGATGGAGCAGCCAACG
Ref ech36	161	GTGTTCACCCTGGATTTGGTTGGACACCAATCGAAAACCCCATCATTAAACAGAATGGCTACAATGTGATCAACGCCGCC
ech36 cloned	161	GTGTTCACCCTGGATTTGGTTGGACACCAATCGAAAACCCCATCATTAAACAGAATGGCTACAATGTGATCAACGCCGCC
Ref ech36	241	TTCCCCGTTATTCTGTCAGATGGCACAGCGTTATGGGAAAACGACATGGCTCCTGACACTCAAGTCGCAACTCCAGCTGA
ech36 cloned	241	TTCCCCGTTATTCTGTCAGATGGCACAGCGTTATGGGAAAACGACATGGCTCCTGACACACAAGTCGCAACTCCAGCTGA
Ref ech36	321	AATGTGTGAGGCTAAAGCAGCTGGAGCCACAATTCTGCTGTCAATTGGAGGTGCTACTGCTGGCATAGATCTCAGCTCCA
ech36 cloned	321	AATGTGTGAGGCTAAAGCAGCTGGAGCCACAATTCTGCTGTCAATTGGAGGTGCTACTGCTGGCATAGATCTCAGCTCCA
Ref ech36	401	GTGCAGTCGCTGACAAGTTCATCGCCACCATTGTACCAATCTTGAAGCAGTACAATTTTGACGGCATTGATATAGACATT
ech36 cloned	401	GTGCAGTCGCTGACAAGTTCATCGCCACCATTGTACCAATCTTGAAGCAGTACAATTTTGACGGCATTGATATAGACATT
Ref ech36	481	GAGACGGGGTTGACCAACAGCGGTAATATCAACACACTTTCCACATCCCAGACCAACTTGATTCGCATCATTGATGGTGT
ech36 cloned	481	GAGACGGGGTTGACCAACAGCGGTAATATCAACACACTCTCCACATCCCAGACCAACTTGATTCGCATCATTGATGGTGT
Ref ech36	561	TCTTGCTCAGATGCCTTCCAACTTCGGCTTGACTATGGCACTTGAGACAGCGTACGTTACAGGCGGTAGCATCACGTATG
ech36 cloned	561	TCTTGCTCAGATGCCTTCCAACTTCGGCTTGACTATGGCACCTGAGACAGCGTACGTTACAGGCGGTAGCATCACGTATG
Ref ech36	641	GCTCTATTTGGGGAGCGTACCTACCTATCATCCAGAAATATGTTCAAAACGGCCGGCTGTGGTGGTTAAACATGCAATAT
ech36 cloned	641	GCTCTATTTGGGGAGCGTACCTACCTATCATCCAGAAATATGTTCAAAACGGCCGGCTGTGGTGGCTAAACATGCAATAT
Ref ech36	721	TACAACGGCGACATGTACGGTTGCTCTGGCGACTCTTACGCAGCTGGCACTGTCCAAGGATTCATCGCTCAGACTGACTG
ech36 cloned	721	TACAACGGCGACATGTACGGTTGCTCTGGCGACTCTTACGCAGCTGGCACTGTCCAAGGATTCATCGCTCAGACTGACTG
Ref ech36	801	CCTAAATGCAGGACTTACCGTCCAAGGCACCACAATCAAGGTTCCGTACGACATGCAAGTACCAGGTCTACCTGCGCAAT
ech36 cloned	801	CCTAAATGCAGGACTTACCGTCCAAGGCACCACAATCAAGGTTCCGTACGACATGCAAGTACCAGGTCTACCTGCGCAAT
Ref ech36	881	CAGGAGCTGGCGGTGGTTATATGAATCCAAGCTTGGTTGGACAAGCATGGGATCACTACAACGGTGCTCTGAAAGGCTTG
ech36 cloned	881	CAGGAGCTGGCGGTGGTTATATGAATCCAAGCTTGGTTGGACAAGCATGGGATCACTACAACGGTGCTCTGAAAGGCTTG
Ref ech36	961	ATGACGTGGTCAATCAACTGGGATGGAGCGGGTAACTGGACACTTGGCGACAATTTGCTCACTCGCATTGGCTAGAAGAA
ech36 cloned	961	ATGACGTGGTCAATCAACTGGGATGGAGCGGGTAACTGGACATTTGGCGACAATTTGCTCACTCGCATTGGCTAGAAGAA
Ref ech36	1041	GATGACGGGAGGAAGATTAGGCGTTGAAGTCAATATATTTCTTGATTTCATACCAGAGATATGCCATGAATTGGTCAATG
ech36 cloned	1041	GATGACGGGAGGAAGATTAGGCGTTGAAGTCAATATATTTCTTGATTTCATACCAGCGATATGCCATGAATTAGTCAATG
Ref ech36	1121	ATATATATACCTTCCAGCAGTATATAGTTGGACTGACTATTTAGTCTATGTACATAAAAAACTAGCTTGATAGCCTTTAA
ech36 cloned	1121	ATATATATACCTTCCAGCAGTATATAGTTGGACTGACTATTTAGTCTATGCACATAAAAAACTAGCTTGATAGCCTTTAA
Ref ech36	1201	TGAATATGGGTA
ech36 cloned	1201	TGAATATGGGTA

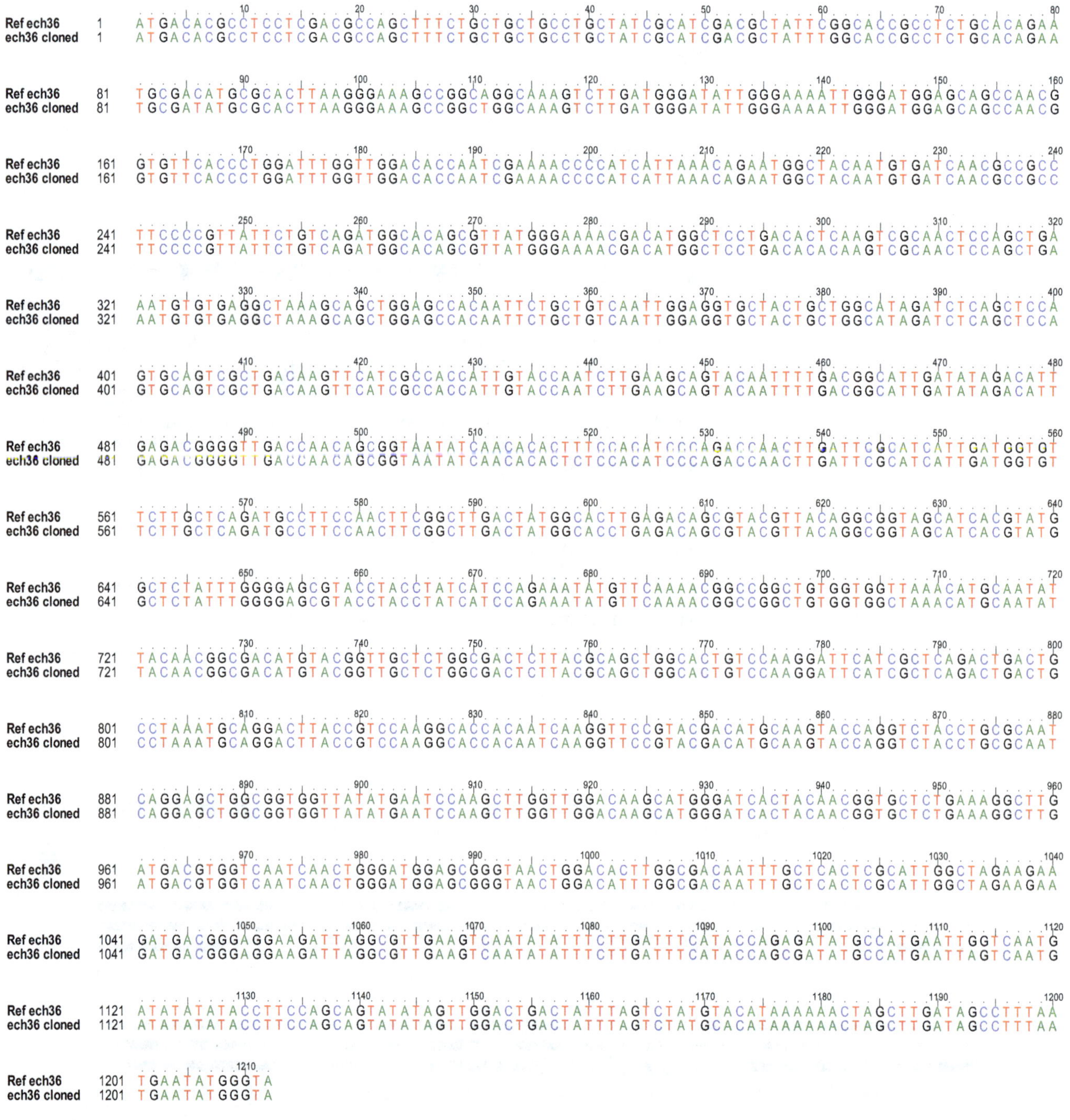

Figure 11. Nucleotid alignment of *ech*36 referance with cloned endochitinase gene (*ech*36) sequences from *T. harzanium* (IABT1042).

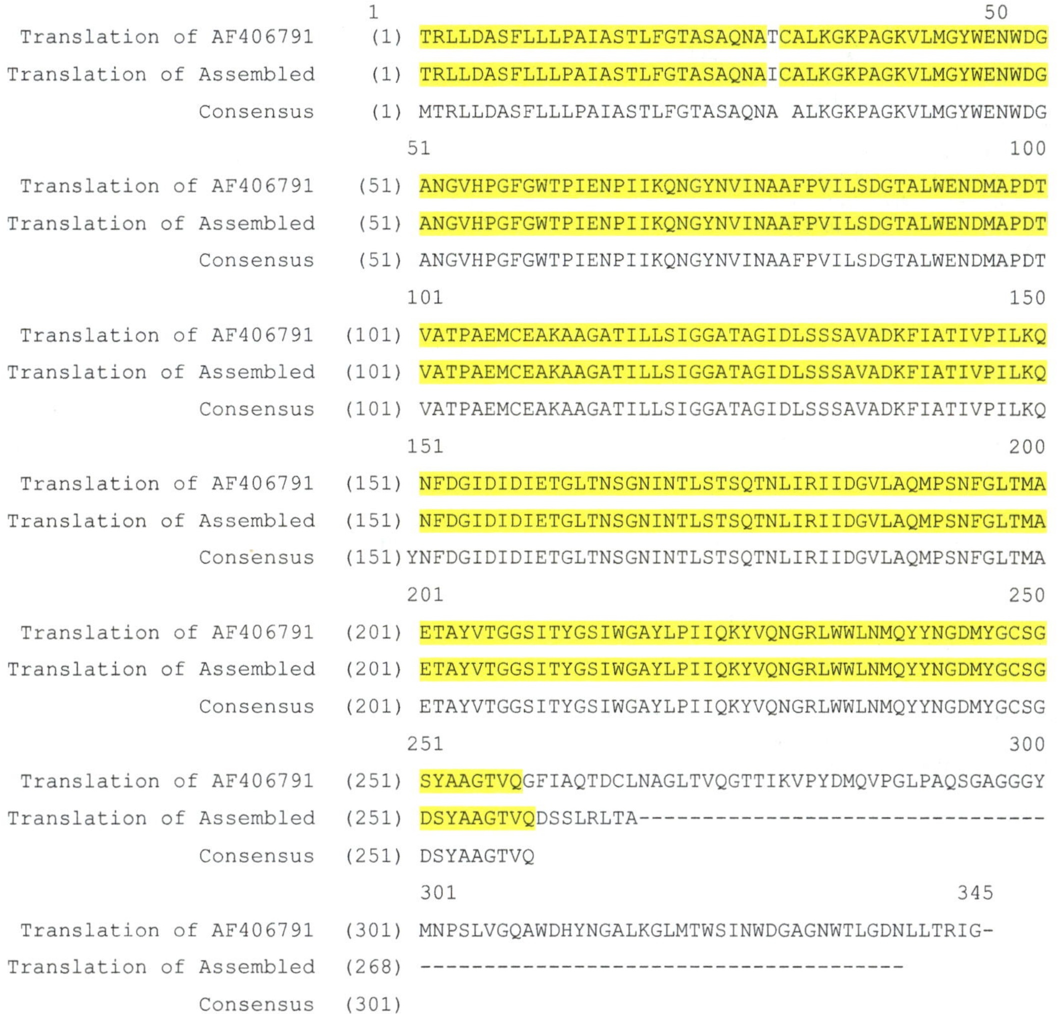

Figure 12. Amino acid alignment of *ech*36 referance with cloned endochitinase gene (*ech36*) sequences from *T. harzanium* (IABT1042).

The cloned *ech*36 in Pbrs-21has unique restriction sites for *Hind*III, *Bgl*II, and *Hae*III at 908, 387, and 692 positions, respectively. The restriction map of the sequence (pBRS-21) is presented in Figure 5. The vector map of pBRS-21 was constructed using the software VECTOR NTI and is presented in Figure 6. The nucleotide sequence of the cloned *ech*36 was subjected for BLASTx, and showed 100% homology with the published *T. harzianum* endochitinase Chit36Y (chit36Y) gene, complete cds (AF406791.1) gene at amino acid levels.

DISCUSSION

The development of resistant traits against pathogen become an important target in plant biotechnology, since the traditional approaches to control of epidemic spread of diseases are no longer sufficient. Successful approaches for enhancing disease resistance of plant were based on the over expression of genes encoding protein that are produced during the natural defense responses of plants. This includes PR proteins of which chitinase is the most important and widely studied one. Thus cloning and characterization of chitinase genes is the first step in development of transgenics resistant to fungal diseases. Therefore, cloning of endochitinase genes and transfering it to plants is a major steps for development of transgenic against resistant to plant pathogen. So in the present study, for the cloning of endochitinase gene from *ech*36 from *T. harzianum* (IABT1042) and sequenced functional analysis were done. The *ech*36 gene had 99% homology with reported endochitinase gene from *T. harzianum* endochitinase Chit36Y (chit36Y) gene, complete cds (AF406791.1) at nucleotide level, and 100% at amino acid level respectively. Similarly 90% homology was observed in *T. asperellum* and *T. viridae* cloned endochitinase gene (Saiprasad et al., 2009).

REFERENCES

Bendtsen JD, Nielsen H, Von Heijne G, Brunak S (2004). Improved prediction of signal peptides: SignalP 3.0. J. Mol. Biol. 340:783-795.

Chet I, Inbar J (1994). Biological control of fungal pathogens. Appl. Biochem. Biotechnol. 48:37-43.

Elad Y, David DR, Levi T, Kapat A, Kirshner B (1999). *Trichoderma harzianum* T-39-mechanisms of biocontrol of foliar pathogens. In: *Modern Fungicides and Antifungal Compounds* II (Eds. H. Lyr, P. E. Russell, H. W. Dehne, and H. D. Sisler). Andover, Hants, UK: *Intercept.,* pp. 459-467.

Hall TA (1999). Bioedit (a user friendly biological sequence 14alignment editor and analysis program for windos 95/98NT. *Nuclic acid. Symp.* Ser 41:95-98.

Haran S, Schickler H, Oppenheim A, Chet I (1996). Differential expression of *Trichoderma harzianum* chitinases during mycoparasitism. *Phytopathol* 86:980-985.

Harman GE, Howell CR, Viterbo A, Chet I, Lorito M (2004). *Trichoderma* species-opportunistic, avirulent plant symbionts. Nature Rev. Microbiol. 2(1):43-56.

Harman GE, Kubicek CP (1998). *Trichoderma* and *Gliocladium*. Taylor and Francis, London. P. 278.

Kapulnik Y, Chet I (2000). Induction and accumulation of PR proteins activity during early stages of root colonization by the mycoparasite *T. harzianum* strain T-203. Plant Physiol. Biochem. 38:863-873.

Liu M, Sun ZZ, Xu T, Harman GE, Lorito M (2004). Enhancing rice resistance to fungal pathogens by transformation with cell wall degrading enzyme genes from *Trichoderma atroviride* J. Zhejiang Uni. Sci. 5(2):133-136.

Lorito M (1998). Chitinolytic enzymes and their genes in : Harman G. E., Kubicek C. P. (Eds). *Trichoderma* and *Gliocladium*, Enzymes, Biological Control and Commercial Application, Taylor and Francis, London UK. pp. 73-99.

Lorito M, Harman GE, Howell CR, Viterbo A, Chet I (2004). *Trichoderma* species-opportunistic, avirulent plant symbionts. Nature Rev. Microbiol. 2(1):43-56.

Lorito M, Mach RL, Sposato P, Strauss J, Peterbauer CK, Kubicek CP (1996). Mycoparasitic interaction relieves binding of Cre1 carbon catabolite repressor protein to promoter sequence of *ech-42* (endochitinase-encoding) gene of *Trichoderma harzianum*. Proc. Nation. Acad. Sci. USA. 93:14868-14872.

Naseby DC, Pascual JA, Lynch JM (2000). Effect of biocontrol strains of *Trichoderma* on plant growth, *Pythium ultimum* population, soil microbial communities and soil enzyme. activities. J. Appl. Microbiol. 88:161-169

Roco A, Perez LM (2001). *In vitro* biocontrol activity of *Trichoderma harzianum* on *Alternaria alternata* in the presence of growth regulators. Electronic J. Biotechnol. 4(2), //www. ejbiotechnology. info/tent/vol4/ issue2/full/1/1. pdf.

Saiprasad GVS, Mythili JB, Anand A, Suneetha G, Rashmi HJ, Naveena C, Ganeshan G (2009). Development of *Trichoderma harzanium* endochitinase gene construct conferring antifungal activity in transgenic tobacco. Indian J. Biotechnol. (8):199-206.

Sambrook J, Russel DW (2001). Molecular Cloning: A Laboratory Manual, Cold Spring Harbour Laboratory, New York. pp. A8:52-55.

Sivasithamparam K, Ghisalberti F (1998). Secondary metabolism Trichoderma and Gliocladium. In : >Trichoderma and Gliocladium. Volume I=. (Eds. C. P. Kubicek and G. E. Harman). Taylor and Francis Ltd. London, pp. 139-191.

Vieira PM, Coelho ASG, Steindorff AS, Siqueira SJL, Silva RN, Ulhoa CJ (2013). Identification of differentially expressed genes from Trichoderma harzianum during growth on cell wall of Fusarium solani as a tool for biotechnological application. BMC Genomics. 14:177.

Zdobnov EM, Apweiler R (2001). InterProScan d an integration platform for the signature-recognition methods in InterPro. Bioinformatics 17:847–848.

8

A review of regeneration and genetic transformation in cowpea (*Vigna unguiculata* L. Walp)

Tie Manman, Luo Qian, Tan Huaqiang, Zhu Yongpeng, Lai Jia and Li Huanxiu

College of Horticulture, Sichuan Agricultural University, Ya'an 625014, Sichuan, China.

Biotechnology techniques including plant tissue culture and recombinant DNA technologies are important tools that can complement traditional breeding. Progress has been slow in cowpea (*Vigna unguiculata* L.), mainly due to its recalcitrant in plant regeneration and genetic transformation. Despite this, some achievements have been obtained. This review presents a consolidated account of explants used, mode of regeneration and genetic transformation in cowpea. Furthermore, it describes the discussion of some existing problems about regeneration and genetic transformation in cowpea. The authors find that organogenesis pathway and Agrobacterium mediated method is widely used in this area in cowpea recent years. Other explants such as anther should be researched in regeneration in cowpea and the affecting factors during transformation process should be explored further.

Key words: Regeneration, organogenesis, somatic embryogenesis, genetic transformation, cowpea.

INTRODUCTION

Cowpea (*Vigna unguiculata* L.), widely grown in Africa, Latin America, Southeast Asia and southwestern regions of North America (FAOSTAT, 2006), is a major source of high-quality dietary protein and energy for local people. It plays a critical role in the lives of millions of people in developing countries of Africa and Asia. In spite of the great importance of this crop, its productivity is low, which is mainly limited by the damage caused by biotic constraints like virus, bacteria, fungus, insects, plants, and nematodes, as well as abiotic stresses such as heat and drought (Singh et al., 1997). In addition, limited genetic diversity in cowpea breeding programs is of special concern because cowpea appears to have lower inherent genetic diversity than other cultivated crops as a result of a hypothesized single domestication event (Fang et al., 2007). Consequently, the transfer of exogenous gene by genetic engineering could potentially accelerate the breeding process. Although some resistance genes to insect pests and fungi have been bred in some International Institute of Tropical Agriculture (IITA) cowpea varieties (Latunde-Dada, 1990), the attempts using conventional breeding methods to introduce the resistance genes into the cultivated cowpea have made little progress for the strong hybrid-incompatibility. Hence, genetic engineering approaches stand out as the most effective alternative strategy to overcome the production constraints (Zaidi et al., 2005). This paper reviews previous work on cowpea tissue culture and transformation.

REGENERATION SYSTEM

There are two pathways in plant regeneration, namely organogenesis and embryogenesis, both of which have been used in cowpea regeneration. Some of the reports affected by many factors, such as appropriate genotype, optimal explants, certain culture medium, specific plant growth regulators, etc.

Organogenesis pathway

Organogenesis pathway means that the regenerative plants differentiate shoot buds directly through eristematic center, thus forming a complete plant eventually. It could be divided into direct organogenesis and indirect organo genesis based on the presence or absence of callus. The successful establishment of a regeneration system is affected by many factors, such as appropriate genotype, optimal explants, certain culture medium, specific plant growth regulators, etc.

Effect of genotype

Genotype is one of the main factors that influence the regeneration in different plant species. Among the different factors affecting cowpea regeneration, the organogenic response varies widely on account of the genotypes. The cotyledons of 36 genotypes of cowpea were cultured by Brar et al. (1999a), 19 among which failed to induce buds. Regeneration percentages (1 to 11%) and the numbers of shoots (4 to 12 per explant) of the other 17 cowpea genotypes showed significant differences. Chaudhury et al. (2007) tested the regeneration capability of seven cultivars of *V. unguiculata* that is, IC-202786, IC-257438, IC-259159, IC-243501, V-240, V-130 and V-585. The frequency of regeneration and number of multiple shoots from the explants were affected by the genotype. Of these, the commercially grown cultivar V-585 produced the highest number of shoots, with an average of 6.8 shoots per explant and an average length of 1.7 cm. Other scholars like Pellegrineschi (1997); Popelk et al. (2006); Manoharan et al. (2008); Solleti et al. (2008a); Raveendar et al. (2009); Bakshi et al. (2012a) also concluded that genotypes have influence on shoot regeneration.

Type and size of explants

The plant regeneration is largely governed by the appropriate choice of the explants. For legume, there are many kinds of explants to choose. Epicotyls, hypocotyls, cotyledons, cotyledonary nodes, shoot tips, plumular apices and shoot meristems (Table 1) are commonly used in regeneration via organogenesis pathway. Both explants types and the size of explants have great influence on regeneration.

Amitha and Reddy (1996) studied the regeneration potential of epicotyl, cotyledon and hypocotyl explants of cowpea. The shoot buds produced from different explants proliferated and regenerated into complete plants. Bao et al. (2006) take 5 parts of cowpea that is, hypocotyls, epicotyls, stem segments, leaves and terminal buds under the same condition for callus and buds induction. All the 5 types of explant are able to induce callus in spite of the obviously different induction rate. On the optimal medium selected for different explants, the callus induced from hypocotyls, epicotyls, stem segments and terminal buds are easy to differentiate adventitious buds, but the leaves cannot produce buds. Li (2011) did a similar test, taking leaves, stem segments, cotyledons and cotyledonary nodes as explants, founding that leaves and stem segments could only induce callus and could not differentiate buds, while the cotyledon and cotyledonary node explants could induce shoots directly, and the shoot induction rate of cotyledonary nodes could reach 100%. Diallo et al. (2008) used three types of explants, that is, cotyledonary nodes with one cotyledon, two cotyledons and without cotyledon. The shoots multiplication rate was influenced by the number of cotyledons. Explants with both entire cotyledons produced the greater number of shoots on the optimal medium, followed by the explants with one cotyledon. The worst results were observed with explants without cotyledons. This is consistent with the research of Chaudhury et al. (2007) and Raveendar et al. (2009). However, researchers such as Chaudhury et al. (2007); Solleti et al. (2008b), Bakshi et al. (2011, 2012a) used cotyledonary nodes without cotyledons for transformation, they thought it will facilitate the transformation of genes and the cleaning of fungus liquid. Cowpea cotyledonary node is widely used in cowpea regeneration recent years due to its efficient transformation regeneration frequency.

In addition to explant types, the cutting mode and size of explants have significant effects on regeneration. Three types of explants were used for cowpea regeneration by Le et al. (2002), that is, longitudinal thin cell layers (0.5×10 mm), transverse thin cell layers (0.3 to 0.5 mm) and longitudinal half axis (5×10 mm). The buds induction frequency of both longitudinal thin cell layers and longitudinal half axis could reach 100%, 32.5 and 23.2 buds were harvested per explants respectively. While the frequency of the transverse thin cell layers was only 52.9%, and the average buds was 11.7 per explant. Muthukumar et al. (1996) studied the regenerative capacity of different cutting methods on cotyledons. Whole cotyledons with intact proximal end without cotyledonary node tissue regenerated well. Regeneration rate of the proximal end of longitudinally sliced halves was higher than the transverse halves containing proximal end.

Effect of basal medium and plant growth regulators

The medium is the basis of tissue culture. Medium containing different ingredients will produce different results in the cultivation of cowpea explants. Three kind of basal medium that is, MS, B5 and MSB (MS salts + B5 vitamins) were commonly used in cowpea tissue culture.

Table 1. *In vitro* plant regeneration in cowpea.

Explants type	Mode of regeneration	Medium+PGRs	References
Shoot apical meristems	Organogenesis	MS+BA+NAA	Kartha et al. (1981)
Immature cotyledons	Somatic embryogenesis	MSB+2,4-D+BA MSB+IAA+KT	Li et al. (1993)
Primary leaves	Organogenesis	B5+BA	Muthukumar et al. (1995)
Cotyledons	Organogenesis	B5+BA	Muthukumar et al. (1996)
Epicotyls, cotyledons and hypocotyls	Organogenesis	MS+BA/IAA	Amitha and Reddy (1996)
Shoot tip	Organogenesis	MS+BA+NAA/2,4-D	Brar et al. (1997)
Hypocotyls and cotyledons	Organogenesis	BM/FBM+BA/ZE/KT+IAA/NAA	Pellegrineschi (1997)
Cotyledons	Organogenesis	MS+BA	Brar et al. (1999a)
Primary leaves	Somatic embryogenesis	MS+2,4-D/ABA	Anand et al. (2000)
cotyledonary node thin cell layer explants	Organogenesis	MSB+IBA/TDZ	Le et al. (2002)
Primary leaves	Somatic embryogenesis	MSB/ B5+2,4-D/ABA	Ramakrishnan et al. (2005)
Embryo axes	Organogenesis	MS+NAA/BA	Odutayo et al. 2005)
Shoot apices	Organogenesis	MSB+BA/GA3/IBA	Mao et al. (2006)
Hypocotyls, epicotyls, stem segments, leaves and terminal buds	Organogenesis	MS+BA/NAA/2,4-D	Bao et al. (2006)
Cotyledonary nodes	Organogenesis	MSB/MS+BA/IBA	Chaudhury et al. (2007)
Embryonic axis explants	Organogenesis	MSB+BA/ZT/GA3/IAA	Raji et al. 2008)
Cotyledonary nodes	Organogenesis	MS/ B5+BA/KT/NAA	Diallo et al. (2008)
Shoot meristems	Organogenesis	MS+BA	Manoharan et al. (2008)
Embryonic axis explants	Organogenesis	MSB+BA	Yusuf et al. (2008)
Cotyledonary nodes	Organogenesis	MSB+BA	Raveendar et al. 92009)
Cotyledonary nodes	Organogenesis	MSB/MS+ BA/KT/GA3/IBA	Solleti et al. (2008b)
Shoot meristems	Organogenesis	MS+BA/NAA/IBA	Aasim et al. (2008)
Shoot tip explants	Organogenesis	MS+TDZ/IBA	Aasim et al. (2009a)
Plumular apices	Organogenesis	MS+BA/NAA/IBA	Aasim et al. (2009b)
Leaves, stem segments, cotyledons, cotyledonary nodes	Organogenesis	MSB+BA/TDZ+IBA	Li 2011)
Cotyledonary nodes	Organogenesis	MSB+BA/TDZ	Bakshi et al. (2011, 2012a)

Chaudhury et al. (2007) used MSB as the basal medium for bud induction while MS for rooting. Diallo et al. (2008) studied the influence of basal medium for cowpea regeneration. More efficient multiple shoots from cotyledonary node explants were obtained on B5 medium. But MS medium was more conducive to shoot elongation. Rooting was favored on half strength MS medium (Bao et al., 2006).

Plant hormones play an important role in the process of cowpea regeneration. The type and concentration of hormone vary at different stages of induction.

In general, higher concentration of cytokinin alone or in combination with a lower concentration of auxin was often used for shoot buds induction. For example, Chaudhury et al. (2007) assessed the effect of different concentration (2.5-10 μM) of BA on shoot buds induction. They observed that BA at 5 μM (1.25 mg/L) was optimal for shoot regeneration as it induced an average of 6.3 shoots (three-fold increase than control) in 80% of the cultures.

Muthukumar et al. (1995) and Solleti et al. (2008a) came to the same conclusion. Basal medium supplemented with 4.4 μM (1 mg/L) BA was used for shoot formation by other scholars (Diallo et al., 2008; Aasim et al., 2009b) from different types of explants (Table 1). Muthukumar et al. (1996) found that when the concentration of BA exceeds 15 μM (3.5 mg/L), the ability of shoot regeneration was inhibited clearly. The regenerated buds were difficult to elongate with the increasing of BA concentration.

Aasim et al. (2009a) compared the effects of different concentrations of TDZ (0.15 to 0.35 mg/L) on shoot regeneration from shoot tip explants. The results presented that shoot regeneration rate raised with increase in TDZ concentration. But higher TDZ concentratlon was observed inhibiting shoot elongation. BA was better than TDZ on shoot regeneration which was reported by Solleti et al. (2008a). High concentration of cytokinins (BA and KT) in conjunction with low concentration of auxins (NAA and 2,4-D) produced higher number of shoots together with higher elongation. Aasim et al. (2009b) and Li (2011) investigated and found that adding low concentration of auxins or not had no clear effect on shoot induction. Many studies showed that KT and GA_3 have apparently stimulative effect on the elongation of adventitious shoots. Solleti et al. (2008a) assessed the effect of KT and GA_3 on shoot elongation.The results showed that 0.5 μM KT enhanced the shoot elongation. Medium containing GA_3 induced non-uniform elongation and the elongated shoots were found thin and lanky, unsuitable for rooting.

Some types of auxin were used for rooting such as IBA (Chaudhury et al., 2007; Diallo et al., 2008; Solleti et al., 2008; Aasim et al., 2009a; Li, 2011), IAA (Mao et al., 2006; Raji et al., 2008) and NAA (Brar et al., 1997;

Diallo et al., 2008) and 2,4-D (Brar et al., 1997). Diallo et al. (2008) explored the influence of different concentration (0.5-5 mg/L) of IBA in comparison with NAA. IBA was found to be more effective for rooting because of the higher root length and activity. Media containing 2.5 μM (0.5 mg/L) IBA were used for rooting (Aasim et al., 2009a; Chaudhury et al., 2007; Solleti et al., 2008a; Li, 2011). Rooting formation disposed with NAA or 2,4-D was described by Brar et al. (1997). NAA was better for rhizogenesis at the concentration of 0.1-0.5 mg/L. Some scholars, however, believed that there was no apparent promoting on rhizogenesis when adding auxins (Mao et al., 2006). Raji et al. (2008) also confirmed that the root induction rate on medium without hormone could reach 100%, same as the effect of low concentration (0.01 mg/L) of IAA. Furthermore, increasing the concentration of IAA would inhibit root formation.

Many studies have found that adding cytokinin to seedling medium have a great impact on regeneration. Amitha and Reddy (1996) preconditioned the seedlings with BA (2 mg/L) and CM (coconut milk) (5%) on MS medium. The regeneration frequency and the shoot buds number presented variety compared to the control group. Brar et al. (1999a) came to a conclusion that explants derived from the seedlings growing on the medium supplemented with 66.6 μM (15 mg/L) BA resulted in the highest regeneration percentage. While the highest number of shoots was observed from those with 155.3 μM (35 mg/L) BA. Le et al. (2002) preprocessed the seedlings with different concentrations of TDZ (0, 1, 5, 10, 20 and 50 μM). The result indicated that the addition of 10 μM/L TDZ was the best for buds induction. Similar result had been reported by Bakshi et al. (2012a). A 10 μM TDZ preconditioning enhanced shoot proliferation ability. Aasim et al. (2009b) observed that seeds pre-cultured with 10 mg/L BA for 5 days could increase the regeneration rate of plumular apices. While Raveendar et al. (2009) drew a conclusion that seeds germinated on MSB medium containing 13.3 μM (4 mg/L) BA for 3 days and cultured on the subsequent optimal medium produced the highest number of shoots (7.2 ± 0.8 per explant), evidently higher than that of control (4.7 ± 0.6 per explant).

Effect of other additives

As a kind of growth regulator, ethylene has an effect on *in vitro* culture. Brar et al. (1999b) studied the regeneration of cowpea from cotyledons with the processing of three ethylene inhibitors respectively, that is, silver nitrate ($AgNO_3$), 2,5-norbornadiene and cobalt chloride ($CoCl_2$). They found the regeneration varied owing to the concentration and type of ethylene inhibitors. The addition of 50 μM $AgNO_3$ or 100 μM 2,5-norbornadiene could improve induction rate of adventitious buds. The greatest percentage of regeneration was observed by adding 60 μM $AgNO_3$ either to both the initiation and regeneration stages, or to only the regeneration stage. $CoCl_2$ at 25 μM or 2,5-norbornadiene at 100 μM enhanced the number of shoots per explant. Either 25 μM $CoCl_2$ or 2,5-norbornadiene was the best for shoot elongation. Mao et al. (2006) found that the addition of 58.8 μM $AgNO_3$ was helpful to reduce callus browning, and this supported the conclusion of Brar et al. (1999b). Aasim et al. (2009b) added 3 g/L active charcoal and 1.25 mg/L polyvinylpyrrolidone (PVP) in the medium. High frequency of callus induction could be obtained due to significantly inhibition of the bleeding of phenolic compounds from the explants.

Somatic embryogenesis

Embryogenesis means that the individual cells experience a series of morphological development stages similar to the formation of zygotic embryo. Embryogenesis is considered better than other in vitro propagation systems as its relatively high genetic uniformity and shorter time in regeneration and propagation. Somatic embryogenesis takes a prominent role in clonal propagation when integrated with a conventional breeding program and molecular and cellular biotechnology. It provides a valuable tool to enhance the pace of genetic improvement of commercial crop species (Stasolla and Yeung, 2003).

Type of explants and plant growth regulators

Explants have been used in cowpea somatic embryogenesis pathway were only immature embryo (Li et al., 1993) and primary leaves (Anand et al., 2000; Ramakrishnan et al., 2005). Li et al. (1993) studied the cowpea somatic embryogenesis using immature embryo which developed 10 to15 days after flowering. The results showed that at the callus induction stage by protoplast, the improved MS medium was more effective than B5 medium.

The former frequency of cell division at the tenth day was 27.7%, while the latter was 20.5%. In their study, MSB was employed as the basal medium in the subsequent culture. 2,4-D and BA was added in callus subculture, while IAA and KT was added in the developmental stages of proembryo. Anand et al. (2000) took leaf segments from young seedlings as explants, used MS as basal medium and added 2,4-D as phytohormone. Maximum proliferation of callus was observed within 15 days on medium containing 6.78 μM 2,4-D. 2.26 μM 2,4-D was the best for embryos development. This conclusion was confirmed by Ramakrishnan et al. (2005). In the process of torpedo-stage somatic embryo transfer to cotyledonary stage, the concentration of 2,4-D shoud be reduced. The presence of ABA was also crucial for the maturity of embryo (Anand et al., 2000; Ramakrishnan et al., 2005). The optimum concentration of ABA is 5 μM. Ramakrishnan et al. (2005) compared the different effects of TDZ and ZE in plantlet conversion. The results indicated that TDZ was crucial for plantlet conversion, while ZE would cause recallusing of cotyledonary-stage embryos. Extension of hypocotyls and complete development of plantlet was achieved on half-strength B5 medium supplemented with 3% maltose, 2500 mg/L potassium nitrate and 0.05 mg/L thidiazuron (TDZ) with 32% regeneration frequency. Nevertheless, Anand et al. (2000) thought that ZE had an enormous effect on plantlet conversion. The maximum percentage of conversion (21.8%) occurred on 1/2 MS semisolid medium containing ZE (2 μM), ABA (5 μM) and mannitol (3%).

Type of the basal medium and other additives

According to Ramakrishnan et al. (2005), MSB had a better effect on embryogenic callus induction and proembryo formation. B5 was essential to the maturity of embryos and the reduction of the number of abnormal embryoid, and 1/2 B5 was very important for cotyledonary-stage somatic embryos to convert into complete plantlets. MS medium was ordinary in the whole process of cowpea somatic embryogenesis without any particular superior effect. They also found that adding 150 mg/l CH and 100 mg/l Gln to medium which contain 1.5mg/l 2,4-D would promote embryogenic callus induction. Among different forms of carbon source, 3% sucrose was better than 3% glucose and maltose in embryogenic callus induction. The frequency of callus induction could reach 100% with 3% sucrose and MSB medium containing 150 mg/L CH and 100 mg/L Gln. However, sugar alcohols such as mannitol and sorbitol were not effective in increasing the frequency of callus induction irrespective of the media.

Other factors

Anand et al. (2000) drew the conclusion that at the stage of suspension-cultured cells of cowpea, a shaking speed of 90 rpm and 0.4 ml packed cell volume per 25 ml medium were found to be optimal for maintaining suspension cultures.

ESTABLISHMENT OF COWPEA GENETIC TRANSFORMATION SYSTEM

There are a variety of methods to establish plant genetic transformation system, but only Agrobacterium mediated and particle bombardment transformation methods have been used in cowpea to date.

Agrobacterium mediated genetic transformation

Garcia et al. (1986, 1987) were the first to employ transformation experiments in cowpea (Table 2). They used an *Agrobacterium*-mediated system to transform cowpea leaf disc explants. Although kanamycin-resistant callus was obtained, the transgenic calli failed to regenerate into mature plants. Perkins et al. (1987) and Filippone (1990) achieved transgenic calli from different explants co-cultivated with *Agrobacterium tumefaciens* respectively, but no intact transgenic plants could be regenerated either. Penza et al. (1991) reported they obtained chimeric transgenic cowpea plants from longitudinal mature embryo slices co-cultivated with *A. tumefaciens*.

The report showed that transformed cells mainly located in the subepidermal regions of the plant stems and there was no evidence of stable integration of the introduced genes. All the experiments above demonstrated cowpea is susceptible to *A. tumefaciens*, however, failed to regenerate transgenic plants.

The first production of transgenic cowpea plants was reported by Muthukumar et al. (1996). They used de-embryonated cotyledons as explants co-cultivating with *A. tumefaciens*, followed by transfer of the explants to a selective medium, and recovered hygromycin-resistant shoots that grew to maturity and set seed. Only one of the plants was confirmed to be transgenic, but the seeds showed failure to germinate and no evidence of transgene transmission to the progeny was obtained. Similarly, Kononowicz et al. (1997) developed an efficient genetic transformation system for cowpea. The introduced gene was detected in T0 cowpea plants but the evidence of transferred gene in the T1 progeny was not presented.

The first success of obtaining transgenic cowpea and transmitting the transgene to their progenies following Mendelian laws was described by Popelka et al. (2006).They obtained transgenic plants with a frequency of transformation of 0.05 to 0.15% which is slightly lower. This technique was improved by Chaudhury et al. (2007) who used cotyledonary node explants as Popelka's group, but they inflicted wounds at the nodal region with

Table 2. Key dates of cowpea genetic transformation.

Date	Methods	Gene	Result	References
1986	*Agbacterium tumefaciens*	Kanamycin resistant	Transgenic calli	Garcia et al. (1986)
1987	*Agbacterium tumefaciens*	Kanamycin resistant	Transgenic calli	Garcia et al. (1987)
1987	*Agbacterium tumefaciens*	Kanamycin resistant	Transgenic calli	Perkins et al. (1987)
1991	*Agrobacterium tumefaciens*	gus	Putative transgenic plant	Penza et al. (1991)
1992	Biolistic	gus	Transient expression	Penza et al. (1992)
1993	Biolistic	gus	Transient expression	Akella and Lurquin (1993)
1996	Agrobacterium tumefaciens	Hygromycin resistant	Transgenic plants, no evidence of transgenic progenies	Muthukumar et al. (1996)
1997	Agrobacterium tumefaciens /Biolistic	α-amylase inhibitor	Transgenic plants, no evidence of transgenic progenies	Kononowicz et al. (1997)
2003	Biolistic	bar	Small proportion of transgenic progenies	Ikea et al. (2003)
2006	*Agrobacterium tumefaciens*	bar	Transgenic progenies	Popelka et al. (2006)
2007	*Agrobacterium tumefaciens*	Kanamycin resistant	Transgenic progenies	Chaudhury et al. (2007)
2008	*Agrobacterium tumefaciens*	gus	Transgenic shoots	Raji et al.,2008
2008	Biolistic	ahas	Transgenic progenies	Ivo et al. (2008)
2008	*Agrobacterium tumefaciens*	nptII and gus	Transgenic progenies	Solleti et al. (2008a)
2008	*Agrobacterium tumefaciens*	α-amylase inhibitor-1	Transgenic progenies	Solleti et al. (2008b)
2011	*Agrobacterium tumefaciens*	cry1Ac	Transgenic plants	Bakshi et al. (2011)
2012	*Agrobacterium tumefaciens*	nptII and gus	Transgenic progenies	Bakshi et al. (2012a)
2012	*Agrobacterium tumefaciens*	pmi	Transgenic progenies	Bakshi et al. (2012b)

sterile needle prior to Agrobacterium infection and generated transgenic plants with an efficiency of 0.76%.

Solleti's group reported much higher transformation efficiencies of 1.64 and 1.67%. They added virulence genes in resident pSB1 vector in Agrobacterium strain LBA4404 and used a regimen of geneticin selection at 45 mg/L, thereby increasing the transformation efficiency and providing a rapid and efficient identification system (Solleti et al. 2008a).

Using the same regimen, they introduced the bean a-amylase inhibitor-1 (aAI-1) gene into an Indian cowpea cultivar, and obtained the transgenic seeds that strongly inhibited the development of *C. maculatus* and *C. chinensis* (Solleti et al., 2008b). Bakshi et al. (2011) improved the method of Agrobacterium-mediated transformation of cowpea. They used both sonication and vacuum infiltration to dispose cotyledonary node explants with *A. tumefaciens*. The stable transformation efficiency of treatments increased by 88.4% compared to the traditional Agrobacterium-mediated transformation in cowpea. Shortly after that, they described how seedling preconditioning in thidiazuron (TDZ) and BA affected the shoot proliferation potential of cotyledonary nodes and transformation process. Explants of TDZ preconditioned seedlings presented significantly higher transient transformation rate as compared to that of explants of BA preconditioned seedlings. Best results were obtained under the condition of explants deriving from seedling preconditioning in 10 μM TDZ for 4 days. The transformation rate enhanced from 0.6 to 2.1% on comparison of in absence and presence of seedling preconditioning (Bakshi et al., 2012a). Besides, they developed a new method for obtaining transgenic cowpea using of *pmi* selection. This system was based on the ability to inhibit shoot organogenesis from non-transformed explants cultured on medium containing mannose as a carbon source, thus obtaining efficient shoot proliferation (Bakshi et al., 2012b).

Genetic transformation by particle bombardment

Penza's and Akella's groups separately applied electroporation method via intact cowpea embryonic tissues, using naked DNA in the presence of protectants such as spermine (Penza et al., 1992; Akella and Lurquin, 1993), but no stably transformed plants were obtained. Ikea et al. (2003) were able to generate transgenic cowpea plants using the particle bombardment process. However, the transgenes were transmitted to only a small proportion of the progeny and there was no further molecular evidence of stable transformation with Mendelian laws. Ivo et al. (2008), employing the biolistic method of gene transfer, were able to generate stable transgenic cowpea plants from bombarded embryonic axes.

Their system is built on combining the use of the herbicide imazapyr to select transformed meristematic cells and a simple tissue culture protocol. The gus gene was used as a reporter gene and their transformation frequency was 0.90%, thereby presenting the first work on the use of this approach of biolistic-mediated for generating transgenic cowpea plants and obtaining the progenies (first and second generations) that co-segregated in a Mendelian fashion.

CONCLUSIONS AND FUTURE PROSPECTS

Significant progress aimed towards regeneration and genetic transformation of cowpea has been made, but still there is a long way to go in this direction. As has been described above, cowpea regeneration depended mainly on direct or indirect organogenesis pathway. Recent years, the most desirable mode of regeneration is via direct organogenesis, and cotyledonary node explants have been preferred. In addition, some researchers have attempted to establish regeneration system via somatic embryogenesis. It is important to note that a liquid suspension protocol is relatively efficient, because a high frequency of somatic embryos can be obtained, and individual embryos can be handled easily for further biotechnological applications (Anand et al., 2000).

In general, two methods including *Agrobacterium*-mediated and particle bombardment have been successfully applied to genetic transformation of cowpea. The continued search has lead to the obtaining of transgenic cowpea and their progenies. Although the transformation frequency keeps increasing, it is still lower compared to other legumes. Furthermore, the efficient recovery of transformed plants depends not only on the mode of regeneration and choice of transformation procedure but also on selectable markers (Chandra and Pental, 2003). Although some results have been obtained with the negative selection protocols, it has not proved an efficient method. The selection of putative transformed shoots on antibiotic or herbicide-supplemented medium lead to regeneration decreasing, growth retardation and low rooting efficiency, all were caused by long-time exposure to stringent selection. Using the 6-phosphomannose isomerase gene as the selectable marker (Bakshi et al., 2011) has been the one and only new attempt for obtaining recovery transgenic cowpea, and further research is needed. In conclusion, developing protocols to enhance regeneration still remains an important goal in cowpea, and the recovery of transgenic cowpea should be the focus of future research.

Abbreviations: **ABA,** Abscisic acid; **BA,** benzylaminopurine; **BM,** basal medium; **CH,** casein hydrolysate; **FBM,** fortified basal medium; **GA3,** gibberellin; **GLN,** L-glutamic acid-5-amide; **IAA,** indoleacetic acid; **IBA**, indole-3-butytric acid; **IITA,** international institute of tropical agriculture; KT, N6-furfuryladenine; **MS,** Murashige and Skoog (1962) medium; **MSB,** Murashige and Skoog (1962) salts and gamborg b5 vitamins (1968) medium; **NAA,** A-naphthlcetic acid; **PGRS**, plant growth regulators; **PVP,** polyvinylpyrrolidone; **TCL,** thin cell layer; **TDZ,** THIDIAZURON; **2,4-D,** (2,4- dichlorphenoxy) acetic acid; **ZE,** zeatin.

REFERENCES

Aasim M, Khawar KM, Ozcan S (2008). In vitro micropropagation from plumular apices of Turkish cowpea (Vigna unguiculata L.) cultivar Akkiz. Sci. Hortic. 122:468-471.

Aasim M, Khawar KM, Ozcan S (2009a). Comparison of shoot regeneration on different concentrations of thidiazuron from shoot tip explants of cowpea on gelrite and agar containing medium. Notulae Botanicae Horti Agrobo. Cluj-Napoca 37(1):89-93.

Aasim M, Khawar KM, Ozcan S (2009b). In vitro micropropagation from plumular apices of Turkish cowpea (Vigna unguiculata L.) cultivar Akkiz. Sci. Hortic. 122:468-471.

Akella,V, Lurquin PF (1993). Expression in cowpea seedlings of chimeric transgenes after electroporation into seed-derived embryos. Plant Cell Rep. 12:110-117.

Amitha K, Reddy TP (1996). Regeneration of plantlets from different explants and callus cultures of cowpea (Vigna unguiculata L.). Phytomorphology 46(3):207-211.

Anand RP, Ganapathi A, Anbazhagan VR, Vengadesan G, Selvaraj N (2000). High frequency plant regeneration via somatic embryogenesis in cell suspension cultures of cowpea vigna unguiculata (L.) walp. *In Vitro* Cell. Dev. Biol. Plant. 36:475-480.

Bakshi S, Sadhukhan A, Mishra S, Sahoo L (2011). Improved Agrobacterium-mediated transformation of cowpea via sonication and vacuum infiltration. Plant Cell Rep. 30:2281-2292.

Bakshi S, Roy NK, Sahoo L (2012a). Seedling preconditionng in thidiazuron enhances axillary shoot proliferation and recovery of transgenic cowpea plants. Plant Cell Tiss.110:77-91.

Bakshi S, Saha B, Roy NK, Mishra S, Panda SK, Sahoo L (2012b). Successful recovery of transgenic cowpea (*Vigna unguiculata*) using the 6-phosphomannose isomerase gene as the selectable marker. Plant Cell Rep. 31:1093-1103.

Bao YH, Bai Y, Wang YM, Huang YY, Xu XJ, Xu QW (2006). The regeneration of cowpea (Vigna unguiculata L.). Agric. Sci. GuangDong 4:31-33.

Brar MS, Al-Khayri, JM, Shambein CE, Mcnew RW, Morelock TE, Anderson EJ (1997). In vitro shoot tip multiplication of cowpea Vigna unguiculata (L.) Walp. *In Vitro* Cell. Dev. Biol. Plant 33:114-118.

Brar MS, Al-Khayri JM, Morelock TE, Anderson EJ (1999a). Genotypic response of cowpea (Vigna unguiculata L.) to in vitro regeneration from cotyledon explants. *In Vitro* Cell. Dev. Biol. Plant. 35:8-12.

Brar MS, Moore MJ, Al-Khayri JM, Morelock TE, Anderson EJ (1999b). Ethylene inhibitors promote in vitro regeneration of cowpea (Vigna unguiculata L.). *In Vitro* Cell. Dev. Biol.Plant 35:222-225.

Chandra A, Pental D (2003). Regeneration and genetic transformation of grain legumes: An overview. Curr. Sci. 84:381-387.

Chaudhury D, Madanpotra S, Jaiwal R, Saini R, Kumar AP, Jaiwal PK (2007). Agrobacterium tumefaciens-mediated high frequency genetic transformation of an Indian cowpea (Vigna unguiculata L. Walp.) cultivar and transmission of transgenes into progeny. Plant Sci. 172:692-700.

Diallo MS, Ndiaye A, Sagna M, Gassama-Dia YK (2008). Plants regeneration from African cowpea (*Vigna unguiculata* L.) variety. Afr. J. Biotechnol. 16:2828-2833.

Fang J, Chao CT, Roberts PA, Ehlers JD (2007). Genetic diversity of cowpea [*Vigna unguiculata* (L.) Walp.] in four West African and USA breeding programs as determined by AFLP analysis. Genet Resour Crop Evol. 54:1197-1209.

FAOSTAT (2006) FAO agriculture database. FAO, Rome. Available at http://faostat.fao.org/faostat

Filippone E (1990). Genetic transformation of pea and cowpea by co-cultivation of tissues with Agrobacterium tumefaciens carrying binary vectors. Cowpea genetic resources, contributions in cowpea exploration, evaluation and research from Italy and IITA 175-181.

Gamborg OL, Miller RA, Ojima K (1968). Nutrient requirements of suspension cultures of soybean root cells. Exp. Cell Res. 50:151-158.

Garcia JA, Hille J, Goldbach R (1986). Transformation of cowpea (Vigna unguiculata) cells with an antibiotic-resistance gene using a Ti-plasmid-derived vector. Plant Sci. 44:37-46.

Garcia JA, Hille J, Vos P, Goldbach R (1987). Transformation of cowpea (Vigna unguiculata) with a full-length DNA copy of cowpea mosaic virus M-RNA. Plant Sci. 48:89-98.

Ikea J, Ingelbrecht I, Uwaifo A, Thotttappilly G (2003). Stable gene transformation in cowpea (Vigna unguiculata L. Walp.) using particle gun method. Afr. J. Biotechnol. 2:211-218.

Ivo NL, Nascimento CP, Vieira LS, Campos FAP, Aragao FJL (2008).Biolistic-mediated genetic transformation of cowpea (Vigna unguiculata) and stable Mendelian inheritance of transgenes. Plant Cell Rep. 27:1475-1483

Kartha KK, Pahl K, Leung NL (1981). Plant regeneration from meristems of grain legumes: soybean, cowpea, peanut, chickpea, and bean. Can. J. Bot. 59:1671-1679.

Kononowicz AK, Cheah KT, Narasimhan ML, Murdock LL, Shade RE, Chrispeels MJ, Filippone E, Monti LM, Bressan RA, Hasegawa PM (1997). Development of transformation system for cowpea (Vigna unguiculata L. Walp). Adv. Cowpea Res. 361-371.

Latunde-Dada AO (1990). Genetic manipulation of the cowpea (Vigna unguiculata [L.] Walp.) for enhanced resistance to fungal pathogens and insect pests. Advances in Agronomy 44:133-154.

Le BV, Carvalho MH, Zuily-Fodil Y, Thi AT, Van KTH (2002). Direct whole plant regeneration of cowpea [(Vigna unguiculata (L.) Walp.] from cotyledonary node thin cell layer explants. J. Plant Physiol.159:1255-1258.

Li XB, Xu ZH, Wei ZM, Bai YY (1993). Somatic embryogenesis and plant regeneration from protoplasts of cowpea (vigna sinensis). Acta. Botnica. Sinica. 35:632-636.

Li XM (2011). Establishment of in vitro high efficient regeneration system of cowpea (Vigna unguiculata L.Walp.) and screening test for Its resistance to kanamycin. Sichuan Agricultural University, China.

Manoharan M, Khan S, James OG (2008). Improved plant regeneration in cowpea through shoot meristem. J. App. Hortic. 10(1):40-43.

Mao JQ, Zaidi MA, Arnason JT, Altosaar I (2006). In vitro regeneration of Vigna unguiculata (L.) Walp. cv. Blackeye cowpea via shoot organogenesis. Plant Cell Tiss. Org. 87:121-125.

Murashige T, Skoog F (1962). A revised medium for rapid growth and bioassays with tobacco tissue cultures. Physiol. Plantarum 15:473-497.

Muthukumar B, Mariamma M, Gnanam A (1995). Regeneration of plants from primary leaves of cowpea. Plant Cell Tiss. Org. 42:153-155.

Muthukumar B, Mariamma M, Veluthambi K, Gnanam A (1996). Genetic transformation of cotyledon explants of cowpea (Vigna unguiculata L. Walp) using Agrobacterium tumefaciens. Plant Cell Rep. 15:980-985.

Odutayo OI, Akinrimisi FB, Ogunbosoye I, Oso RT (2005). Multiple shoot induction from embryo derived callus cultures of cowpea (Vigna unguiculata L.Walp.). Afr. J. Biotechnol. 4(11):1214-1216.

Penza R, Lurquin PF, Filippone E (1991). Gene transfer by cocultivation of mature embryos with Agrobacterium tumefaciens-application to cowpea (Vigna unguiculata Walp). J. Plant Physiol. 138:39-43.

Penza R, Akella V, Lurquin PF (1992). Transient expression and histological localization of a gus chimeric gene after direct transfer to mature cowpea embryos. Biotechniques 13:576-580.

Pellegrineschi A (1997). *In vitro* plant regeneration via organogenesis of cowpea (Vigna unguiculata L.). Plant Cell Rep. 17:89-95.

Perkins EJ, Stiff CM, Lurquin PF (1987). Use of *Alcaligenes eutropus* as a source of genes for 2,4-D resistance in plants. Weed Sci. 35:12-18.

Popelka JC, Gollasch S, Moore A, Molvig L, Higgins TJV (2006). Genetic transformation of cowpea (Vigna unguiculata L.) and stable transmission of the transgenes to progeny. Plant Cell Rep. 25:304-312.

Raji AAJ, Oriero E, Odeseye B, Odunlami T, Ingelbrecht IL (2008). Plant regeneration and Agrobacterium-mediated transformation of African cowpea [Vigna unguiculata (L.) Walp] genotypes using embryonic axis explants. J. Food Agric.Environ. 6:350-356.

Ramakrishnan K, Sivakumar PR, Manickam A (2005). In vitro somatic embryogenesis from cell suspension cultures of cowpea (Vigna unguiculata L.Walp). Plant Cell Rep. 24:449-461.

Raveendar S, Premkumar A, Sasikumar S, Ignacimuthu S, Agastian P (2009). Development of a rapid, highly efficient system of organogenesis in cowpea Vigna unguiculata (L.). Walp. S. Afr. J. Bot. 75:17-21.

Singh BB, Chambliss OL, Sharma B (1997). Recent advances in cowpea breeding. Advances in Cowpea Research. pp 30-50.

Solleti SK, Bakshi S, Sahoo L (2008a). Additional virulence genes in conjunction with efficient selection scheme and compatible culture regime enhance recovery of stable transgenic plants in cowpea via Agrobacterium tumefaciens-mediated transformation. J. Biotechnol 135:97-104.

Solleti SK, Bakshi S, Purkayastha J, Panda SK, Sahoo L (2008b). Transgenic cowpea (Vigna unguiculata) seeds expressing a bean α-amylase inhibitor 1 confer resistance to storage pests, bruchid beetles. Plant Cell Rep. 27:1841-1850.

Stasolla C, Yeung EC (2003). Recent advances in conifer somatic embryo genesis: improving somatic embryo quality. Plant Cell Tiss. Org. 74:15-35.

Yusuf M, Raji AA, Ingelbrecht I, Katung MD (2008). Regeneration efficiency of cowpea [Vigna unguiculata (L.) Walp.] via embryonic axes explants. Afr. J. Plant Sci. 2(8):105-108.

Zaidi MA, Mohammadi M, Postel S, Masson L, Altosaar I (2005). The Bt gene cry2Aa2 driven by a tissue specific ST-LS1 promoter from potato effectively controls Heliothis virescens. Transgenic Res. 14:289-298.

Microwave assisted hot air drying of papaya (*Carica papaya* L.) pretreated in osmotic solution

Alireza Yousefi, Mehrdad Niakousari and Mehdi Moradi

Department of Food Science and Technology, Shiraz University, Shiraz, Iran.

In this study, mathematical modeling of microwave (MW) assisting hot air drying of thin-layer papaya (*Carica papaya* L.) slices with 5 ± 1 mm thickness in an experimental drying process is presented. The osmosis solution comprised 50% sucrose + 2% NaCl solutions. The osmosis dehydration characteristics obtained by solid gain (SG), water loss (WL) and weight reduce (WR) parameters. The drying air velocity (0.9 ± 0.1 m/s) and temperatures (40, 50 and 60°C) were examined in drying papaya slices from initial moisture content of 700 ± 2% (d.b) to moisture content of 20 ± 1% (d.b). The slices were subjected to 10 s 540 W MW power at each 15 min interval as an auxiliary drying method. Ten thin-layer drying models were fitted to drying experimental data of papayas, implementing non-linear regression analysis techniques. The statistical analysis using correlation coefficient (R^2), chi-square (χ^2) and root mean square error (RMSE) concluded that, the best model in terms of fitting performance for MW assisted hot air drying of papaya pretreated in osmosis solution at 40 and 50°C was page model while that at 60°C was two-term model.

Key words: Papaya, mathematical modeling, microwave (MW), cabinet drier.

INTRODUCTION

Drying is a technique of conservation that consists of the elimination of large amount of water present in a food by the application of heat under controlled conditions, with the objective to diminish the chemical, enzymatic and microbiological activities that are responsible for the deterioration of foods (Barnabas et al., 2010). Air-drying is one of the traditional method which used for food dehydration by means of some kinds of driers such as cabinet drier, fluidized-bed drier, but recent studies pointed out the efficiency in water removal when air drying is combined with microwaves (MW) heating (Momenzadeh et al., 2010; Funebo and Ohlsson, 1998; Drouzas et al., 1999). Fruit dehydration by immersion in osmotic solutions has been of rising interest during the last decades since it can improve food quality when combined with other type of dehydration method (Mauro and Menegalli, 2003). To carry out osmotic dehydration, fruit pieces were immersed in a concentrated solution containing one or more solutes.

In the osmotic dehydration, because of high concentration in solutes (sugar and salt), promoting two simultaneous flows in counter current, a water outflow from the food, and an inflow of solutes from the solution to the food, due to the establishment of gradients of chemical potential of water and solutes (Petchi and Manivasagan, 2009). In the process, more water than solute is usually transfers due to the deferential permeability of cellular membranes (Mauro and Menegalli, 2003). MW drying is a rapid dehydration technique that can be applied to specific foods. Compared to conventional hot air drying, MW drying is rapid, more uniform and energy efficient and includes space savings and energy drying. In MW, processing the energy is transferred directly to the sample producing a volumetric heating (Abbasi and Mowla, 2008; Oliveira and Franca, 2002). Momenzadeh et al. (2010) studied drying characteristics of shelled corn (*Zea mays* L) in a fluidized bed dryer assisted by MW heating. Their results showed that, increasing the drying air temperature resulted in up to 5% decrease in drying time while in the MW -assisted fluidized bed system, the drying time decreased dramatically up to 50% at a given and corresponding drying air temperature at each MW energy level.

The drying kinetics of food is a complex phenomenon and requires dependable models to predict drying behavior (Sharma et al., 2003). Kingsly and Sing (2007) studied thin-layer drying of pomegranate arils in a cabinet drier at drying temperatures of 50, 55, and 60°C. They reported that, the page model satisfactorily represented the drying characteristics of pomegranate arils than other models. This present article focuses on mathematical modeling of papaya during drying using MW assisted hot air drying pretreated in osmotic solution (osmosis-cabinet- MW) process as a new kind of combined drying method in different hot air drying temperatures (40, 50 and 60°C). The MR (moisture ratio) for each drying temperature was obtained and a suitable thin-layer drying model to describe the drying process was developed.

MATERIALS AND METHODS

Sample preparation

Papayas (*Carica papaya L.*) were provided by a producer in Sistan and Blouchestan province- Iran. Papayas were stored at 4°C prior to use in the drying experiment. After one-day storage in a refrigerator, it was taken and reached to room temperature (24 ± 1°C) 1 h before the start of the experiment. For all experiments papayas were peeled and sliced (5 ± 1 mm thickness) with a stainless steel knife in approximately 5 × 2 cm^2. The initial moisture content of papaya was 700 ± 2 % (d.b.). The pH and Brix of fresh papaya was 5.1 and 11.4, respectively.

Preparation of osmosis solution

Osmotic solution was prepared with sucrose 50 and 2% NaCl, which were obtained from Merck Company, Germany. The product to solution ratio was prepared 1:10 (weight basis) (Antonio et al., 2004). Temperature controlled mixing tank was used for osmotic operation.

Cabinet dryer and microwave setup

Cabinet dryer with controllable airflow and temperature system and air humidity monitoring system utilized for hot air drying process. The absolute humidity and the hot air velocity for all of drying temperatures were 0.6 ± 0.02 g/kg dry air and 0.9 ± 0.1 m/s, respectively.

Combination oven MW system (LG, Korea) with power control dial (power output 180 to 900 W) was used in combination with hot air drying method.

Microstructure analysis

SEM imaging of papaya was carried out to exhibit the surface properties of the samples. The thin layers prepared from the untreated and dried papaya coated with gold using an ion sputter (Fisons Instruments, UK). The coated samples were viewed and photographed using the scanning electron microscope (model 5526, Cambridge, UK) at 20 kV.

Experimental procedure

Osmosis treatment

Papaya slices were weighed and placed into an osmotic solution in a dynamic conditions provided by agitation (150 rpm) at room temperature (24 ± 1°C) for 4 h. The product to solution ratio was 1:10 (w/w). The samples were then removed from the solution and were blotted in order to remove excess solution and then dried with filter paper for almost 5 min (Antonio et al., 2004). It was weighed prior to placing in the cabinet drier. For each treatment, water loss (WL), solid gain (SG) and weight reduction (WR) were evaluated based on following equations and the results were expressed in g/100g of initial fresh fruit weight:

$$WL = \frac{(ww_0) - (w_t - ws_t)}{(ws_0 + ww_0)} \quad (1)$$

$$SG = \frac{(ws_t - ws_0)}{(ws_0 + ww_0)} \quad (2)$$

$$WR = WL - SG \quad (3)$$

Where, ww_0 is the weight of water in initial sample, ws_0 is the weight of solids initially present in the fruit (g); w_t and ws_t are the weight of the fruit (g) and the weight of solids at the end of treatment (g), respectively (Petchi and Manivasagan, 2009; Mujica-Paz et al., 2003; Lazarides et al., 1995). The changes in SG as well as WL determined consecutively each 30 min after the process for 4 h. All the experiments were done in triplicate.

Microwave assisted hot air drying

One layer of samples after 4 h osmosis dehydration placed in the cabinet dryer at 3 temperature of 40, 50, and 60°C for hot air drying process and weight – time data were recorded consecutively using a digital weighter (....) until 20 ± 1% (d.b) moisture content was achieved from initial moisture content of 700 ± 2% (d.b). In connection with using MW energy as an accessory drying method which combined with hot air drying, selected samples were exposed to MW energy following convection hot air drying. The MW power output was set to 540W. Fruits were subjected to 10 seconds micro power at each 15 min interval. A sample for each treatment was placed in an oven set to 105°C to obtain the solid content (total moisture content). The MR vs. drying time curve was obtained for each treatment.

Mathematical modeling

Ten models of thin layer drying described in Table 1 were investigated to find the most suitable drying model for the drying process of papaya. In these models, the moisture ratio was simplified to M/M_0 instead of $(M - M_e)/(M_0 - M_e)$ as the value of M_e is relatively small compare to M or M_0 (Doymaz, 2004). Correlation coefficient (R^2) was one of the primary criteria to select the best

Table 1. Mathematical models for thin-layer drying.

Model name	Model equation	References
Newton	$MR = \exp(-kt)$	Westerman et al. (1973)
Page	$MR = \exp(-kt^n)$	Guarte (1996)
Modified Page	$MR = \exp(-kt)^n$	Yaldiz et al. (2001)
Henderson and Pabis	$MR = a.\exp(-kt)$	Yagcioglu et al. (1999)
Logarithmic	$MR = a.\exp(-kt) + c$	Yaldiz et al. (2001)
Two-term	$MR = a.\exp(-k_0 t) + b.\exp(-k_1 t)$	Rahman et al. (1997)
Exponential two-term	$MR = a.\exp(-kt) + (1-a)\exp(-kat)$	Yaldiz et al. (2001)
Wang and Sing	$MR = 1 + at + bt^2$	Ozdemir and Devres (1999)
Approximation of diffusion	$MR = a\exp(-kt) + (1-a)\exp(-kbt)$	Yaldiz and Ertkin (2001)
Midilli et al.	$MR = a\exp(-kt^n) + bt$	Sacilik et al. (2006)

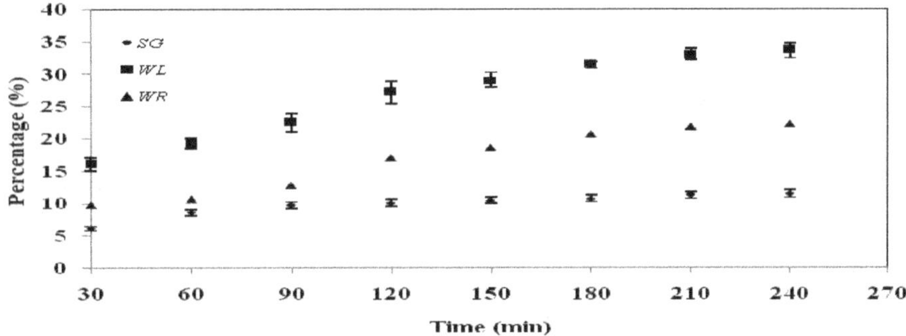

Figure 1. The changes in SG, WL and WR during osmosis dehydration.

model. Other statistical parameters such as chi-square (χ^2) and root mean square error (RMSE) were used to determine the quality of the fit. These parameters can be calculated as given equations: (Singh and Sodhi, 2000; Togrul and Pehlivan, 2002).

$$R^2 = \frac{\sum_{i=1}^{N}(MR_{\exp,i} - \overline{MR}_{\exp})^2 (MR_{pre,i} - \overline{MR}_{pre})^2}{\sum_{i=1}^{N}(MR_{\exp,i} - \overline{MR}_{\exp})^2 \sum_{i=1}^{N}(MR_{pre,i} - \overline{MR}_{pre})^2} \quad (4)$$

$$RMSE = \left[\frac{1}{N}\sum_{i=1}^{N}\left(MR_{exp,i} - MR_{per,i}\right)^2\right]^{1/2} \quad (5)$$

$$\chi^2 = \frac{\sum_{i=1}^{N}\left(MR_{exp,i} - MR_{per,i}\right)^2}{N - z} \quad (6)$$

The model is said to be good if R^2 value is high, χ^2 and RMSE values are low (Sarsavadia et al., 1999; Togrul and Pehlivan, 2002).

RESULTS AND DISCUSSION

Osmosis treatment

The moisture content of the fresh papaya was 700 ± 2% (d.b). The effect of osmosis treatment on dehydration of the samples pointed through the changes in parameters of the process (SG, WL, and WR) during dehydration that are shown in Figure 1. The results showed that, the WL, SG and WR of papaya increased with increasing immersion time in the osmotic solution. These results were in agreement with that of Petchi and Manivasagan (2009) and Heng et al. (1990).

In the most intense processing condition (after 4 h immersion), water loss and weight loss attained to 33.64 and 22.19 g/100 g of initial fresh fruit, respectively. Figures 1 shows that, at the end of the osmosis treatment especially at 210 and 240 min after immersion, the changes in WL were very low that the slope of the changes was near to zero. This is to say that, the rate of mass transfer by increasing the immersion time will be lower because of the decreasing in driving force between the solution and the product that hinders the mass transfer. It means that, the system comes to equilibrium.

Microwave assisted hot air drying

Figure 2 shows the drying curves (moisture ratio versus time) obtained under specific process conditions. As

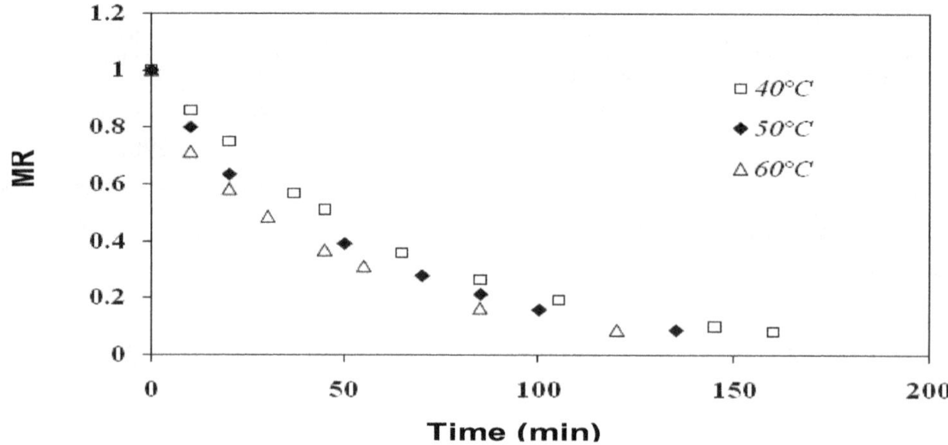

Figure 2. Effect of osmosis-cabinet- MW drying time on moisture ratio at 40, 50 and 60°C.

Table 2. Regression analysis coefficients of mathematical drying models for osmosis-cabinet-microwave drying process at 40°C

Model name	Model coefficients	R^2	RMSE	χ^2
Newton	K = 0.015	0.999	0.007	5.05E-05
Page	K = 0.013 n =1 .030	0.999	0.004	2.38E-05
Modified page	K = 0.015 n =1.027	0.999	0.007	5.68E-05
Henderson and pabis	K = 0.015 a =1.007	0.999	0.006	4.52E-05
Logarimetric	K = 0.015 a =1.015 c = -0.011	0.999	0.005	4.28E-05
Two-term	K_0 = 1.015 K_1 = 0.015 a = -0.015 b = 1.015	0.999	0.005	4.24E-05
Exponential two-term	K = 0.017 a = 1.392	0.999	0.005	2.71E 05
Approximation of diffusion	K = 0.307 a = -0.016 b = 0.051	0.999	0.005	3.49E-05
Wang and Sing	A = -0.012 b = 4.36E-05	0.997	0.019	0.000
Midilli et al.	K = 0.029 n = 0.810 a = 1.021 b = -0.000	0.995	0.021	0.001

expected, increasing air temperature reduced the drying time. At higher temperature due to the quick removal of moisture, the drying process was shorter. The decrease in drying time with increase in drying temperature may be due to increase in water vapor pressure within the papaya pieces, which increased the migration of moisture especially as the drying occurs only in falling rate period. Similar observation was reported for apple purees (Vergara et al., 1997).

The moisture ratio of papaya reduced exponentially as the drying time increased. Continuous decrease in moisture ratio indicates that, diffusion governed the internal mass transfer (Haghi and Amanifard, 2008).

Mathematical modeling

Figure 2 shows drying curves of osmosis-cabinet- MW process which exhibits the effect of drying time on moisture ratio. The MR values were fitted against the drying time for each temperature by applying the non-linear regression analysis technique. The best model for each treatment obtained using comparison of statistical parameters of R^2, RMSE and χ^2. The characteristics of the curves fitting attributed to 40, 50 and 60°C of osmosis-cabinet- MW drying process are shown in Tables 2 to 4, respectively.

As the results show, Page model demonstrated the best curve fitting results (highest R^2, lowest RMSE and χ^2) at 40 and 50°C while Two-term model represented the best characteristics for the thin-layer drying of papaya at 60°C. These models can be used as the best ones to predict the drying behavior. Figure 3 shows the coincidence between experimental MR and predicted MR obtained from the best model related to each drying condition, which banded around the straight line *(X = Y)*; that proved the feasibility of the selected models in describing the drying behavior of *Carica papaya*.

Microstructure analysis

Figure 4 shows the scanning electron microscopy images of fresh papaya. Papaya at the end of osmosis

Table 3. Regression analysis coefficients of mathematical drying models for osmosis-cabinet-microwave drying process at 60°C.

Model name	Model coefficient	R^2	RMSE	χ^2
Newton	k = 0.023	0.990	0.033	0.001
Page	k = 0.048 n=0.800	0.998	0.011	0.000
Modified page	k = 0.027 n = 0.851	0.990	0.033	0.001
Henderson and pabis	k = 0.021 a = 0.949	0.988	0.027	0.002
Logarithmic	k = 0.025 a = 0.899 c = 0.066	0.991	0.023	0.002
Two-term	K_0 = 0.018 K_1 = 0.896 a = 0.859 b = 0.140	0.999	0.003	3.96E-05
Exponential two-term	k = 0.128 a = 0.148	0.998	0.011	0.000
Approximation of diffusion	k = 0.025 a = 0.152 b = 0.863	0.990	0.033	0.002
Wang and Sing	a = -0.018 b = 9.36E-05	0.971	0.053	0.004
Midilli et al.	k = 0.0617 n = 0.711 a = 0.998 b = -0.0006	0.999	0.006	9.14E-05

Table 4. Regression analysis coefficients of mathematical drying models for osmosis-cabinet-microwave drying process at 50°C

Model name	Model coefficient	R^2	RMSE	χ^2
Newton	k = 0.018	0.999	0.007	8.11E-05
Page	k = 0.022 n = 0.952	0.999	0.002	6.4E-06
Modified Page	k = 0.019 n = 0.955	0.999	0.007	9.46E-05
Henderson and Pabis	k = 0.018 a = 0.989	0.999	0.006	6.34E-05
Logarithmic	k = 0.019 a = 0.097 c = 0.022	0.999	0.004	3.91E-05
Two-term	K_0 = 0.017 K_1 = 0.967 a = 0.974 b = 0.025	0.999	0.003	2.71E-05
Exponential two-term	k = 0.025 a = 0.530	0.999	0.003	1.97E-05
Approximation of diffusion	k = 0.017 a = 0.933 b = 4.004	0.999	0.003	2.56E-05
Wang and Sing	a = - 0.014 b = 6.030	0.993	0.027	0.001
Midilli et al.	k = 0.036 n = 0.808 a =1.017 b =-0.0005	0.998	0.012	0.000

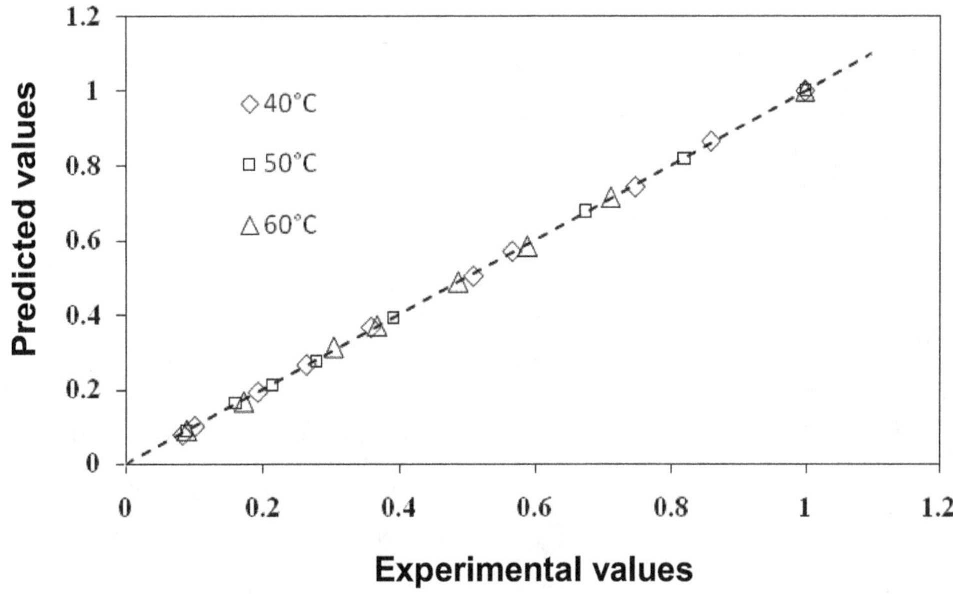

Figure 3. Comparison of the experimental and predicted MR (by the best resulting model).

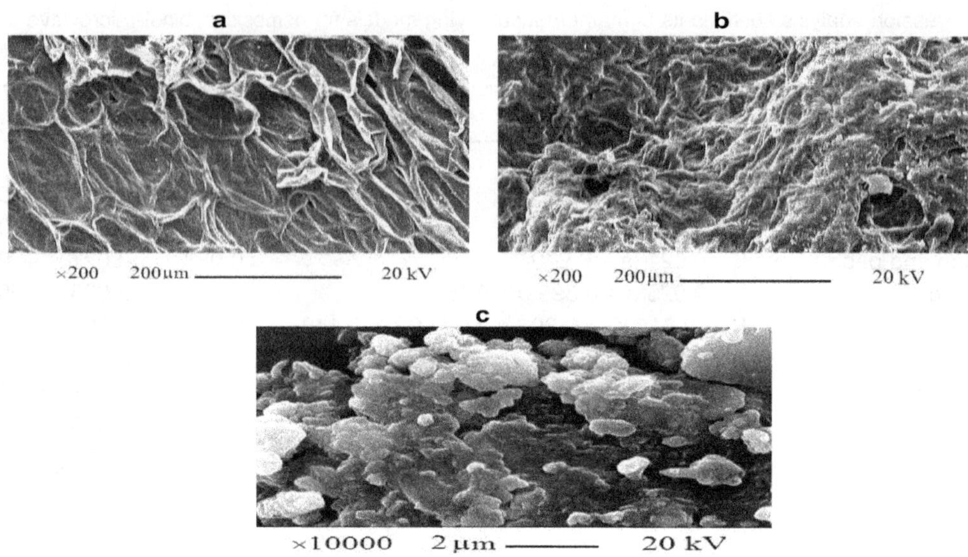

Figure 4. SEM images of a: fresh papaya, b: dried papaya at 50°C using osmosis-cabinet- MW process, c:osmosis treated papaya for 4 h.

dehydration (after 4 h) and papaya sample at the end of osmosis-cabinet- MW drying process at 50°C. Figures 4b and 4c clearly exhibit the sucrose particles that migrated to sample's surface. These particles are responsible for increasing the SG during 4 h immersion in the 50% sucrose + 2% NaCl solution. Comparsion of Figures 4a with 4b obviously shows that, the external porosity of the samples obtained from osmosis-cabinet- MW drying process was less than that of the fresh sample. The similar results reported for garlic and carrot slices (Lozano et al., 1980).

Conclusion

As the results, Page model showed the best curve fitting results for the experimental moisture ratio values for osmosis-cabinet- MW drying process at two temperatures of 40 and 50°C. In addition, Two-term model represented the best fitting traits to predict the drying characteristics of papaya slices during the drying process at 60°C. So, the introduced models can be used to predict moisture content of papaya fruit slices during drying at each mentioned temperature. The less porosity obtained for dried papayas using osmosis-cabinet- MW drying process which observed from scanning electron microscopy pictures causes more easily transportation of their slices because of lesser bulk volume.

Nomenclature: a, b, c, n, Empirical coefficients in drying models; **k,** k_0, k_1, empirical constants in drying models; **M,** moisture content at any time of drying; M_e, **MR,** equilibrium moisture content; M_0, initial moisture content; moisture ratio; $MR_{pre,i}$, ith predicted M.R; $MR_{\exp,i}$, ith experimentally observed M.R; **N,** number of observations; **z,** number of constants in models; **R^2,** coefficient of determination; **RMSE,** root of mean square error; χ^2, reduced chi-square; **ww$_0$,** Weight of water in initial sample; **ws$_0$,** Weight of solids initially present in the fruit; **ws$_t$,** weight of solids at the end of osmosis treatment; **w$_t$,** weight of the fruit.

REFERENCES

Abbasi Souraki B, Mowla D (2008). Experimental and theoretical investigation of drying behavior of garlic in an inert medium fluidized bed assisted by microwave. J. Food Eng. 88:438–449.

Antonio GC. Azoubel,PM, Alves DG, El-Aouar AA, Murr FEX (2004). Osmotic dehydration of papaya (*carica papaya* l.): influence of process variables. Proceedings of the 14th International Drying Symposium, Sao Paulo, Brazil, C, pp. 1998-2004.

Barnabas M, Siores E, Lamb A (2010). Non-thermal microwave reduction of pathogenic cellular population. Int. J. Food Eng. 6:1-18.

Doymaz I (2004). Pretreatment effect on sun drying of mulberry fruits (*Morus* alba L.). J. Food Eng. 65:205–209.

Drouzas AE, Tsami E, Saravacos GD (1999). Microwave/vacuum drying of model fruit gels. J. Food Eng. 39:117-122.

Funebo T, Ohlsson T (1998). Microwave-assisted air dehydration of apple and mushroom. J. Food Eng. 38:353-367.

Guarte R (1996). Modeling the drying behavior of copra and development of a natural convection dryer for production of high quality copra in the Philippines. Ph.D. dissertation, Hohenheim University, Stuttgart, Germany. P. 287.

Haghi AK, Amanifard N (2008). Analysis of heat and mass transfer during microwave drying of food products. Brazilian J. Chem. Eng. 25:491-501.

Heng W, Guilbert S, Cuq JL (1990). Osmotic dehydration of papaya: influence of process variables on the quality. Sciences des Aliments. 10:831-848.

Kingsly ARP, Singh DB (2007). Drying kinetics of pomegranate arils. J. Food Eng. 79:741–744.

Lazarides HN, Katsanidis E, Nickolaidis A (1995). Mass transfer kinetics during osmotic preconcentration aiming at minimal solid uptake. J. Food Eng. 25:151-166.

Lozano JE, Rotstein E, Urbicain MJ (1980). Total porosity and open-pore porosity in the drying of fruits. J. Food Sci. 45:1403-1407.

Mauro MA, Menegalli FC (2003). Evaluation of water and sucrose diffusion coefficients in potato tissue during osmotic concentration. J. Food Eng. 57:367-374.

Momenzadeh L, Zomorodian A, Mowla D (2010). Experimental and theoretical investigation of shelled corn drying in a microwave-assisted fluidized bed dryer using Artificial Neural Network. *Food and Bioproducts Processing*. DOI:10.1016/j.fbp.2010.03.007.

Mujica-Paz H, Valdez-Fragoso A, Lopez-Malo A, Palou E, Welti-Chanes J (2003). Impregnation and osmotic dehydration of some fruits: effect of the vacuum pressure and syrup concentration. J. Food Technol. 57:305-314.

Oliveira MEC, Franca AS (2002). Microwave heating of foodstuff. J. Food Eng. 53:347-359.

Ozdemir M, Devres Y (1999). The thin layer drying characteristics of hazelnuts during roasting. J. Food Eng. 42:225-233.

Petchi M, Manivasagan R (2009). Optimization of osmotic dehydration of radish in salt solution using response surface methodology. Int. J. Food Eng. 5:1-11.

Rahman MS, Conrad OP, Caroline T (1997). Desorption isotherm and heat pump drying kinetics of peas. Food int. Res. 30:485-491.

Sacilik K, Keskin R, Elicin A (2006). Mathematical modeling of solar tunnel drying of thin layer organic tomato. J. Food Eng. 73:231-238.

Sarsavadia PN, Sawhney RL, Pangavhane DR, Singh SP (1999). Drying behaviour of brined onion slices. J. Food Eng. 40:219–226.

Sharma GP, Prasad S, Datta AK. (2003). Drying kinetics of garlic cloves under convective drying conditions. J. Food Sci. Technol. 40:45–51.

Singh H, Sodhi NS (2000). Dehydration kinetics of onions. J. Food Sci. Technol. 37:520–522.

Togrul IT, Pehlivan D (2002). Mathematical modeling of solar drying of apricots in thin layers. J. Food Eng. 55:209–216.

Vergara F, Amezaga E, Barcenas ME, Welti J (1997). Analysis of the drying processes of osmotically dehydrated apple using the characteristic curve model. Drying Technol. 15:949–963.

Westerman PW, White GM , Ross IJ (1973). Relative humidity effect on the high temperature drying of shelled corn. Trans. Am. Soc. Agric. Eng. 16:1136-1139.

Yagcioglu A, Degirmencioglu A, Cagatay F (1999). Drying characteristics of laurel leaves underdifferent drying conditions. In proceedings of the 7th international congress on agricultural mechanization and energy 26-27 May, Adana, Turkey pp. 565-569.

Yaldiz O, Ertekin C (2001). Thin layer solar drying some different vegetables. Drying Technol. 19:583-596.

Yaldiz O, Ertekin C, Uzun HI (2001).Mathematical modeling of thin layer solar drying of Sultana grapes. Energy 42:167-171.

10

Nutrient potential of Almond seed (*Terminalia catappa*) sourced from three states of Eastern Nigeria

Mbah B. O.*, Eme P. E. and Eze C. N.

Department of Home Science, Nutrition and Dietetics, University of Nigeria Nsukka, Enugu State, Nigeria.

The study was carried out to evaluate the proximate and nutrient contents of Almond (*Terminalia catappa*) seeds obtained from three states of eastern Nigeria, namely: Imo, Enugu and Anambra states. Hundred seeds were handpicked from each state. The physical characteristics (length, circumference, whole weight, weight of kernels (nuts) and the hull) of the fruits were measured. Proximate, physiochemical and anti-nutritional content analysis were performed on triplicate samples of raw or roasted kernels across the locations after sun drying. The treatment means (location x processing) were compared using Steel and Torrie (1960) procedure. Location significantly ($p<0.05$) affected the protein, fibre and carbohydrate (CHO) contents of the seeds. Protein content of *T. catappa* seeds was highest (24.2%) in Anambra and lowest (14.5%) in Imo. Roasting decreased the CHO content from 35.6 to 33%. Roasted seed is advantageous since it increased protein, fat and fibre content of *T. catappa*. Calcium, iron, zinc and tannin were influenced by location. Seeds from Enugu had the least content of these nutrients. However, those from Imo were the largest in size and had the highest kernel and seed weights. As judged by the results the concentration of macro and micronutrients in this lesser known legume have promising potentials to alleviate the risk of some nutritional deficiency diseases in Nigeria.

Key words: Nutrient, potentials, phytochemicals, anti-nutrient content, Almond seeds.

INTRODUCTION

Underutilized foods could be used to meet world food security demands when properly processed and prepared for consumption. In Africa, deficiency disease due to lack of proteins is common. Under-nutrition in Nigeria is mainly due to poverty, inadequate energy intake as well as protein and micronutrients owing to poor nutrition education. Animal and animal products are very expensive as source of nutrients in developing countries (Obizoba, 1983). Discovery of alternative protein sources is a major need in Africa and Nigeria in particular. The food seeds rich in protein particularly legumes could effectively reduce the level of malnutrition (Tropilab Incorporated, 2005).

The tropical Almond (*Terminalia catappa*) is one of the lesser known legumes found in the tropics and in Nigeria ecosystem. Almond is a large deciduous tree that thrives as an ornamental tree. The leaves are arranged in close spirals. The leaf blade is simple broadly obviate, the leaf top is rounded and blunt, gradually tapering to a narrowing substrate base. The tree is slightly deciduous during dry season, and in some environments may lose their leaves twice in a year (Thamson and Evans, 2006).

Due to the phytochemical properties of *T. catappa*, the leaves, bark and fruits are useful in the treatment of dysentery, rheumatism, cough and asthma. The fruits are also to treat leprosy and headaches. The leaves are specifically used to treat intestinal parasites, eye problems, wounds and liver ailments (Kirtikar and Basu, 1999; Corner, 1997). The seed is edible and highly cherished by children. It is also used by many rural dwellers in southern Nigeria to fortify the local complimentary foods, which are usually low in protein. The *T. catappa* tree produces fruits whose pulp is fibrous, sweet and edible when ripe. The fruit is widely eaten by children as forage snack with the nuts and seeds often discarded. The thrust of this study was to determine and compare the nutrient and antinutrient contents of almond seeds, in the south eastern states of Nigeria.

*Corresponding author. E-mail: biomambah@yahoo.com.

Table 1. Physical characteristics of *T. catappa* seed in three states of south eastern Nigeria.

Local	Circumference (cm)	Length (cm)	Kernel wt. (g)	Seed shell wt. (g)	Whole seed wt. (g)
Anambra	9.6	5.7	12.5	78.8	91.2
Enugu	8.2	5.5	10.5	67.0	75.5
Imo	10.1	5.7	13.0	80.8	93.8
LSD (0.05)	0.2	0.1	2.0	7.1	7.7

MATERIALS AND METHODS

Collection and preparation of samples

T. catappa fruits from three states in South Eastern Nigeria (Imo, Anambra and Enugu) were gathered from various homes.

The seeds were sun dried to prevent rancidity of the kernel and to facilitate dehulling. The dried seeds were dehulled by cracking along the margins with a piece of pebble to obtain the brown spindle-shaped kernels. The kernels were dried at 60°C for six hours to 5% moisture level. Hundred spindle-shaped kernels as obtained from each location were divided in groups of 10 seeds per location. The length and circumferences of each individual kernel were measured and the weight of each replicate was obtained.

The kernels and the dry shell from each state were packed in separate cellophane bags and stored in refrigerator until used for various analysis.

Laboratory analysis

In the analytical laboratory of the Department of Crop Science, University of Nigeria, Nsukka, nutrient composition of *T. catappa* seeds was analyzed using the standard procedures of the Association of Official Analytical Chemists (AOAC, 1990). The kernels were analyzed to determine the moisture, ash, crude fat, crude fibre, crude protein, carbohydrate, antinutrients and mineral contents of *T. catappa*. Determination of fat was carried out by soxhlet procedure method which was also employed for estimation of crude protein. The percentage carbohydrate content was obtained by difference after summation of the values for moisture, ash, crude protein, crude fat and crude fibre, and subtracting the sum from 100%. Fibre was determined using the AOAC (2005) which involves hydrolyzing the protein, starch, fat and other digestible carbohydrates from the sample. Tannins extraction was by the modified methanol procedure of Prince et al. (1977). The photometric method by Onwuka (2005) was used to determine the phytate content.

Statistical analysis

The data collected were subjected to the analysis of variance using Steel and Torrie procedure (1960). F.LSD was used to separate the significant differences in the treatment mean (Obi, 2002).

RESULTS

The result presented in Table 1 is the physical characteristics of *T. catappa* seeds sourced from Anambra, Enugu and Imo State. The circumference of the seeds differed signification ($P<0.05$) across the locations. Enugu had the least value (8.2 cm) followed by Anambra (9.6 cm), while Imo had the largest sized seeds (10.1 cm). The seed length from Anambra and Imo were higher than that from Enugu. The weight of the kernels also varied significantly ($p<0.05$) ranging from 10.5 to 13.0. The kernel weight from Imo state was higher than those from Anambra and Enugu. Those from Enugu were the least (10.5 g). The seed shell weight (residues) varied significantly ($p<0.05$) ranging from 67.0 to 80.8 g. Seeds from Enugu had the least shell weight 67.0 g, and the value for Anambra and Imo were statistically similar.

The whole seed weights differed significantly ($p<0.05$) across the locations. The seeds from Anambra and Imo had no significant difference in weights of 91.2 and 93.8 g respectively, and the values were significantly ($p<0.05$) heavier than that of Enugu.

Data on Table 2 shows the effects of location and processing on proximate composition of the Almond kernels. The main effects and the interaction were in most cases non-significant. The moisture content varied from 9.5 to 10% with the kernels from Anambra having the least moisture content. The seeds (kernels) from Enugu state had the highest moisture content (10.7%) followed by that of Imo (10.3%). There was no difference in moisture contents of the raw and roasted seeds (kernels) across the location. The seed protein value differed significantly across the location. The protein seeds from Anambra had the highest protein (24.0%) followed by those from Enugu (18.4%), and the least was that of Imo (14.5%).

Roasting of the seeds sparingly increased the protein value when compared with the raw kernels obtained from Anambra and Enugu. The fat and ash contents of the sample where neither influenced by location nor processing. However, it appeared that the samples from Anambra and Enugu had higher values of fat and ash than that of Imo. The fibre content of the food differed across the locations. Seed from Anambra had the least fibre content (6.4%). Both Enugu and Imo seeds (kernels) had nearly similar fibre content values of 13.3 and 12.1% respectively.

Roasting non-significantly increased the fibre content of the kernels across the locations. The carbohydrate contents of the seeds from the three locations varied significantly ($P<0.05$). Seeds (kernels) from Imo state had the highest carbohydrate value (37.8%). Values from Anambra and Enugu state were statistically similar (33.8 and 31.3%, respectively). Roasting significantly ($P<0.05$)

Table 2. Effect of location, processing and local x processing on proximate composition of almond *(T. catappa)* seeds.

Location	Processing	Moisture (%)	Protein (%)	Fat (%)	Ash (%)	Fibre (%)	CHO (%)
	Raw	9.5	23.4	22.0	4.1	6.4	34.6
Anambra	Roasted	9.6	24.7	22.6	3.8	6.5	32.9
	Mean	9.6	24.0	22.3	3.9	6.4	33.8
	Raw	10.8	18.2	21.6	3.7	12.6	33.1
Enugu	Roasted	10.5	18.6	23.8	3.7	14.0	29.5
	Mean	10.7	18.4	22.7	3.7	13.3	31.3
	Raw	10.1	14.5	21.8	3.5	10.9	39.2
Imo	Roasted	10.5	14.5	21.7	3.6	13.3	36.5
	Mean	10.3	14.5	21.8	3.5	12.1	37.8
L.SD for comparing:							
Location		NS	1.8	NS	NS	2.3	3.3
Location × processing		NS	NS	NS	NS	NS	NS
Processing							
Raw		10.2	18.7	21.8	3.8	10.0	35.6
Roasted		10.2	19.2	22.7	3.7	11.2	33.0
L.SD (0.05)		NS	NS	NS	NS	NS	2.7

NS, Non-significant.

Table 3. Effect of location, processing, location x processing on mineral and anti-nutient content of *T. catappa* seeds (mg/100 g sample).

Location	Processing	Ca	P	Fe	Zn	Tannin	Phytate
	Raw	20.7	17.2	0.7	1.0	0.3	0.1
Anambra	Roasted	24.3	17.0	0.7	1.0	0.3	0.3
	Mean	22.5	17.1	0.7	1.0	0.3	0.2
	Raw	23.5	16.7	0.8	1.2	0.3	0.3
Enugu	Roasted	22.8	16.2	0.8	1.2	0.3	0.2
	Mean	23.2	16.4	0.8	1.2	0.3	0.2
	Raw	25.8	16.1	1.4	0.9	0.3	0.2
Imo	Roasted	24.4	16.0	1.2	0.8	0.4	0.3
	Mean	25.1	16.1	1.3	0.8	0.4	0.2
L.SD for comparing:							
Location		2.1	NS	0.1	0.1	0.04	NS
Location x processing		2.9	NS	0.1	0.1	NS	NS
Processing							
Raw		23.3	16.7	1.0	1.0	0.3	0.2
Roasted		23.9	16.4	0.9	1.0	0.3	0.3
L.SD (0.05)		NS	NS	NS	NS	NS	NS

NS, Non-significant.

decreased the carbohydrate content of the kernels. The raw almond kernel had 35.6%, while the carbohydrate content of the roasted kernel is 33.0%.

Table 3 presents the effects of location, location x processing, and the main effect of processing on the mineral and anti-nutrient content of *T. catappa* seeds (kernels). The mineral and anti-nutrient content of the seeds were in most cases not significantly influenced by processing. There was significant location effect on most of the mineral and anti-nutrient contents studied. Imo seeds (kernels) had the highest calcium (25.1 mg) followed by that of Enugu state (23.2 mg). Roasting slightly increased the calcium content from 23.3 to 23.9 mg. The phosphorus content of the seeds ranged from

16.0 to 17.2 mg. Although the location effect was not significant, the Anambra seeds had the highest (17.1 mg) phosphorus content. The difference in phosphorus content was only 0.3 mg between Enugu and Imo seeds (kernels). Roasting seemingly decreased the phosphorus content of the seeds. Anambra had the least iron content 0.7 mg while seeds from Imo had the highest content of iron (1.3 mg). There was no significant difference between the iron contents of the raw and the processed kernels. Zinc content of the seed differed significantly ($P<0.05$) across the locations. Imo had the least Zinc content (0.8 mg) followed by that of Anambra (1.0 mg) while Enugu had the highest value (1.2 mg). Roasting caused no change to the zinc content when compared with the raw seeds. Tannins content was significantly high in the seeds collected form Imo state. Roasting had no effect on the tannin content of the seeds. The phytate contents of the seeds were neither influenced by location nor was processing (roasting), though the least value obtained from raw seeds from Anambra state. Roasting seemingly increased the kernel phytate content from 0.2 to 0.3 mg in seeds obtained from Imo State.

DISCUSSION

The variation in moisture of the Almond seed (kernel) might be attributed to varietal differences as well as soil and other environmental variables. There was no difference in moisture between the raw and the roasted samples; this suggests that roasting has little or no effect on the moisture content of the seeds (kernel). The higher protein content of seeds from Anambra as compared to those from Enugu and Imo state could be attributed to accessional differences and or soil and climatic visibilities between the states. The increase in protein due to roasting suggests that roasting is an important domestic food processing technique to increase protein in almond seeds (kernel). Many processing methods have been shown to improve the nutritive value of plant foods for human nutrition (Obizoba, 1994). The similarity in fat content of the seeds from the three locations indicates that the seeds (kernels) had the same fat concentration regardless of location.

The slight increase in fat in the roasted seeds (kernel) could be due to the fact that at high temperature, the complex organic compound of seed (kernel) was disintegrated to release more free fat molecules. The similarity in ash content across the states indicates that location and or soil differences had no effect on ash content of the seeds (kernels). The slight decrease in ash content of the seeds due to roasting suggest that roasting had adverse effect on ash content of the seeds (kernels). The decrease in ash content might be due to the loss of vegetable part of the seed (kernels) during roasting.

The smaller seed size (lower circumference, seed length, kernel and residue) from Anambra and Imo states had an advantage on physical characteristic, hence large sized. The higher fibre content of the roasted seeds (kernels) suggest that roasting is a good food processing method to improve the fibre content of almond seeds (kernels). Fibre is advantageous in human nutrition, since soluble fibre seems to reduce the level of low density lipoprotein (that is, cholesterol) in the blood (King and Burgess, 2000). The higher amount of carbohydrate in the seeds collected from Imo and Anambra state which are closer to the forest zone may be explained by the more available moisture which is necessary for optimum rate of Photosynthesis. Seeds from Enugu had the lowest carbohydrate content since it is closer to the drier Northern Nigeria. The amount of rainfall decrease as one move from the southern guinea savannah further north (Ugese et al., 2008).

The slightly higher calcium in almond seeds (kernels) from Imo might be associated with concentrations of the nutrients and the differences in location. The higher phosphorus of seeds from Anambra (17.1) indicates that Anambra seed had an edge over those of Enugu and Imo state.

The decrease in phosphorous, iron, zinc and tannins contents of the roasted seeds might be associated with vegetative loss during processing. This is however advantageous since at low concentration tannins play an anti-inflammatory role (Nwosu et al., 2008). Nevertheless, the increase in phytate of rosted seeds has some nutritional implications. The current information is that phytate lowers serum cholesterol as well as decreases risk of cancer (Turner, 2002).

Conclusion

The consequences of malnutrition in the under developed or developing countries like Nigeria cannot be over emphasize; this tragedy is mostly found in the rural areas. This can be attributed to mere ignorance of food trees around them. Approximately 33 g of protein are lost each day by the average adult male and can be replaced in the diet. The body has no means of storing amino acids, but the reserves are depleted in only a few hours (Ukoha, 2003). Also growing children need more protein than adults per unit weight since more protein is needed for growth than for maintenance (Chilaka, 2004). In this regard, attention should be drawn to cheap sources of protein like *T. catappa*. *T. catappa* among other leguminous plants has been proven to be edible, available, affordable and contains most of the nutritional requirements in large proportion (Paterson, 2002). Seeds rich in proteins particularly those of leguminous plants like *T. catappa* could reduce the level of malnutritions in most impoverished countries of Africa. The seed is hard to crack and often breaks into pieces in the course of kernel extraction. This is a major hindrance to the full commercialization of *T. catappa* kernels.

RECOMMENDATION

T. catappa should be used in food supplements among the rural dwellers. It can be used raw or roasted to improve the food nutrient content.

REFERENCES

AOAC (1990). Official method of analysis of the Association of official Analytical Chemist, 5th ad. AOAC Press, Arlington, Virginia, USA.
AOAC (2005). Official method of analysis of the Association of official Analytical Chemist, 5th ad. AOAC Press, Arlington, Virginia, USA.
Chilaka FC (2004). Biochemistry for Beginners. AP-Express Publishers Ltd. Nsukka. pp.113-131.
Corner EJH (1997). In Wayside Trees of Malaya, 4th Edn, The Malayan Nature Society, Malaya 1:217.
King FS, Burgess A (2000). Nutrition for developing Country. pp.10-12.
Kirtikar M, Basu BD (1999). In India Medicinal Plants, Vol. Periodical Exparts Books Agency, New Deihi, pp.10-16.
Nwosu FO, Dosumu OO, Okocha JOC (2008). The Potential of *T. catappa* (Almond) and Hyliaene thebaica (Dum palml) fruits as raw materials for Livestock feed. Afr. J. Biotechnol. 7(24):4576-4580.
Obi IU (2002). Statically Mathematic of detecting differences between treatment means. A P-Express Publishers Ltd. Nsukka. pp. 48-52.
Obizoba IC (1994). Plant Food for Human Nutrition. Kluver Academic Publishers 45:23-34.
Obizoba IC (1983). Nutritive value of Cowpea-bambaranut Groundnut-rice Mixture in Adult rats, Nig. J. Nutr. Sci. 4:35-40.
Onwuka GI (2005). Food Analysis and Instrumentation Theory and Practice. Napthali Prints, Lagos: Nigeria. pp. 142-143
Paterson J. (2002). The Need for Copper and zinc supplementation in Montana. In Beef: Questions and Answers, Bozeman, M.T. Motana State University 8(3):23-26.
Prince ML, Hagerman AE, Butter LG (1977). Tannin content of cowpea, Pigeon and Mung Beans. J. Agric Food Chem. 28:459.
Steel RD, Torrie JH (1960) Principles and procedures of statistics with special reference to the biological sciences. MC Graw-Hill Book Company, New York, London and Toronto.
Thamson J, Barry E (2006). Species profiles Pacific Island Agroforestry. Plant Science Publishers. pp.1-17.
Tropilab Incorporated (2005). Terminalia catappa- Tropical Almond. Retrieved on June 5th, 2005 from http://www.tropilab.com/terminalia-cat.html.
Turner L (2002). The top 10 anti oxidants rich foods 1 power houses. Better nutrition. Biofactor 12:5-11.
Ugese FD, Baiyeri PK, Mbah BN (2008). Mineral content of the pulp of Shea Butter fruit (vitellaria paradoxa C.F. Garthn) sourced from seven locations in the savannah Ecolopgy of Nigeria. Fruits 63:69-70.
Ukoha AI (2003). Foundation Biochemistry in Basic Biological Sciences. Niger Publishers Ltd., Nsukka. pp. 121-131.

Effect of calcium fortification and steaming on chemical, physical and cooking properties of milled rice

Abid Ali Lone[1], Qazi Nissar Ahmed[1], Shaiq A. Ganai[1], Imtiyaz A Wani[2], Rayees A. Ahanger[3], Hilal A. Bhat[3] and Tauseef A. Bhat[4]

[1]Division of Post Harvest Technology, Sher-e-kashmir University of Agricultural Sciences and Technology of Kashmir, Shalimar, Srinagar 191121, India.
[2]Division of Pomology, Sher-e-kashmir University of Agricultural Sciences and Technology of Kashmir, Shalimar, Srinagar 191121, India.
[3]Division of Plant Pathology, Sher-e-kashmir University of Agricultural Sciences and Technology of Kashmir, Shalimar, Srinagar 191121, India.
[4]Division of Agronomy, Sher-e-kashmir University of Agricultural Sciences and Technology of Kashmir, Shalimar, Srinagar 191121, India.

Milled rice of two varieties, Jehlum and Shlimar Rice-1 were subjected to seven chemical treatments and five steaming treatments for calcium fortification. Physic-chemical and organoleptic evaluations of each treatment were carried out before and after fortification. Jehlum proved superior in whiteness (L - value), starch and sugar content, volume expansion, elongation ratio and minimum cooking time, while Shalimar Rice-1 proved better in crude protein and calcium content. Fortification treatments resulted in an increase in yellowness (*b* value), water uptake, calcium content and calcium retention in cooked rice. However, significant decrease was noticed in whiteness (L- value) and redness (*a*- value) with increasing concentration of calcium in soaking solution. The highest calcium content (184.45 mg/100 g) was recorded in T_3 (Ca lactate at 0.8% a.i. Ca) followed by 184.04 mg/100 g recorded in T_6 (Ca gluconate at 0.8% a.i. Ca) samples, while the minimum (8.05 mg/100 g) was observed in T_0 (control-water dip) samples. Steaming of rice resulted in an increase in total sugar, moisture and yellowness (*b*- values), as well as calcium retention in cooked rice, while as a significant decrease was observed in colour (L and *a* values), crude protein and starch content, volume expansion, elongation ratio and water uptake.

Key words: Calcium, chemical properties, cooking quality, fortification, physic-chemical properties, steaming and milled rice.

INTRODUCTION

Rice (*Oryza sativa* L.) is the most popular cereal worldwide serving as a staple food for nearly half of the world's population and accounts for about 22% of the total energy intake (Chukwu and Osch, 2009). At least 90% of the world's rice farmers and consumers are in Asia, where rice provides up to 75% of the dietary energy and protein to 2.5 billion people (Gnanamanickam, 2009). Although milled rice is superior to brown rice in palatability and digestibility yet milling process results in loss of various nutrients in brown rice (Misaki and Yasumatsu, 1985). Depending upon the extent of milling process, two to five times of reduction have been

Table 1. Recommended dietary allowance (RDA) of calcium in India and in USA (Harinarayan et al., 2007).

Group	India (mg/day)	USA (mg/day)
Adults	400	800-1000
Infants	500	500-750
Children	500	800-1300
Pregnant and lactating women	100	1200-1300

reportedly observed in nutrients such as thiamine, fat, fibre, niacin, phosphorus, potassium, iron, calcium, sodium and riboflavin at 10 to 12% degree of milling (Pillaiyar, 1988).

Furthermore, mineral malnutrition is a serious threat to the health and productivity of more than two billion people worldwide particularly to women and children (Nagpal et al., 2005). Every night 800 million people go to bed hungry and about 842 million suffer from malnutrition for one or another reason (Toenniessen, 2002). Low energy and protein intakes as well as micronutrient deficiencies are common nutritional problems for people in rice consuming countries like India, Indonesia, Myanmar, Nepal, the Philippines, Bangladesh, Sri Lanka, and Vietnam (Dexter, 1998). It can be addressed by various programmes including supplementation and fortification of commonly eaten foods with required or missing micronutrients (Venkatesh and Sankar, 2004; Aquilanti et al., 2012). Furthermore, increasing intake of a variety of calcium-containing foods might be safer than calcium supplementation because the ingestion of large quantities of calcium concentrated in tablet form may suppress bone remodelling as well as bone resorption (Niewoehner, 1988). Calcium (Ca) is an essential nutrient required for critical biological functions and an adequate intake of calcium has been demonstrated to reduce the risk for chronic diseases such as osteoporosis, hypertension and possibly colon cancer, as well as a number of other disorders associated with aging. Table 1 shows the recommended dietary allowance (RDA) of calcium in India recommended by Indian Council of Medical Research (ICMR). It is 400 mg/day for adults, 500 mg/day for infants, 500 mg/day for children and 1000 mg/day for pregnant and lactating mothers, much lower than the RDAs in the United States of America, which are 800-1000, 500-750, 800-1300 and 1200-1300 mg/day, respectively (Harinarayan et al., 2007). The mean daily intake of calcium is very low in different parts of India (Joshi and Harinarayan, 2009). In Kashmir, the mean daily intake of calcium has been reported to be 230 ± 63 mg/day for men and 178 ± 45 mg/day for women (Zargar et al., 2007). Teotia and Teotia (2008) reported that in 22 States of India, 52% of the populations have nutritional bone diseases, 40.6% have dietary calcium deficiency in critical years of growth and 8% have calcium deficiency in adults and post menopause. Thus there is an urgent need to improve calcium intake through dietary and fortified sources. Therefore, enrichment and fortification of rice are of continued interests to consumers and food processors. Among various calcium sources, calcium lactate has been frequently selected for fortification because of its non-metallic and pleasant mouth feel (Lee et al., 1995). Keeping in view the above facts, the present study on the fortification of rice with calcium was undertaken on two local varieties with the objectives of assessing the impact of calcium fortification on physic-chemical and cooking properties of milled rice.

MATERIALS AND METHODS

Two rice varieties, viz. Jehlum and Shalimar Rice-1 released by Sher-e-kashmir University of Agricultural Sciences and Technology of Kashmir (SKUAST-K) were used in the present study. Paddy was supplied by Seed Processing Centre, Division of Plant Breeding and Genetics, SKUAST-K, Shalimar. Paddy (14% MC) was cleaned and milled to rice, and samples were stored in plastic containers prior to analysis.

Method of fortification

The method employed was essentially based on the technique of fortification developed by Lee et al. (1995). The rice samples (1.5 kg of each variety) were separately soaked for three hours in calcium lactate solution or calcium gluconate solution using three concentrations (0.4, 0.6 and 0.8% a.i. Ca) for each treatment. Rice soaking solution ratio was kept constant at 1:0.75 for each treatment and soaking was done under ambient conditions (20 to 25°C). Soaking in distilled water (0% Ca) served as control. The solution/water was drained off by placing the soaked product on an 18-mesh sieve for 5 min. Each treatment sample was spread (<0.5 inch depth) onto cheese cloth in shallow stainless steel basket (9 × 11 inch in size, having 0.25 inch diameter holes at the bottom). The baskets with rice samples of each treatment were autoclaved in a retort at four different pressures and durations viz. 10 psi for 10 min (S_4), 10 psi for 5 min (S_3), 5 psi for 10 min (S_2) and 5 psi for 5 min (S_1). The samples of each fortified treatment concentrations were also maintained as un-steamed (S_0). The rice was removed from retort and all the samples including non-steamed samples were dried under ambient conditions (20 to 25°C) under shade to moisture content of 11 to 12%. Each treatment was replicated thrice in completely randomized design with 100 g of rice as a unit of replication.

Physical characteristics

Colour was determined by using colorimeter (Make: ColourTec

PCM/PSM, New Jersey). The parameters L, *a* and *b* were determined for each sample. L (Luminosity or brilliance) varies from black (0) to white (100), *a* from green (-a*) to red (+a*) and *b* from blue (-b*) to yellow (+b*).

Chemical parameters

Samples were ground to flour prior to analyses. Crude protein was determined by the method described in official methods of analysis (AOAC, 1995). Starch and total sugars were determined according to the methods described by Nagi et al. (2007). Calcium content of both cooked and un-cooked rice was determined by official methods of analysis (AOAC, 1995). About 5 g flour samples were ashed at 550 to 600°C for 8 to10 h in a muffled furnace. The residue was treated with 5 ml concentrated HNO_3 and heated on a steam bath for 5 min and filtered. The filtrate was made up to 50 ml using one N HNO_3 and aspirated into the flame generated in an atomic absorption spectrophotometer (Make: Electronic Corporation of India Limited). To avoid interference, a few drops of 1% lanthanum chloride were added to each sample.

Cooking properties

To determine the elongation ratio, 5 g rice sample was taken in a 50 ml graduated centrifuge tube and 15 ml of water was added to each tube initially and cooked for 20 min on a water bath. The cooked rice was placed on a blotter paper and the length of 10 kernels was recorded randomly before and after cooking (DRR, 2000). The elongation ratio was calculated by using Equation (1):

$$\text{Elongation ratio} = \frac{\text{Kernel length after cooking (mm)}}{\text{Kernel length before cooking (mm)}} \quad (1)$$

To calculate volume expansion, five grams of head rice were weighed into a perforated basket (height = 7 cm and internal diameter = 4.2 cm). The basket was suspended in a long-type 150 ml beaker containing 100 ml of deionized water and allowed to stand for 10 min. The beaker was transferred into a rice cooker (RC-100H, Toshiba Corp., Tokyo) with the inner pan containing 200 ml of deionized water, and then steam-cooked for 30 min. The beaker with the cooked sample was taken out and excessive water was allowed to drain for five minutes. The height of the cooked rice in the perforated basket was taken at three points with a sliding steel tape. The average of the three measurements was used in calculating cooked rice volume. Volumetric expansion was expressed as the quotient of cooked rice volume over rice sample weight (Patindol et al., 2008). Samples were cooked by using the method described by Gujral and Kumar (2003). Two grams of rice was taken in a test tube from each sample and cooked in 20 ml distilled water at 90°C in a water bath. The weight of each sample was recorded using an analytical balance and water uptake was calculated by using Equation (2):

$$\text{Water uptake (ml/g)} = \frac{\text{WC - WUC}}{\text{WUC}} \quad (2)$$

Where, WC and WUC represent the weight of cooked and uncooked rice samples respectively. The rice samples were cooked for minimum cooking time as described by Batcher et al. (1956). Two grams of rice was taken in a test tube from each sample and cooked in 20 ml distilled water at 90°C in a water bath. The minimum cooking time was determined by removing a few kernels at regular time intervals of one minute during cooking and pressing between two glass plates till no white core was left. The samples were considered to be cooked when no white core was left.

Statistical analysis

Analysis of variance (ANOVA) was carried out and critical difference (CD) test was used to describe means with 95% confidence. Analysis of variance was calculated by SPSS statistical software (SPSS Inc., Chicago, Illinois, USA) at a probability level of $p < 0.05$. Each test was performed in triplicates and the unit of replication was 100 g rice. The experimental design was completely randomized design (factorial).

RESULTS AND DISCUSSION

Physical properties

It was observed that raw milled rice of Jehlum and SR-1 exhibited L (whiteness) values of 79.62 and 77.81, *a* (redness) value of 0.28 and 0.30, and *b* (yellowness) values of 16.96 and 16.90, respectively (Table 2). The differences in colour values between the test varieties could be governed by the genetic variability of the test varieties. However, soaking of milled rice in water or fortificant solution resulted in the reduction of L and *a* values, while as *b* values increased in response to soaking. Ca fortified rice of Jehlum variety exhibited significantly higher L, *a* and *b* values compared to that of SR-1 variety. Fortification treatments differed significantly from one another with regard to all the three colour values (Table 3). It was also observed that T_0 (control) samples exhibited the highest mean L (72.80) and *a* (-0.45) values, while the lowest (67.84 and -0.90) were recorded in T_3 (Ca lactate at 0.8% a.i. Ca) samples. In case of *b* values (yellowness), mean values were recorded highest (17.45 and 17.46) in statistically at par T_6 and T_3 treatments and minimum (17.23 and 17.25) in statistically at par T_5 and T_1 as against 17.14 value recorded in T_0 samples. The results are in agreement with the findings of Lee et al. (1995) and Chitpan et al. (2005). Steaming treatments differed significantly from one another with regard to their mean L, *a* and *b* values (Table 4). There was a gradual decrease in mean L and *a* values with increasing steaming intensity from S_0 (no steaming) to S_4 (10 psi pressure for 10 min), while as an increasing trend was noticed in mean *b* values from S_0 to S_4 steaming intensity. S_0 samples provided significantly higher L (74.71) and *a* (-0.33) values compared to the rest of the steaming treatments (S_1 to S_4), which showed 65.92 to 72.84 L values and -1.26 to -0.45 *a* values. In contrast, *b* values were recorded maximum (17.84) in S_4 and minimum in S_0 (17.09) samples. Similar effects on colour have been reported by Islam et al. (2002) and Lamberts et al. (2006) in response to steaming, and by Lee et al. (1995) and Chitpan et al. (2005) in response to Ca fortificants. The discolouration and intensity of colour has been reportedly attributed to Maillard type of non-enzymatic browning reaction and processing conditions during parboiling (Lamberts et al., 2006).

Table 2. Effect of variety on physical, cooking and chemical properties of milled rice.

Variety	Physical properties			Cooking properties				Chemical properties				
	L	A	b	Volume expansion (cm^3/g)	Elongation ratio	Water uptake (g/g)	Cooking time (min)	Protein (%)	Starch (%)	Total sugars (%)	Calcium in un-cooked rice (mg/100 g)	Calcium retained on cooking (mg/100 g)
Jehlum (Raw milled)	79.62	0.28	16.96	5.10	1.72	2.42	18.33	8.42	70.26	0.74	5.36	4.70
Jehlum(Fortified)	71.59	-0.67	17.36	4.75	1.51	2.42	21.22	8.14	67.65	1.00	122.57	93.96
Shalimar Rice-1 (Raw milled)	77.81	0.30	16.90	4.96	1.67	2.44	21.33	8.65	69.76	0.71	12.46	11.70
Shalimar Rice-1 (Fortified)	69.92	-0.72	17.28	4.57	1.43	2.41	22.97	8.33	66.74	0.94	124.10	95.61
C.D. ($P \leq 0.05$)	1.26	0.01	0.02	0.01	0.01	NS	0.21	0.04	0.90	0.02	0.21	NS

*L=luminosity, varies from 100- white to black-0, b = varies from – (negative) blue to + (positive) yellow, a = varies from - (negative) green to + (positive) red.

Table 3. Effect of fortification treatments on physical, cooking and chemical properties of milled rice.

Treatment		Physical properties (Colour)			Cooking properties				Chemical properties				
Chemical	Conc. (a.i. % calcium)	L	a	b	Volume expansion (cm^3/g)	Elongation ratio	Water uptake (g/g)	Cooking time (min)	Protein (%)	Starch (%)	Total sugars (%)	Calcium in un-cooked rice (mg/100 g)	Calcium retained on cooking (mg/100 g)
Control (T_0)	0.0	72.80	-0.45	17.14	4.66	1.43	2.30	23.03	8.32	67.15	0.95	8.05	7.96
Calcium lactate (T_1)	0.4	71.64	-0.60	17.25	4.66	1.47	2.37	22.47	8.24	67.20	0.96	108.95	82.55
-do- (T_2)	0.6	70.93	-0.73	17.36	4.66	1.48	2.44	22.13	8.25	67.27	0.97	135.63	101.51
-do- (T_3)	0.8	67.84	-0.90	17.46	4.67	1.47	2.50	21.73	8.24	67.17	0.98	184.45	141.50
Calcium gluconate (T_4)	0.4	71.77	-0.55	17.23	4.66	1.47	2.36	22.13	8.23	67.15	0.96	105.54	81.12
-do- (T_5)	0.6	68.91	-0.69	17.34	4.66	1.47	2.43	21.80	8.23	67.22	0.95	136.03	98.07
-do- (T_6)	0.8	69.70	-0.85	17.45	4.65	1.48	2.51	21.40	8.24	67.18	0.98	184.04	144.79
Mean		70.51	-0.68	17.32	4.66	1.47	2.41	22.98	8.25	69.19	0.96	123.24	93.93
C.D. ($P \leq 0.05$)		2.37	0.12	0.04	NS	NS	0.01	0.40	NS	NS	NS	0.39	4.54

Chemical properties

SR-1 variety possessed significantly higher crude protein content in raw milled as well as water soaked rice as compared to Jehlum variety which could be attributed to its inherently higher protein content (Table 2). Crude protein content was observed to decrease from 8.65% in raw milled rice to 8.60% in water soaked rice of SR-1 variety and from 8.42 to 8.37% in case of Jehlum variety. Irrespective of variety, it was observed that chemical treatments (fortificant solutions) had no effect on crude protein content of Ca fortified rice Table 3). However a significant decrease was observed in protein content with the increase in steaming intensity of fortified rice samples (Table 4). The decrease in protein content may be probably due to leaching of non-protein nitrogen, albumin and proteinaceous substances (Raghavendrarao and Julaino, 1970; Otegbayo et al., 2001). Jehlum variety possessed significantly (higher starch content in both raw milled as well as water soaked rice kernels in comparison to SR-1 variety. However, on soaking, starch content of milled rice decreased from 70.26 to 70.00% in Jehlum variety and 69.76 to 69.50% in SR-1 variety. Chemical treatments had non-significant

Table 4. Effect of steaming treatments on physical, cooking and chemical properties of milled rice.

Steaming treatment	Physical properties			Cooking properties				Chemical properties				
	L*	*a**	*b**	Volume expansion (cm^3/g)	Elongation ratio	Water uptake (g/g)	Cooking time (min)	Protein (%)	Starch (%)	Total sugars (%)	Calcium in un-cooked rice (mg/100 g)	Calcium retained on cooking (mg/100 g)
No steaming (S_0)	74.71	-0.33	17.09	4.97	1.63	2.55	18.76	8.46	69.76	0.75	123.32	74.34
5 psi for 5 min (S_1)	72.84	-0.45	17.27	4.83	1.55	2.48	20.59	8.36	68.17	0.83	123.30	86.36
5 psi for 10 min (S_2)	70.62	-0.63	17.44	4.71	1.46	2.41	21.93	8.25	67.05	0.93	123.36	98.28
10 psi for 5 min (S_3)	69.67	-0.81	17.66	4.54	1.39	2.34	23.90	8.13	65.86	1.08	123.22	104.70
10 psi for 10 min (S4)	65.92	-1.26	17.84	4.26	1.34	2.30	25.29	8.03	63.91	1.23	123.48	111.24
Mean	70.75	-0.70	17.46	4.66	1.47	2.41	22.09	8.25	66.95	0.96	123.34	94.98
C.D. ($P \leq 0.05$)	2.00	0.01	0.03	0.02	0.02	0.01	0.34	0.06	1.53	0.03	NS	3.84

effect on starch content of Ca fortified rice. The present study further indicated that steaming had a significant influence on mean starch content of Ca fortified rice, which decreased gradually with the increase in steaming intensity. Changes in starch content on steaming could be due to gelatinization of starch and some losses might have occurred during soaking and steaming process (Chukwu and Osch, 2009). Depending on the pressure, starch gets solubilized during parboiling process (Pillaiyar, 1988), which might have leached away with the soaking solution. Raghavendrarao and Juliano (1970) have attributed similar observations to appreciable decrease in molecular size of the starch in response to parboiling. Total sugar content of raw milled as well as calcium fortified rice varied significantly with respect to varieties. The mean value for total sugars in raw milled and calcium fortified rice was 0.74 and 1.00% in Jehlum and 0.71 and 0.94% in SR-1, respectively (Table 2). Chemical (fortification) treatments had non-significant impact on the level of total sugars in Ca fortified rice. Total sugars increased with soaking and the increasing intensity of steaming. Maximum mean total sugar (1.23%) was observed in S_4 (10 psi pressure for 10 min) and minimum (0.75%) was observed in S_0 (no steaming) which is shown in Table 4. Ali and Bhattacharya (1980) concluded that paddy grains are subjected to enzymatic action during the soaking stage, reducing sugars being produced from sucrose. The present findings are completely in consonance with the observations of Priestly (1976) and Lamberts et al. (2006). SR-1 proved significantly superior to Jelum variety with regard to Ca content of raw milled as well as fortified rice. SR-1 possessed significantly higher (12.46 mg/100 g) Ca content in raw milled rice as compared to Jehlum (5.36 mg/100 g) variety. On soaking in distilled water, Ca content of milled rice decreased to 11.25 and 4.65% in respective varieties, thus indicating a loss of 7.62 and 14.36% in Ca content, respectively. Irrespective of the source of Ca, fortified rice samples of both the test varieties showed a significant increase in Ca content with increasing Ca concentration in soaking solution. Irrespective of chemical and steaming treatments, Ca fortified samples of SR-1 variety exhibited significantly higher Ca content (124.10 mg/100 g) as compared to Jehlum (122.57 mg/100 g), which could be attributed to its inherently higher Ca content.

Irrespective of varieties, treatment T_3 (Ca lactate at 0.8% a.i. Ca) proved to be significantly superior by providing a maximum mean Ca content of 184.45 mg/100 g followed by 184.04 mg/100 g in T_6 (Ca gluconate at 0.8% a.i. Ca) as compared to 8.05 mg/100 g recorded in control (T_0). Soaking milled rice in the test solutions had increased the Ca content of fortified rice by 1211.05 to 2191.30% with a minimum increase recorded in samples subjected to Ca gluconate at 0.4% a.i. Ca (T_4) and a maximum increase in samples of T_3 (Ca lactate at 0.8% a.i. Ca). Varieties varied significantly in the retention of Ca in cooked raw milled rice, which was in accordance with the Ca content of raw milled rice in the respective variety. Before fortification, cooked rice of SR-1 had significantly higher Ca content (11.70 mg/100 g) than Jehlum (4.70 mg/100 g) variety. After soaking in distilled water, Ca content of cooked and 10% of Ca was lost during cooking and soaking in each variety. Almost similar findings have been reported by Cheigh et al. (1977) and Hayakawa and Igaue (1979) in response to combined washing and cooking of rice. However, test varieties showed a non-significant effect on Ca retention in fortified rice in the present study. Cooked samples showed a significant increase in retention of Ca in response to increasing Ca concentration in soaking solution as well as increasing steaming intensity. Irrespective of chemicals, treatment concentrations differed significantly from one another in Ca retention. The higher retention in T_3 and T_6 could be attributed to higher levels of Ca in these samples before cooking as a result of fortification.

Steaming had a significant effect on retention of Ca in cooked Ca fortified rice (Table 4). Maximum mean Ca content (111.24 mg/100 g) was recorded in S_4 (10 psi pressure for 10 min) while minimum (73.34 mg/100 g) was recorded in S_0 (no steaming) cooked rice samples. Ca is a stable nutrient which reportedly gets lost by leaching (Chitpan et al., 2005). In the present study, about 10% of Ca was lost in S_4 samples, while as losses to the extent of 42% were recorded in S_0 samples. More or less similar losses have been reported by Hettiarachchy et al. (1996). The retention of Ca in fortified rice might be due to physical entrapment of Ca compound in the rice grain (Lee et al., 1995). Furthermore, Ca salt may possibly form complexes with amylose of rice starch during its contact period with the grain and subsequent heat treatment (Hettiarachchy et al., 1996).

Cooking properties

Steaming treatments had a significant influence on elongation ratio, volume expansion and water uptake of Ca fortified rice, which decreased significantly with the increase in steaming intensity. The present findings are in agreement with the observations of Raghavendrarao and Juliano (1970) and Sowbhagya and Ali (1991). Maximum and minimum mean elongation ratio of 1.63 and 1.34, volume expansion of 4.97 and 4.26 cm^3/g, and water uptake of 2.55 and 2.30 ml/g were recorded in S_0 (no steaming) and S_4 (10 psi pressure for 10 min) samples, respectively (Table 4). The decrease in water uptake on parboiling is well supported by earlier findings of Mahedevappa and Desikachar (1968), Bhattacharya and Subbarao (1968), Ali and Bhattacharya (1972), Arai (1975), Anand (1983) and Patindol et al. (2008). The decrease in water uptake with increasing soaking and steaming time has been attributed to modification of starch granules by heating and parboiling process (Parnsakhorn and Noomhorm, 2008). While as Swasidisevi et al. (2010) concluded, the occurrence of partial pre-gelatinization in pre-steamed rice might have assisted in reduction of fissures within grain kernels and thus reduced water absorption. Chemical treatments exhibited non-significant impact on elongation ratio and volume expansion (Table 3). However, with the increase in Ca concentration of soaking solution, mean water uptake increased significantly from the minimum of 2.30 ml/g observed in T_0 (water dip- control) samples to maximum of 2.50 and 2.51 ml/g recorded in statistically at par T_3 and T_6 (Ca lactate or Ca gluconate at 0.8% a.i.Ca), respectively. Almost similar results have been observed by Lee et al. (1995) in Ca fortified rice and by Sudha and Leelavathi (2008) in Ca fortified wheat flour dough. Ca fortified rice developed from Jehlum variety exhibited significantly higher mean elongation ratio (1.51) as well as volume expansion (4.75 cm^3/g) compared to 1.43 ratio and 4.57 cm^3/g recorded in samples of SR-1 variety (Table 2). The differences in elongation ratio among rice varieties have also been reported by Adu-kwarteng et al. (2003), Singh et al. (2005) and Sidhu et al. (1975). However, the test varieties had a non-significant effect on water uptake of Ca fortified rice. Although chemical treatments showed varying influences on minimum time required for rice cooking yet significantly more cooking time (23.03 minutes) was required for T_0 (water dip-control) samples as against a minimum mean time of 21.40 min required for T_6 (Ca gluconate at 0.8% a.i.Ca) samples (Table 3). A minimum time required for cooking of fortified rice samples was observed to increase significantly with the increase of steaming intensity (Table 4). A minimum mean cooking time of 18.76 min required for S_0 (no steaming) samples was followed by 20.59 min for S_1 (5 psi pressure for 5 min) samples, which increased to a maximum of 25.29 min required for S_4 (10 psi pressure for 10 min). The increase in cooking time might have been due to increased kernel hardness during the steaming process (Islam et al., 2002). Ca fortified rice developed from SR-1 took significantly longer time (22.97 min) to cook as compared to 21.22 min recorded in samples of Jehlum variety (Table 2). The present findings are completely in accordance with the findings of Kaur et al. (1991), Islam et al. (2002) and Patindol et al. (2008).

Conclusion

From the present investigation, it can be concluded that, fortifying milled rice of Jehlum (V_1) or SR-1 (V_2) variety with 0.8% a.i. Ca through either Ca lactate (T_1) or Ca gluconate (T_4) resulted in enhanced levels of Ca in fortified rice as compared to raw milled rice. Steaming helped in retention of calcium in cooked rice. The fortified rice thus produced may be consumed as such and may also find applications in rice based food systems.

REFERENCES

Aquilanti L, Kahraman O, Zannini E, Osimani A, Silvestri G, Ciarrocchi F, Garofalo C, Tekin E, Clementi F (2012). Response of lactic acid bacteria to milk fortification with dietary zinc salt. Int. Dairy J. 25(1):52-59.

Adu-kwarteng E, Ellis WO, Oduro I, Manful JT (2003). Rice grain quality: a comparison of local varieties with new varieties under study in Ghana. Food Control. 14(7):507-514.

Ali SZ, Bhattacharya KR (1980). Changes in sugars and amino acids during parboiling of rice. J.Food Biochem. 4(3):169-179.

Ali SZ, Bhattacharya KR (1972). Hydration and amylose solubility behaviour of parboiled rice. Libensmittel Wissenscaft and Technology. 5: 207-12.

AOAC (1995). *Official Methods of Analysis.* Association of Official Agricultural Chemists, Washington D. C.

Anand AK (1983). Effect of pretreatments on the milling and cooking quality of some high yielding varieties of rice. M. Sc. Thesis submitted to Punjab Agricultural University, Ludhiana, India. pp. 65-67.

Arai K, Raghavendrarao SN, Desikachar HSR (1975). Studies on the effect of parboiling on japonica and indica rice. Japan Journal of Tropical Agriculture 19: 7-14.

Bhattacharya KR, Subbarao PV (1968). Effect of processing conditions on quality of parboiled rice. Central Food Technology Research Institute Reporter 476: 479-484.

Cheigh HS, Ryu JS, Jo JS, Kwon TW (1977). A type of post harvest loss: nutritional losses during washing and cooking of rice. Korean J.Food Sci.Technol. 1977(9):229-233.

Chitpan M, Chavasit V, Kongkachuichai R (2005). Development of fortified dried broken rice as a complementary food. Food. Nutr. Bull. 26(4):376-384.

Chukwu O, Osch FJ (2009). Response of nutritional contents of rice (*Oryza sativa*) to parboiling temperatures. American-Eurasian J. Sustain. Agric. 3(3):381-387.

Dexter PB (1998). Rice fortification for developing countries: opportunities for micronutrient interventions. *OMN1/USAID* 1-12.

Gnanamanickam SS (2009). Rice and its importance to human life. Progress Biological Control. 8:1-11.

Harinarayan CV, Ramalakshmi T, Prasad UV, Sudhakar D, Srinivasarao PV, Sarma KV, Kumar EG (2007). High prevalence of low dietary calcium, high phytate consumption and vitamin D deficiency in healthy south Indians. Am. J. Clin. Nutr. 85 (4):1062-1067.

Hayakawa T, Igaue I (1979). Studies on washing of milled rice: scanning electron microscopic observation and chemical study of solubilized materials. Nippon Nogali Kagaku Kaishi. 53(10):321-327.

Hettiarachchy NS, Ganasambandam R, Lee MH (1996). Calcium fortification of rice: distribution and retention. J. Food Sci. 61(1):195-197.

Islam MR, Roy P, Shimizu N, Kimura T (2002). Effect of processing conditions on physical properties of parboiled rice. Food Sci. Technol. Res. 8 (2):106-112.

Joshi SR, Harinarayan CV (2009). Vitamin D status in India – its implications and remedial measures. J. Assoc. Physic. India 57:40-48.

Lamberts L, Debie E, Derycke V, Veraverbeke WS, Deman W, Decour JA (2006). Effect of processing conditions on colour change of brown and milled parboiled rice. Cereal Chem. 83(1):80-85.

Lee MH, Hettiarachchy NS, Mcnew RW, Ganasambandam R (1995). Physiochemical properties of calcium fortified rice. Cereal Chem. 72:352-355.

Mahadevappa M, Desikachar HSR (1968). Expansion and swelling of raw and parboiled rice during cooking. J. Food Sci. and Technol. 5: 59-62.

Misaki M, Yasumatsu K (1985). Rice enrichment and fortification. In *Rice: Chemistry and Technology.* 2nd edition. American Association of Cereal Chemistry, St. Paul, Minnesota, USA. pp. 354-358.

Nagi, HPS, Sharma S, Sekhon KS (2007). *Handbook of Cereal Technology*. Kalyani Publishers, India. pp.111-155.

Nagpal S, Na S, Rathnachalam R (2005). Non-calcemic actions of vitamin D receptor ligands. Endocrinol. Rev. 26:662-667.

Niewoehner C (1988). Calcium and osteoporosis. Cereal Foods World. 33(7-8):784-787.

Otegbayo BO, Osamuel F, Fashakin JB (2001). Effect of parboiling on physico-chemical qualities of two local rice varieties in Nigeria. The J. Food Technol. Afr. 6(4):130-132.

Parnsakhorn P, Noomhorm A (2008). Changes in physico-chemical properties of parboiled brown rice during heat treatment. Agricutlural Engineering International: The CIGR E-Journal 10:1-20.

Patindol J, Newton J, Wang YJ (2008). Functional properties as affected by laboratory-scale parboiling of rough rice and brown rice. J. Food Sci. 73: 370-377.

Pillaiyar P (1988). *Rice Postproduction manual*, Wiley Eastern, New Delhi, India.

Priestly RJ (1976). Studies on parboiled rice. Part I. Comparison of characteristics of raw and parboiled rice. Food Chem. 1(1): 5-14.

Raghavendrarao SN, Juliano BO (1970). Effect of parboiling on some physicochemical properties of rice. J. Agric. Food Chem. 18(2):289-294.

Singh N, Kaur L, Sodhi NS, Sekhon KS (2005). Physico-chemical, cooking and textural properties of milled from different Indian rice cultivars. Food Chem. 89:253-259.

Sidhu SJ, Gill MS, Bains SG (1975). Milling of paddy in relation to yield and quality of rice of different Indian varieties. J. Agric. and Food Chem. 23(6) : 1183-1185.

Sudha ML, Leelavathi K (2008). Influence of micronutrients on rheological characteristics and bread-making quality of flour. Inter. J. Food sci. and Nutri. 59: 105-115.

Swasidisevi T, Sriviyakula W, Tia W, Soponronnaret S 2010. Effect of pre-steaming on production of partially parboiled rice using hot air fludization technique. Journal of Food Engineering 96: 455-462.

Teotia SPS, Teotia M (2008). Nutritional bone disease in Indian population. Indian J.Med. Res. 127:219-228.

Toenniessen GH (2002). Crop genetic improvement for enhanced human nutrition. J. Nutr. 132(9):29435-29465.

Venkatesh MG, Sankar R (2004). Micronutrient fortification of foods – rationale, application and impact. Indian J.Pediat. 71(11):997-1002.

Zargar AH, Ahmad S, Masoodi SR, Wani AI, Bashir MI, Laway BA, Shah ZA (2007). Vitamin D status in apparently healthy adults in Kashmir valley of Indian subcontinent. Postgraduate Med. J. 83 (985):713-716.

12

Effect of blending fresh-saline water and discharge rate of drip on plant yield, water use efficiency (WUE) and quality of tomato in semi arid environment

D. D. Nangare[1], K. G. Singh[2] and Satyendra Kumar[3]

[1]Central Institute of Post Harvest Engineering and Technology, (CIPHET), Abohar, Punjab, India.
[2]Department of Soil and Water Engineering, PAU, Ludhiana, Punjab, India.
[3] Central Soil Salinity Research Institute, Karnal, Haryana, India.

The use of alkali ground water constitutes a major threat to irrigated agriculture in semiarid parts of India. The entire arid and semiarid region in India is characterized by low rainfall and has the problems either of water scarcity or poor quality ground water and it can be better utilized for irrigation through drip irrigation system. An experiment was conducted on tomato crop at Central Institute of Post Harvest Engineering and Technology (CIPHET) Abohar, Punjab to study the effect of blending fresh and saline irrigation water on yield and quality. The good quality canal water (EC of 0.38 dS/m) and ground water (EC 19.5 dS/m) were mixed in ratio of 100% Fresh (F), 75:25 (Fresh: saline; F:S) and 50:50 (F:S). The irrigation was done through drip system with three discharge rates (1.2, 2.4 and 4.2 lph) at three irrigation levels of 0.6, 0.8 and 1.0. The plant yield decreased significantly with increase in salinity levels of irrigation water (that is, increase in proportion of saline water). The maximum plant yield (3.55 kg/plant) was recorded with fresh water irrigation while 50% saline water blending in irrigation produced the lowest yield (2.64 kg/plant). The average yield decreased significantly when the discharge rate of emitters increased from 1.2 to 2.4 lph. The quality of tomato is observed inferior in saline water treatment compared to fresh water treatment. The TSS and acidity of tomato fruits increased with increase in the saline water ratios of irrigation water. As compared to 100% fresh water treatment, the mixing of 75% fresh and 25% saline water reduced tomato yield by 11% and gave a better quality tomato fruits at the discharge rate 2.4 lph and irrigation level 0.8. Hence, saline water can be utilized through drip system for sustainable yield and quality tomato production in water scarce area having poor quality ground water.

Key words: Saline water, drip irrigation, tomato, quality.

INTRODUCTION

Water scarcity is becoming one of the major limiting factors for sustainable agriculture in the semi-arid regions of the world. In India the entire arid and semiarid regions have been characterized by low rainfall and have the problems either of water scarcity or poor quality ground water. The use of alkaline ground water possesses a major threat to plant growth and health, which is mostly observed in semiarid parts especially, south Asia (Minhas and Bajwa, 2001). In India, the regions identified for poor quality water are major parts of Rajasthan, Gujrat, Haryana, North Western UP and South Western parts of Punjab. Around 30 to 50% such type of lands are found

Table 1. Physio-chemical properties of experimental soil.

Depth (cm)	Sand (%)	Silt (%)	Clay (%)	Initial EC (dS/m)	pH	Bulk density (/cm^3)
0-15	76.57	8.02	15.41	0.15	8.52	1.69
15-30	77.92	7.69	15.39	0.13	8.63	1.71
30-45	78.21	7.35	14.44	0.13	8.70	1.82
45-60	76.96	8.36	14.68	0.14	8.70	1.79
60-90	78.28	8.18	13.54	0.15	8.63	1.76

in semiarid parts, where annual rainfall is 500 to 700 mm whereas, it is most intensively cultivated areas of the Indo-Gangetic plains. Oster and Jaiwardhane (1998) reported that soil properties such as permeability and availability of soil nutrients are adversely affected by irrigation with sodic water in these regions. Poor quality water constitutes 32 to 84% of ground water surveyed in different parts of India is related either saline or alkali (Minhas, 1996). Farmers of these regions are compelled to use poor quality water to irrigate their crops due to inadequate availability of good quality water. In South western Punjab, the quality of underground water is marginal and unfit for irrigation. About 22% ground water is fit, 31% marginal and 47% water is unfit for irrigation due to poor quality. Brackish groundwater with high EC (0.2 to 12.6 dSm^{-1}) and RSC ranging from 0.3 to 35.1 mel^{-1} has been observed in this zone (Jain and Kumar, 2007). Saline water up to 11 dS m^{-1} has been used successfully for commercial irrigation for a number of crops globally (Rhodes et al., 1992). However, in order to assure maximum yields from crops irrigated with saline water, it is necessary to develop special management procedures (Pasternak and De Malach, 1994). Presently, drip irrigation is widely regarded as the most promising irrigation system to use saline water (Meiri and Plantz, 1985). Several factors contribute to the good results obtained with saline water irrigation using drip irrigation (Dasberg and Or, 1999): (i) less water use (high application efficiency) results in less salt deposited on the field, (ii) avoidance of leaf burn, (iii) high frequency drip irrigation prevents the soil from drying out between irrigation events, thereby avoiding peaks in salt concentration and concomitant high osmotic potentials and (iv) salts are continuously leached out from the wetted section and accumulate at the wetting front away from the active root zone.

Keeping the aforementioned facts in view, the present investigation was planned with the aim to develop appropriate management practices for using saline water in conjunction with fresh water through drip system to grow tomatoes successfully in semi arid environment.

MATERIALS AND METHODS

Experimental site

A field experiment was conducted in sandy loam soil during 2008 to 2009 at research farm of Central Institute of Post Harvest Engineering and Technology (CIPHET), Abohar, Punjab, India. Abohar is located at southwestern part of Punjab (30° 4′ N and 74° 21′ E and mean sea level of 185 m). The climate of Abohar is semi-arid with severe summer and winter having the average annual rainfall is 300 to 400 mm. The soil type at the experimental site is sandy loam. The physico-chemical properties of soil at different depths are given in Table 1. The field capacity (1/3 atm) and wilting point (15 atm) of soil at 0 to 30 cm depth were found to be 13.30 and 3.52%, respectively on dry weight basis.

The canal (fresh) and tubewell (saline) water were both available at experimental site. The fresh water was used from the water tank constructed at CIPHET, Abohar farm. The groundwater in the CIPHET farm was saline in nature. The same saline water was pumped for mixing with fresh water and made three different salinities of water. The EC of Fresh canal water and 100% saline water was 0.38 and 19.5 dS/m, respectively. The quality of water after blending fresh canal and saline water in different ratios is given in Table 2. One tank of 2000 lit capacity was used for preparing mixture of fresh and saline water as per given ratio and then this mixed water was transferred to storage tank. The Na^+ and Cl^- ions present in fresh water were 1.6 and 1.0 meq/l, respectively. The Na^+ and Cl^- ions present in 100% saline water were 29.12 and 107.0 meq/l, respectively. The sodium adsorption ratio (SAR) of fresh irrigation water and 100 % saline water was 1.46 and 5.02, respectively.

Nursery raising and treatments

Nursery of tomato (*Lycopersicon esculentum* Cv GC 1500) was raised by sowing the seeds on raised beds in a plastic greenhouse (to protect seedlings from cold weather). Transplanting was done after 45 days and seedlings were transplanted after hardening in 5 m^2 plots with the spacing of 75 × 30 cm. Cultural practices (pest and disease control, weeding and fertigation) were carried out as per the guidelines given from Punjab Agricultural University (PAU), Ludhiana. Fertilizer application per hectare was done at the rate of 50 kg N, 62.5 kg P_2O_5 and 62.5 kg K_2O. Fifty percent of N and full dose of P and K were applied before the transplanting and the remaining 50% nitrogen fertilizers applied in 10 splits through fertigation.

The experimental plots were irrigated with three levels of saline water which was prepared by mixing different ratios of fresh (F) and saline (S) water. The resulted irrigation water's EC of different mixing ratios was found to be 0.38 (T_1), 6.3 (T_2) and 9.1 (T_3) dS/m for the mixing ratios of 100% F, 75:25 (F:S) and 50:50 (F:S), respectively. Initially, after transplanting equal quantities of fresh water were applied to all treatment plots in respect of irrigation levels for proper establishment of seedlings. After establishing the seedlings, fresh water was applied to all plots in three irrigation levels 0.6 (I_1), 0.8 (I_2) and 1.0 (I_3) IW/CPE ratios with three discharge rates of 1.2 (Q_1), 2.4 (Q_2) and 4.2 (Q_3) lph. To avoid water stress on plants due to high salinity of water and cold weather condition in early stage after transplanting, the irrigation treatments

Table 2. Characteristics of water used for irrigation.

Blending water ratio (F:S)	Ec (dS/m)	pH	HCo_3^-	Cl^-	Ca^{2+}	Mg^{2+}	Na^+	K^+	SAR
			me/l						
100 : 0	0.38	7.51	2.00	1.00	1.60	0.80	1.60	0.17	1.46
75 : 25	6.30	7.66	3.00	30.50	7.00	12.00	12.24	0.49	3.97
50 : 50	9.10	7.77	3.00	46.50	9.20	19.80	14.40	0.64	3.78
0 : 100	19.50	7.79	4.00	107.00	20.60	46.60	29.12	1.09	5.02

F= Fresh canal water; S = tube well saline water.

Figure 1. View of experimental field with low tunnels during winter.

with mixed water was started at the end of winter season after 45 days of transplanting. During the winter, the plants were covered with polyethylene film making as low tunnel to avoid the plants from frost injury during January to mid February (Figure 1) and polyethylene film was removed during daytime when sunshine occurs (Figure 2). The volume of irrigation water applied per plant through drip irrigation was estimated on the basis of plant spacing, pan evaporation, pan factor and crop coefficient. The irrigation was scheduled when cumulative pan evaporation (CPE) was 10 mm. Based on discharge capacity of emitters, the drip system was operated for determined time to apply given volume of water per plant in each irrigation level. Plant height was recorded regularly at 30 days interval. The yield and quality parameters were worked out during harvesting stage.

Statistical analysis

The experimental data was statistically analyzed using AgRes Software statistical package (Agres Statistical Software Version 3.01, Pascal International Software Solutions, USA) for getting best treatment.

RESULTS AND DISCUSSION

Growth parameters

Plant height

The recorded data from field investigation is presented in Figure 3; it is clear that plant height was decreased with increasing salinity levels. The maximum height (77.8 cm) was recorded in plots irrigated with fresh water (T_1) followed by salinity treatments T_2 and T_3, respectively. The highest salinity levels reduced the plant height significantly compared to fresh water treatment. The obtained results are in agreement with the findings reported by Malash et al. (2008). The highest tested salinity (T_3) reduced plant height (7.71%) compared to T_1 (fresh water) after 120 days after transplanting (DAT). Cetin and Demet (2008) reported that canopy cover of tomato crop increased and reached maximum (70-100%) value at the middle of crop growth period and afterwards it remain constant until end of growing season. On the other hand the obtained results showed that the different discharge rates significantly affected the plant height. The highest plant height was recorded in irrigation applied with discharge of Q_3 followed by Q_2 and Q_1.

Plant yield

Plant yield as affected by different salinity levels, irrigation levels and discharge rate of emitters is shown in Table 3. It was found that the plant yield decreased with the increase of salinity levels in irrigation water, that is, from T_1 to T_3 significantly. The maximum plant yield (3.91

Figure 2. View of the experimental field after removing plastic cover during day time.

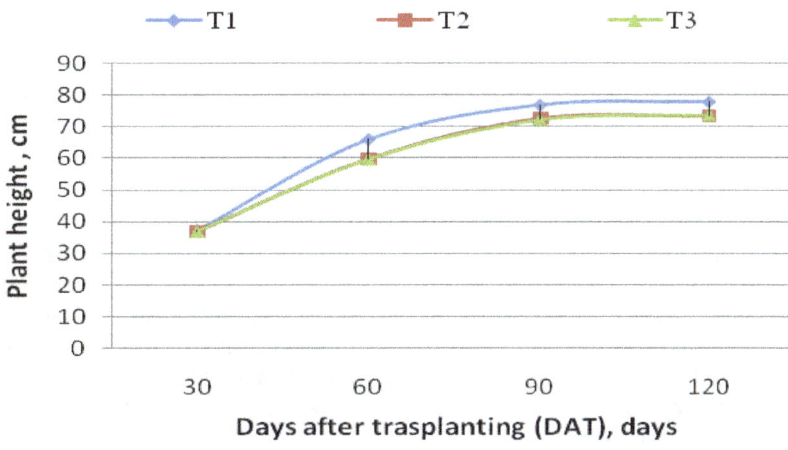

Figure 3. Variation in plant height at different salinity level.

kg/plant) was recorded in the treatment $I_3Q_2T_1$ followed by $I_3Q_1T_1$ while, the minimum plant yield (2.39 kg/plant) was recorded in the treatment $I_1Q_3T_3$. The irrigation levels I_2 and I_3 showed a significant increase in plant yield as compared to I_1. On the other hand, the increase in the discharge rate of emitters significantly decreased the plant yield. The average plant yield decreased by 11.51, and 25.84% in saline water treatment T_2 and T_3, respectively as compared to T_1 (fresh water). The interaction of irrigation levels and salinity treatment also showed a significant effect on plant yield. The tomato plants recorded the maximum yield under the discharge rate of Q_1 followed by Q_2 and Q_3 in each salinity treatment (T_1 to T_3). This finding has close proximity with Badr and Taalab (2007) who reported that the maximum yield obtained with surface drip irrigation applied at 2 lph and lowest yield obtained when water applied at 8 lph with surface drip irrigation in saline water. Yield reduction due to saline water caused mainly due to reduction in fruit weight which in turn is directly proportional to fruit size. This is in accordance with the findings of Reina-Sa´nchez et al. (2005) and Malash (2008) who reported that the plant yield decreased with increasing salinity probably due to reduction in fruit weight. The reduction in water

Table 3. Tomato yield under different irrigation levels, discharge rate and fresh and saline water ratio (salinity levels).

Treatment	I_1			I_2			I_3		
	Q_1	Q_2	Q_3	Q_1	Q_2	Q_3	Q_1	Q_2	Q_3
T_1	3.32	3.27	3.09	3.80	3.55	3.52	3.82	3.91	3.76
T_2	3.10	3.00	2.76	3.54	3.32	3.32	3.33	3.28	2.72
T_3	2.93	2.85	2.39	2.80	2.55	2.47	2.67	2.64	2.47

LSD(0.01) T= 0.231, IxT = 0.401; SE T = 0.086, IxT = 0.149.

Table 4. Water use efficiency (WUE) of tomato under different irrigation levels, discharge rate and fresh and saline water ratio (salinity levels) on crop.

Treatment	I_1			I_2			I_3		
	Q_1	Q_2	Q_3	Q_1	Q_2	Q_3	Q_1	Q_2	Q_3
T_1	1.53	1.56	1.55	1.28	1.20	1.15	0.98	0.89	0.86
T_2	1.44	1.44	1.27	1.09	1.14	1.00	0.83	0.82	0.88
T_3	1.36	1.37	1.23	0.93	0.88	0.89	0.72	0.71	0.66

LSD(0.01): T = 0.039, IxQxT = 0.119; SE: T = 0.0149, IxQxT = 0.044.

uptake when irrigated with saline water was also main cause of yield reduction. This finding has close proximity with Romero-Aranda et al. (2000) who reported that that tomato plants irrigated with a saline solution transpire less water than irrigated with fresh water.

The variation in yield with the different combination of treatments was probably due to the variation in presence of soluble salts. The increasing EC value for different treatments was due to the increase in Na^+ and Cl^- ions concentration. The increase in concentration of these ions might have developed more negative osmotic potential in soil solution and caused reduced water and nutrient uptake thereby growth and yield affected adversely. Grattan and Grieve (1999) observed that presence of Na^+ reduces K^+ uptake while Cl^- reduces NO_3^- uptake by the plant. Irshad et al. (2009) also reported maize leaf and root dry matter yield was reduced significantly by saline irrigation water.

Water use efficiency (WUE)

The water use efficiency under different salinity levels, discharge rates and irrigation levels are presented in Table 4. The data showed that the water use efficiency decreased with the increase in salinity levels of irrigation water. A significant difference was found between salinity treatments T_1, T_2 and T_3. The average maximum WUE (1.22 t/ha-cm) was found in T_1 followed by T_2 (1.10 t/ha-cm) and T_3 (0.97 t/ha-cm). The irrigation levels and discharge rate of emitters had a significant effect on WUE of tomato crop. The WUE decreased with increasing irrigation levels from I_1 to I_3. Also, the average WUE decreased with the increase in the discharge rate of emitters under given treatment combination. The interaction of IxT also showed the significant effect on WUE. The maximum and minimum WUE of 1.54 and 0.69 t/ha-cm was found in treatment combination of I_1T_1 and I_3T_3, respectively. The similar findings reported by Reina-Sa´nchez et al. (2005) that WUE reduced as salinity in nutrient solution increased. Romero-Aranda et al. (2002) also reported lower WUE in saline water than control condition.

Quality parameters of tomato

The effect of water salinity and discharge rate on the quality of tomato fruits such as Total soluble solids (TSS), acidity and ascorbic acid were studied and discussed under following heads.

Total soluble solids (TSS)

The data on the effect of irrigation levels, discharge rate of emitters and ratios of fresh and saline water on TSS of tomato fruits is shown in Table 5. The significant variation was found in TSS of tomato fruits with the treatments of T_1, T_2 and T_3. The reduced water uptake in plants irrigated by saline water led to increase in solute concentrations (particularly sugars) and hence increased TSS contents. TSS increased with increase in salinity of irrigation water, that is, T_1 to T_3 under all discharge rates and irrigation levels. The maximum TSS of 6.2 was found in I1Q1T3 treatment. The minimum TSS was found in tomato irrigated with 100% fresh water in all irrigation levels. Yureseven et al. (2005) showed that increase of

Table 5. TSS (°brix) of tomato fruits under different irrigation levels, discharge rate and fresh and saline water ratio (salinity levels).

Treatment	I_1			I_2			I_3		
	Q_1	Q_2	Q_3	Q_1	Q_2	Q_3	Q_1	Q_2	Q_3
T_1	5.8	5.7	5.5	5.5	5.3	5.0	5.4	5.3	5.2
T_2	6.1	6.0	5.8	5.9	5.9	5.6	5.9	5.8	5.6
T_3	6.2	6.2	6.2	6.1	6.2	5.9	6.2	5.9	5.8

LSD(0.01): T = .443, Q = NS, I = NS; SE: T = 0.165.

Table 6. Acidity (percent) of tomato fruits under different irrigation levels, discharge rate and fresh and saline water ratio (salinity levels).

Treatment	I_1			I_2			I_3		
	Q_1	Q_2	Q_3	Q_1	Q_2	Q_3	Q_1	Q_2	Q_3
T_1	0.695	0.674	0.651	0.684	0.668	0.658	0.715	0.683	0.689
T_2	0.708	0.710	0.668	0.733	0.698	0.687	0.742	0.723	0.705
T_3	0.735	0.727	0.697	0.755	0.752	0.720	0.758	0.738	0.726

LSD(0.05): T = 0.0393, Q = NS, I = NS; SE: T = 0.0196.

Table 7. Ascorbic acid (mg/100 g) of tomato fruits under different irrigation levels, discharge rate and fresh and saline water ratio (salinity levels).

Treatment	I_1			I_2			I_3		
	Q_1	Q_2	Q_3	Q_1	Q_2	Q_3	Q_1	Q_2	Q_3
T_1	29.23	30.18	31.55	31.48	32.36	24.78	30.78	32.89	28.88
T_2	32.76	31.33	31.85	33.23	32.76	30.21	37.16	35.80	35.06
T_3	34.98	35.10	32.63	34.54	33.15	32.13	37.41	37.23	36.75

LSD(0.01): T = 4.169, Q = NS, I = NS; SE: T = 1.559.

TSS in tomato fruits with increasing salinity levels from 0.25 to 10 dS/m. Malash et al. (2005, 2008) also reported increasing TSS content of tomato fruits with increasing saline ratio of irrigation water. The average value of TSS was found maximum in Q_1 treatment followed by Q_2 and Q_3 treatments but the differences were statistically insignificant. In irrigation levels, average TSS was found maximum in I_1 followed by I_2 and I_3. This trend supports findings obtained by Hanson and May (2004) and Hanson et al. (2006) who reported that soluble solids increased with decreasing applied water.

Acidity

The data presented in Table 6 show that acidity of tomato fruits increased with the increase in saline water ratio. A significant difference was found in acidity of tomato among T_1, T_2 and T_3. However, no significant difference was in the acidity between T_1 and T_2 treatments. Among the irrigation levels and discharge rate of emitters, there was no significant difference was found in acidity of tomato fruits while the average acidity increased with the increase in irrigation levels and decreased with the increase in the discharge rate of emitters. The maximum acidity (0.758%) was recorded in $I_3Q_1T_3$ treatment and the minimum (0.651%) in $I_1Q_3T_1$ treatment. The similar trend was found by Magan et al. (2008) and Mitchell et al. (1991) who reported that the titrable acidity increased with increasing EC of the nutrient solution and tomato fruits grown under salt stress showed higher organic acid contents and titrable acidity than fruits grown with fresh water.

Ascorbic acid

The effect of salinity levels, irrigation levels and discharge rate of emitters on ascorbic acid of tomato fruits is presented in Table 7. The ascorbic acid increased significantly with the increasing in salinity levels of irrigated water. The results are in concurrence with Sandra et al. (2006) who reported that rising salinity levels in nutrient solution significantly increased Vitamin C, lycopene, β carotene in fresh fruits. The average value of ascorbic acid was minimum (24.78 mg/100 g) in treatment $I_3Q_3T_1$ and the maximum (37.41 mg/100 g) under $I_3Q_1T_3$ treatment. The significant difference was found between of the treatments T_1 and T_3. However, no significant difference was found between T_2 and T_3 treatments. It was observed that there was no significant difference in ascorbic acid was found between the tested discharge rate of emitters or irrigation levels. Although, it was observed that ascorbic acid was decreased with the increase in discharge rate of emitters non-significantly. The maximum ascorbic acid was found with discharge rate of Q_1 followed by Q_2 and Q_3.

Conclusions

It can be concluded from the present field investigation that the maximum yield can be obtained by applying fresh water with discharge rate of 1.2 lph at IW/CPE ratio of 0.8. The increase in salinity ratio increased the quality of tomato but reduced the yield. The mixing of 75% fresh and 25% saline water with EC up to 6.3 dS/m can be applied with discharge rate of 2.4 lph and IW/CPE ratio of 0.8 improve the quality of tomato with slightly reduction in fruit yield of 11% as compared to fresh water. So, keeping in mind the availability of both fresh and saline water, yield reduction and quality of tomato, the discharge rate and IW/CPE ratio was recommended.

REFERENCES

Badr MA, Taalab AS (2007). Effect of Drip Irrigation and Discharge Rate on Water and Solute Dynamics in Sandy Soil and Tomato Yield. Australia J. Basic Appl. Sci. 1(4):545-552.

Cetin O, Demet U (2008). The effect of drip line spacing, irrigation regime and planting geometries of tomato on yield, irrigation water use efficiency and net return. Agric Water Mangt. 95:949-958.

Dasberg S, Or D. (1999). Drip Irrigation. Springer-Verlag, Berlin, P. 162.

Grattan SR, Grieve CM (1999). Salinity-mineral nutrient relations in horticultural crops. Scientia Horticulturae 78:127-157.

Hanson BR, Hutmacher RB, May DM (2006) Drip irrigation of tomato and cotton under shallow saline ground water conditions. Irrig. Drain. Syst. 20:155-175.

Hanson B, May D (2004) Effect of subsurface drip irrigation on processing tomato yield, water table depth, soil salinity and profitability. Agric Water Mangt. 68:1-17.

Irshad MA, Eneji EA, Khattak RA, Khan A (2009). Influence of Nitrogen and SalineWater on the Growth and Partitioning of Mineral Content in Maize. J. Plant Nutr. 32:458-469,

Jain AK, Kumar R (2007). Water management issues – Punjab, North-West India. Indo-US Workshop on Innovative E-technologies for Distance Education and Extension/Outreach for Efficient Water Management, March 5-9, 2007, ICRISAT, Patancheru/Hyderabad, Andhra Pradesh, India

Magan JJ, Gallardo M, Thompson RB, Lorenzo P (2008). Effect of salinity on fruit yield and quality of tomato grown in soil-less culture in greenhouse in Mediterranean climate conditions. Agric Water Mangt. 95:1041-1055.

Malash N, Flowers TJ, Ragab R (2005). Effect of irrigation systems and water management practices using saline and non-saline water on tomato production Agric Water Mangt. 78:25-38.

Malash, NM, Flowers TJ, Ragab R (2008). Effect of irrigation methods, management and salinity of irrigation water on tomato yield, soil moisture and salinity distribution. Irrig. Sci. 26:313–323.

Meiri A, Plantz Z (1985). Crop production and management under saline conditions. Plant soil. 89:253-271.

Minhas PS (1996). Review-Saline water management for irrigation in India. Agric. Water. Mangt. 30:1-24.

Minhas PS, Bajwa MS (2001). Use and management of poor quality waters for the rice-wheat based production system. J. Crop Prod. 4:273–305.

Mitchell JP, Shenna C, Grattan SR, May DM (1991). Tomato fruit yield and quality under water deficit and salinity. J. Am. Soc. Horti. Sci. 116:215-221.

Oster JD, Jaiwardhane NS, (1998). Agricultural management of sodic soils. In: summer, M.E. Naidu, R. (Eds) sodic soils: Distribution, processes, management and environmental consequences oxford university. pp.125-147.

Pasternak D, De Malach Y. (1995). Crop irrigation with saline water. In: Pessarakli, M. (Ed.),Handbook of Plant and Crop Stress. Marcel Dekker, Inc., NY, pp. 599-622.

Reina-Sa´nchez A, Romero-Aranda R, Cuartero J (2005) Plant water uptake and water use efficiency of greenhouse tomato cultivars irrigated with saline water. Agric. Water Mangt. 78:54-66.

Rhodes JD, Kandiah A, Mashak AM. (1992). The use of saline waters for crop production. Irrigation and Drainage, FAO, Rome Italy, 48:150.

Romero-Aranda, R, Soria T, Cuartero J (2000) Tomato plant-water uptake and plant-water relationships under saline growth conditions. Plant Sci. 160:265-272.

Romero-Aranda, R, Soria T, Cuartero J, (2002) Greenhouse mist improves yield of tomato plants grown under saline conditions. J. Am. Soc. Hort. Sci. 127:644-648.

Sandra K, Wilfried H. Schnitzler, Johanna G, Markus W (2006). The influence of different electrical conductivity values in a Simplified re-circulating soilless system on inner and outer fruit quality characteristics of tomato. J. Agric. Food Chem. 54 (2):441–448.

Yureseven E, Kesmez G D,Unlukara A (2005) The effect of water salinity and potassium levels on yield, fruit quality and water consumption of native central anatolian tomato species (Lycopercicon esculantum). Agric. Water Mangt. 78:128-135.

Effect of pruning on carbohydrate dynamics of herbal and medicinal plant species: Prospects leading to research on the influence of pruning on productivity and biochemical composition of bush tea (*Athrixia phylicoides* D.C.)

Nyembezi Marasha[1], Irvine Kwaramba Mariga[2], Wonder Ngezimana[1] and Fhatuwani Nixwell Mudau[1]

[1]Department of Agriculture and Animal Health, College of Agriculture and Environmental Sciences, University of South Africa, Private Bag X6, Florida, 1710, South Africa.
[2]Department of Soil Science, Plant Production and Agricultural Engineering, University of Limpopo, Private Bag X1106, Sovenga, 0727, South Africa.

Pruning is an essential agronomic practice and it has been shown to be the most important operation, next to plucking, which directly determines the productivity and quality of tea bushes. In tea, pruning helps stimulate vegetative growth and prevent reproductive growth phase. In addition, pruning leads to enhanced branching and hence it rejuvenates the tea plants resulting in a greater number of tender leaves for healthier and better quality tea plants. This review illustrates the effects of pruning on yield and quality of herbal teas. Through the review, it is worthy investigating the effect of pruning and pruning time on yield and chemical compositions. It will also shed light on the effects of resting period before pruning. Lastly the review focuses on the effect of different pruning intervals on productivity of tea.

Key words: Carbohydrate dynamics, bush tea, productivity, pruning, quality.

INTRODUCTION

Athrixia phylicoides DC. belongs to the Asteraceae family and is commonly known as bush tea or bushman's tea. Botanically, bush tea is an attractive, aromatic, perennial, leafy shrub of up to 1 m in height with woolly white stems (Fox and Young, 1982). Leaves are simple, alternate, linear to broadly lanceolate, dark-green, glossy or shiny above and woolly white below (Roberts, 1990; van Wyk et al., 2009), with margins entirely or slightly revolute (Mudau et al., 2007a). The inflorescence head is sessible or subsessible and terminates auxillarily in large subcorymbose panicles. Flowers are daisy-like, with pink to purple petals and bright yellow centres (Van Wyk and Gericke, 2000). The fruits of bush tea consist of narrow, cylindrical and thin achenes that are approximately 0.01 to 0.06 mm wide. The seed is 4 mm in length and has 2 pappuses which act as parachutes during seed dissemination (Araya, 2005).

Bush tea has been used for many decades as a health tea and medicinal beverage in South Africa (Maudu, 2010; Rampedi, 2010) by different ethnic groups, which are Vha-Venda, Zulu, Lobedu, Southern Sotho and the Xhosa (Roberts, 1990; van Wyk et al., 2009). It is believed to have aphrodisiac, diuretic, laxative and anti-helmintic properties. It has been used by many generations to treat a wide range of ailments such as headaches, high blood pressure, diabetes, heart disease (Rampedi and Olivier, 2005; Rampedi, 2010), vomiting, and for washing and as a lotion on boils or skin eruptions (Roberts, 1990; Mudau et al., 2007b). It is also used as a cough remedy, and a purgative, blood purifier and the relief of sore feet (Olivier and Rampedi, 2008).

It has been shown that bush tea leaves contain 5-hydroxy-6,7,8,3',4',5'-hexamethoxy flavon-3-ol which is considered to be a new flavonoid, 3-0-demethyldigicitrin, 5,6,7,8,3',4'-hexamethoxyflavone and quecertin, total polyphenols, tannins and total antioxidants (Mashimbye et al., 2006; Rampedi and Olivier, 2008; Padayachee, 2011; Nchabeleng et al., 2012), which are major quality parameters for medicinal tea. Pharmacological evaluation of leaf extracts confirmed that the plant has anti-inflammatory, anti-hypertensive, narcotic and analgesic properties (Moller et al., 2006). The fact that *A. phylicoides* has high contents of polyphenols, lacks caffeine and cytotoxic effects, demonstrates its commercial development potential as a health beverage (Rampedi and Olivier, 2005; Joubert et al., 2008). van Wyk and Gericke (2000) and Mudau et al. (2007a) reported the suitability of bush tea for domestication and development as a commercial health beverage. However, the success of domestication and commercialization of bush tea hinges on maintenance and/or enhancement of quality of bush tea as a herbal beverage.

Research reports by Mudau et al. (2007b) reveals that herbal tea quality is one of the critical factors in commercialization that would determine the price of tea for sale and export. Cultural practices, is one of the factors that has been shown to significantly influence growth, yield and quality of herbal teas (Owour et al., 1990, 2000; Venkatesh et al., 2007). Pruning is an important cultural practice that has been shown to enhance both productivity and quality of *Camellia sinensis*. A study by Yilmaz et al. (2004) has shown that pruning affected the composition of tea leaves which determines quality. This finding is consistent with prior work by TRIT (2006) and recent studies by Tockclai (2008). Based on their studies they concluded that pruning is an essential agronomic practice and it is the most important operation, next to plucking, which directly determines the productivity and quality of tea bushes.

Pruning removes substantial amounts of leaves and branches which result in a drastic reduction in photosynthesis. Barua (1960) has showed that the peak photosynthetic efficiency of maintenance leaf remained for six months, then gradually declined and drops off due to senescence after 12 to 18 months. Barbora (1994) pointed out the need to replenish and maintain foliage for better yield and crop distribution. Plants have a functional equilibrium between their above ground parts (leaves) and below ground parts (roots) (Zeing, 2003). If plants are pruned the starch reserves in the roots are utilised for shoot growth to maintain the equilibrium. The speed of recovery from pruning of a tea bush depends on the plant's starch reserves in the roots (Ndunguru, 2004).

The main objective of this paper is to review the effect of pruning on productivity of tea bushes. More specifically, the review aims to investigate the relationship between leaf growth and root carbohydrate reserves after pruning; investigate the effect of aging of the pruning cycle on productivity and variation in concentration of secondary metabolites (which determines aroma, colour, and taste) with time from pruning. The review also will investigate the change in growth and productivity from skiffing as a method to prolong the pruning cycle and improve yield and again evaluate the effect of pruning and time from pruning on biochemical and quality parameters of bush tea.

This foundation will also assist in determining the role of pruning on yield, chemical composition and bioactivity of herbal teas. Understanding these functions can lead to proper agronomic practices to assess commercial viability of bush tea growth on large commercial scale in a well established orchard.

IMPORTANCE OF PRUNING IN MEDICINAL AND HERBAL PLANTS

Tockclai (2008) and TRIT (2006) provides insights into the importance of pruning as an agronomic practices and it has been shown to be the most important operation, next to plucking, which directly determines the productivity and quality of tea bushes. In tea, pruning helps stimulate vegetative growth and prevent reproductive growth phase. Ravichandran (2004) and Yilmaz et al. (2004) have shown that pruning leads to enhanced branching and hence it rejuvenates the tea plants resulting in a greater number of tender leaves for healthier and better quality tea plants. Satyanarayana et al., (1994) also reported that pruning leads to enhanced branching and hence a greater number of tender leaves. Tea plants are pruned to obtain a given table form and height thereby maintaining tea bushes in a manageable condition for plucking and to eliminate unnecessary and diseased branches (Ravichandran, 2004). In some areas pruning may also help to protect the plants from moisture stress. Tockclai (2008) pointed out that one of the main objectives of pruning is to divert stored energy to production of growing shoots. Although a lot of research has been carried out on the effect of pruning on productivity and chemical composition of black tea, little is known about its effect on bush tea.

CARBOHYDRATE DYNAMICS IN MEDICINAL PLANTS

A. phylicoides has been shown to contain different phenolic acid and flavonoid components (Mashimbye et al., 2006; Maudu, 2010; Maudu et al., 2010; Padayachee, 2011). Currently, the effect of dietary phenolics is of great interest due to their antioxidative and possible anticarcinogenic activities (Moller et al., 2006). Phenolic acids and flavonoids also function as reducing agents, free radical scavengers and quenchers of singlet oxygen formation.

In plants, carbohydrates in excess of growth requirements are used to synthesise carbon based secondary metabolites (CBSM) (Haukioja et al., 1998). According to the carbon-nutrient balance hypothesis, carbohydrates accumulated in excess of growth requirements will be allocated to carbon-based secondary metabolism, but does not explain how this carbon is distributed among different pathways and compounds. Nen et al. (1999) is of the opinion that different branches of the biosynthetic pathway of phenolic compounds may compete for substrates, and such internal metabolic trade-offs may explain the differential accumulation of the compounds. Studies done have shown the effect of pruning on total polyphenols and non structural carbohydrate but did not show how the carbon is distributed among different pathways and compounds. Moreover, the concentration of one compound may increase at the expense of another; thus, changes in the amounts of individual compounds are not necessarily correlated with overall changes in total secondary metabolism (Herms and Mattson, 1992). Reichardt et al. (1991) challenges the above theory and suggests that increased concentrations of secondary chemicals may, in some cases, be masked by rapid metabolic turnover. Against this background, there is need to investigate how CBSM are allocated to different pathways and compounds in leaves and roots in response to pruning and also to assess the patterns of concentration of primary and secondary metabolites in response to pruning.

EFFECT OF LUNG PRUNING ON BUSH RECOVERY AND YIELD

Through continued photosynthesis on the lungs, starch reserves are accumulated which help the tea bush to recover after pruning (Bore, 2001; Bore et al., 2003). However for it to be effective, lungs should be left on the bushes until the pruned portions have fully recovered. In contrast, Nwaka (1997) argues that lungs may direct some reserves for their own use and this may retard recovery of the tea bushes after pruning. It is therefore crucial to ensure that lungs are removed at an optimum time when they have helped the bush to recover but before they start retarding regrowth of the bushes.

The lung branches may continue to be plucked and would therefore help to sustain cropping before the pruned bush recover. In studies carried out in Kenya, Sri Lanka and north east India, it was found that tea yields increase in the year of prune when lungs were retained on the bushes until tipping-in time (TRI, 2001). The size of lung branches greatly contributed to the overall yield, with yield in the pruned year increasing with lung size. However, variations on the size of lungs left on the bushes, their contribution to yield has varied from non-significant to significant. In studies where lungs have been plucked, it has been shown that retaining lungs beyond tipping -in time would lead to yield reduction, probably because of retarding effects on regrowth (Mphangwe, 2012). Although a lot of research has been done on the effect of lung pruning on bush recovery and yield of *Camellia sinensis*, no research has been done in bush tea. Thus there is need for researchers to investigate the effect of lung pruning on bush tea recovery and yield after pruning.

EFFECT OF PRUNING LEVEL ON PRODUCTIVITY

Studies conducted in Sri Lanka by on the effect of different pruning levels, revealed that top pruning significantly increased the number of plucking points, fresh and dry weight of leaves. All the four pruning levels did not significantly affect the growth rate. However results revealed that top pruning was found to be the best among the remaining different pruning levels. Down-pruning caused more plant deaths and reduced subsequent yields (Magambo, 1980). The same results were observed by Maudu et al. (2010) who concluded that basal pruning is not viable in bush tea because the treatment showed highly reduced growth.

EFFECT OF PRUNING ON QUALITY

It had been demonstrated that variation in growth rate is expected to cause some changes in the green leaf constituents and hence the quality of made tea. Similarly, later research demonstrated that the quality of black tea improves with time from pruning. Thus, according to Ravichandran (2004) pruning as such leads to adverse effects on tea quality. Although a lot has been done on the effect of pruning height in *Camellia sinensis*, little has been done on the effect of pruning bush tea on productivity. Studies by Maudu et al. (2010) on the effect of pruning height on growth (yield) and quality (chemical composition) of bush tea showed that pruning reduced yields and quality. The results also showed that pruning at different heights have no favourable effect on quality of bush tea. Total polyphenols remained higher in unpruned tea plants, no significant differences were observed in tannin and total antioxidant content in pruned tea at different heights. Unpruned tea plants remained the tallest plants, with higher number of branches, bigger leaf area and larger biomass than pruned tea at different heights. Research including that of Yilmaz et al. (2004), reported the same trend of less yields in the first year, with yields increasing in the subsequent second and third years. It has also been shown that during the first year of the pruning cycle, yields are generally low, because bushes have to recover from the prune before plucking can resume and because it takes time for shoot numbers to build up (TRIT, 2006). It is recommended that there is need to investigate the impact of pruning bush tea in the long run stretching over two and or more years to allow for accumulation of starch reserves in stems and roots. Mudau and Mariga (2012) pointed out that pruning has a direct effect on regrowth and recommended that it should be tried at field scale on cultivated bush tea to assess the vigour of growth, productivity and quality.

EFFECT OF AGING OF THE PRUNING CYCLE ON PRODUCTIVITY OF TEA

In the first year of the pruning cycle, yields are generally low, because bushes have to recover from the prune before plucking can resume. It takes time for shoot numbers to build up and a ground cover to be established. As the cycle progresses, shoot numbers build up, average shoot size and weight diminish and growth rates slow down (TRIT, 2006). The net effect is that 2nd year yields tend to be the highest, with the crop falling away in subsequent years (Satyanarayana and Sharma, 2004). Prior studies by Barua (1989) are consistent with the above observation and showed that productivity in *C. sinensis* decreases 2 to 3 years after pruning. However, the length of, and productivity during, the pruning cycle, vary greatly among tea varieties, and environmental and management conditions (Mphangwe, 2012).

Yilmaz et al. (2004) observed that polyphenol contents, which are one of the most significant quality traits of tea, gradually declined until the pruning age of 5. Tea yields started to decrease after four harvest years following pruning. Therefore, after 3rd or 4th year of pruning yield starts to gradually decline and this leads to another pruning operation (Satyanarayana and Sharma, 2004). Skiffing is normally light pruning and involves removal of the green wood at about 15 cm above the pruning height and has been found to prolong the pruning cycle in Sri Lanka (Nissanka et al., 2004). Light skiffing at the 5th year after pruning has been shown to enhance productivity by 42%.The effect of age after pruning on productivity and possible ways to prolong the pruning cycle has not been studied in *A. phylicoides*. Therefore, there is need to investigate the impact of aging of the pruning cycle on physiological and yield parameters and investigate the change in growth and productivity from skiffing as a means of prolonging the pruning cycle.

EFFECTS OF PRUNING TIME ON TEA REGROWTH

The time at which pruning is done is important in determining regrowth and performance of tea (Mphangwe, 2012). A lot of research has been done in an attempt to find the best time of pruning for maximum regrowth and these attempts have yielded varying results that could be attributed to complex factors prevailing during pruning time (Mphangwe, 2012). However, no work has been done on the effect of pruning time on tea regrowth and productivity.

Root starch reserves play an important role in recovery of tea after pruning (Wijeratne et al., 2002). The speed of recovery from pruning of a bush depends on plant's starch reserves in the roots (TRIT, 2006). The more starch reserves there are, the faster will be the recovery from the prune. Potassium deficiency, in particular, has been shown to have a very marked effect on recovery.

A study was conducted in Sri Lanka to monitor the variation of root reserves in relation to yield over a 3-year pruning cycle of *Camellia sinensis* and to assess the influence of root reserves on recovery after pruning (Wijeratne et al., 2002).Three periods of resting before pruning (0, 1 and 2 months) were tested on tea clones. The results showed that those 1 to 2 months of resting before pruning increased root reserves. The level of root reserves declined rapidly soon after pruning. The replenishment of root reserves commenced only after two months from pruning and complete replenishment of root reserves occurred about 10 months after pruning. Further analysis of results showed a clear positive relationship between the level of root reserves at pruning and the degree of recovery following pruning. This finding shows that, to sustain the productivity of tea bushes, the importance of building up of adequate root reserves before pruning of tea is emphasized.

Studies done by Baktir (1987) on the effect of pruning ages (1, 2, 3 and 4 years) on root starch content and production of seedling tea have shown that the older the pruning age, the higher the starch content. Maximum starch content of 15.30% was reached 3 years after pruning and at the pruning age of 4 years starch content decreased to 13.58. Hence, a combined reduction of both sink strength and source capacity during the fourth year could have brought the significant reduction in tea yields. It is recommended that pruning should be done when carbohydrate reserves are not limiting. No work has been done on the effect of resting period and pruning time on quality and yield of bush tea. There is need to investigate the relationship between leaf growth and root carbohydrate reserves after pruning. Carbohydrate reserves are utilised in times of adverse seasonal conditions. The timing, amount and location of their deposits should be investigated as a possible indicator to predict vigour and yield (Bore et al., 2003). Pruning should be done when the root reserves are high. Pruning and or tipping have been shown to check root growth and development. Carbohydrate reserves are enhanced by agronomic practices.

EFFECTS OF RESTING PERIOD BEFORE PRUNING ON PRODUCTIVITY

Studies conducted by Watson and Gunasereka (1986) in Sri Lanka, showed that resting period had a significant effect only on regrowth at pruning. Resting tea bushes for 11 months before pruning lowered yields compared to resting tea for eight months; moreover, the unrested control had the highest yields, and hence, no advantage of resting period was observed (Watson and Gunasereka, 1986). This result indicates that more dry matter accumulated during the resting period and that resting period before pruning might have an effect on replenishment of carbohydrates used in regrowth of tea after pruning (Bore, 2001). On the contrary, bushes that were plucked continuously had better bud-breaks. This finding is consistent with prior observation by Wadasinghe and Gunasereka (1986) who observed an increase in carbohydrate concentration with increased resting period. However, no significant effect of resting period on yields was obtained in the 1997 pruning cycle. Bore et al. (2003) claim that the above observation is due to the ``carry-over" effect. Based on the aforementioned discussion, it is hypothesised that, treatment effects may not be realised within the year of experimentation, but may be carried into subsequent years.

CONCLUSIONS

It is evident from the above discussion that extensive academic research has been explored on the effects of pruning on productivity and quality in tea. However, little has been done in bush tea. The effect of phenolics and flavonoids is of great interest due to their antioxidative activities, possible as anticarcinogenic reducing agents, free radical scavengers and quenchers of singlet oxygen formation. Therefore, there is need to carry out extensive research to address the fore mentioned objectives in bush tea. Knowledge of the importance of pruning and its effects on productivity and quality of *A. phylicoides* is necessary for future research which will lead to proper agronomic practices for the commercialization of bush tea. With the increasing demand for herbal teas worldwide, South African bush tea can be an important shrub to supply herbal tea.

REFERENCES

Araya HT (2005). Seed germination and vegetative propagation of *Athrixia phylicoides* (Bush tea). MSc. Thesis, Department of Plant Production and Soil Science, University of Pretoria.

Baktir FH (1987). Effect of pruning ages and ways of plucking on root starch content and production of seedling tea (*Camellia sinensis* L. O. Kuntze). Warta BPTK 13:37-44.

Barbora BC (1994). Practices adopted in North-East India to Enhance Tea Productivity. Proceedings of the International Seminar on "Integrated Crop Management in Tea: Towards Higher Productivity" Colombo Sri Lanka. April 26-27, 1994.

Barua DN (1960). Effect of age and carbon-dioxide concentration on assimilation of and by detached leaves of tea and sunflower. J. Agric. Sci. 55:413-421.

Barua DN (1989). Effect of age and carbon-dioxide concentration on assimilation of and by detached leaves of tea and sunflower. J. Agric. Sci. 55:413-421.

Bore JK (2001). Effects of pruning time, lungs, and resting period on total non-structural carbohydrate reserves, regrowth and yield of tea (*Camellia sinensis* [L] O. kuntze). MSc. Thesis. Egerton University. Njoro, Kenya.

Bore JK, Isutsa DK, Itulya FM, Ng'etich WK (2003). Effects of pruning time and resting period on total non-structural carbohydrates, regrowth and yield of tea (*Camellia sinensis* L.). J. Hort. Sci. Biotech. 78:272-277.

Fox FW, Young NM (1982). Food from the veld. Edible Wild Plants of Southern Africa. Delta Books (Pty) Ltd., Johannesburg, South Africa, pp. 119-120.

Haukioja E, Ossipov V, Koricheva J, Honkanen T, Larsson S, Lempa K (1998). Biosynthetic origin of carbon-based secondary compounds: cause of variable responses of woody plants to fertilization. Chemoeco 8:133-139.

Herms DA, Mattson WJ (1992). The dilemma of plants: to grow or defend. Quart. Rev. Biol. 67:283-335.

Joubert E, Gelderblom WCA, Louw A, de Beer D (2008). South African herbal teas: *Aspalathus linearis*, *Cyclopia* spp. and *Athrixia phylicoides*—A review. J. Ethnopharmacol. 119:376-412.

Magambo MJS (1980). Influence reducing bush (down prune) on recovery from pruning and yield. Tea 1:1-3.

Mashimbye MJ, Mudau FN, Soundy P, Van Ree T (2006). A new flavonol from *Athrixia phylicoides* (Bush tea). S. Afr. J. Chem. 59:1-2.

Maudu M (2010). Chemical profiles of Bush tea (*Athrixia phylicoides* D.C.) at different phenological stages as influenced by pruning and growth regulators. MSc. Mini-dissertation. University of Limpopo, Mankweng. South Africa.

Maudu M, Mudau FN, Mariga IK (2010). The effect of pruning on growth and chemical composition of cultivated bush tea (*Athrixia phylicoides* D.C). J. Med. Plants Res. 4:2353-235.

Moller A, du Toit ES, Soundy P, Olivier J (2006). Morphology and ultra and nonglandular trichomes on the leaves of *Athrixia* phylicoides (Asteraceae). S. Afr. J. Plant Soil 23:302-304.

Mphangwe NIK (2012). Lung Pruning: A Review of Practice. Tea Research Foundation of Central Africa (TRFCA) News. pp. 18-23.

Mudau FN, Mariga IK (2012). Bush tea as a herbal beverage and medicinal plant. Tea in Health and Disease Prevention. Reedy VR (ed), 1st Edition. Elsevier. Academic press, Oxford. UK, pp. 182-192.

Mudau FN, Soundy P, du Toit ES (2007a). Effects of nitrogen, phosphorus and potassium nutrition on total polyphenols content of bush tea (*Athrixia phylicoides* L.) in a shaded nursery environment. Hort. Sci. 42:334-338.

Mudau FN, Araya HT, du Toit ES, Soundy P, Olivier J (2007b). Bush Tea (*Athrixia phylicoides* DC.) as an alternative herbal and medicinal plant in Southern Africa: Opportunity for commercialization. Med. Aromatic Plant Sci. Biotech. 1:70-73.

Nchabeleng L, Mudau FN, Mariga IK (2012). Effects of chemical composition of wild bush tea (*Athrixia phylicoides* DC.) growing at locations differing in altitude, climate and edaphic factors. J. Med. Plants Res. 6:1662-1666.

Ndunguru BJ (2004). Tea pruning and tipping. Module 6.Tea Research Institute of Tanzania. Available online at http://www.trit.or.tz/Training%20modules/MODULE%20No.%206%20 Pruning.pdf Accessed on 16/06/2013.

Nen MK, Julkunen-Tiitto R, Mutikainen P, Walls M, Ovaska J, Vapaavuori E (1999). Trade-offs in phenolic metabolism of silver birch: Effects of fertilization, defoliation, and genotype. Ecology 80:1970-1986.

Nissank SP, Coomaraswamy AA, Seneviratne CK (2004). Change in growth and productivity from pruning and skiffing as a method to prolong pruning cycle. Proceedings for the 4th International Crop

Science Congress, Brisbane, Australia, 26 September – 1 October 2004.

Nwaka E (1997). Tea pruning for yield improvement. Tea Board of Kenya 18:144-148.

Olivier J, Rampedi IT (2008). The *Athrixia phylicoides* research project: summary of research findings. Poster presented at a Conference organised by the Indigenous Plant Use Forum (IPUF) held at Rhodes University, Grahamstown, Eastern Cape province (27-30/06/2005).

Owour PO, Munavu RH, Muritu JW (1990). Plucking standard effects and the distribution of fatty acids in the tea (*Camellia sinensis* L.) leaves. J. Food Chem. 37:27-35.

Owour PO, Ng'etich KW, Obanda M (2000). Quality response of clonal black tea to nitrogen fertilizer, plucking interval and plucking standard. J. Sci. Food Agric. 70:47-52.

Padayachee K (2011). The phytochemistry and biological activities of *Athrixia phylicoides*. MSc Thesis. Faculty of Health Sciences, University of Witwatersrand. South Africa.

Rampedi IT (2010). Indigenous plants in the Limpopo province: potential for their commercial beverage production. PhD Thesis. University of South Africa. South Africa.

Rampedi IT, Olivier J (2005). The use and potential commercial development of *Athrixia phylicoides*. Acta Acad. 37:165-183.

Rampedi IT, Olivier J (2008). Mountain tea: a new herbal tea from South Africa. J. Inst. Food Sci. Tech. 22:47-49.

Ravichandran R (2004). The impact of pruning and time from pruning on quality and aroma constituents of black tea. Food Chem. 84:7-11.

Reichardt PB, Chapin FSI, Bryant JP, Mattes BR, Clausen TP (1991). Carbon: nutrient balance as a predictor or plant defence in *Alaskan balsam* poplar: Potential importance of metabolic turnover. Oecologia 88:401-406.

Roberts M (1990). Indigenous healing plants. Southern Book publishers: Half way House, South Africa. pp. 51-52.

Satyanarayana N, Sharma VS (2004). Pruning and harvesting practices in relation to tea productivity in South India. Proceedings of the International Seminar on "Integrated Crop Management in Tea: Towards Higher Productivity" Colombo, Sri Lanka, April 26-27, 2004.

Satyanarayana N, Sreedhar CH, Cox S, Sharma VS (1994).Response of tea to pruning height. Journal of Plantation Crops, 22.81–86.

Tea Research Institute (TRI) (2001). Pruning Tea. Tea Research Institute Advisory circular. September 2001. Serial no 3/01.

Tea Research Institute of Tanzania (TRIT) (2006). Pruning and tipping. Module 6. August 2004. Available online at http://www.trit.or.tz/Training%20modules/MODULE%20No.%206%20 Pruning.pdf. Accessed on 21/03/2013.

Tockclai tea research association (Tockclai) (2008). Pruning and skiffing of mature tea. Available online at http://www.tocklai.net/Cultivation/young_tea.aspx. Accessed on 21/03/2013.

Van Wyk BE, Gericke M (2000). People's Plants. A guide to useful plants of Southern Africa. Briza Publications, Pretoria, South Africa, pp. 102-103.

Van Wyk BE, Van Oudtshoorn B, Gericke N (2009). People's Plants. A guide to useful plants of Southern Africa. 2nd Edition. Briza Publications, Pretoria, South Africa, pp. 56-57.

Venkatesh P, Jaiprakash M, Prasad, P, Pillai B, Sadhale PP, Sinkare VP (2007). Flavonoid biosynthesis in tea (*Camellia sinensis*). Original Research Report. [PDF]. Available online at http://www.teascience.org/pdf/chapter-2-6-2. Accessed on 21/03/2013.

Wadasinghe G, Gunasereka C (1986). Resting of tea bushes prior to pruning and its effects on recovery and growth at Palm Garden State Plantation, Ratnapura. Annual report of the Tea Research Institute of Sri Lanka. P. 33.

Watson M, Gunasereka C (1986). Effects of resting tea after pruning on frame development and yield at the low country station, Ratnapura-1984. Annual Report of the Tea Research Institute of Sri Lanka. P. 32.

Wijeratne MA, Premathunga A, Karunaratne WRMM (2002). Variation of root starch reserves of tea and its impact on recovery after pruning. J. Plant. Crops 30:33-37.

Yilmaz G, Kandemir N, Kinalioglu K (2004). Effects of different pruning intervals on fresh shoot yield and some quality properties of tea (*Camellia sinensis* L. Kuntze) in Turkey. Pak. J. Biol. Sci. 7:1208-1212.

Zeing B (2003). Functional equilibrium between photosynthetic and above ground non-photosynthetic structures of plants: Evidence from a pruning experiment with three subtropical tree species. Acta. Bot. Sinica 45:152-157.

Assessment of heavy metals and aflatoxin levels in export quality Indica rice cultivars with different milling fractions

Muhammad Asim Shabbir, Faqir Muhammad Anjum, Moazzam Rafiq Khan, Muhammad Nadeem and Muhammad Saeed

National Institute of Food Science and Technology, University of Agriculture, Faisalabad, Pakistan.

The present study was designed to characterize the Pakistani rice varieties (Basmati and coarse), along with their milling fractions intended for export in Europe and other regions for heavy metals and aflatoxins content. The content of heavy metals, that is, arsenic, cadmium, lead and mercury varied significantly among the rice varieties. The highest content of cadmium (0.63 mg kg^{-1}) and lead (0.27 mg kg^{-1}) was found in the rice variety IRRI-6, whereas the highest mercury level was found in KS-282 (0.075 mg kg^{-1}). The content of lead in IRRI-B, IRRI-W and brown rice of Basmati 2000 was higher than that allowed by European Union (EU) legislation, whereas the contents in milling fractions SB-B, SB-W KS-B, KS-W and B2-B were within the EU prescribed limits. The highest arsenic content was found in the rice variety IRRI-6 while Basmati 2000 had the lowest contents. The content of aflatoxins, that is, B1, B2, G1 and G2 varied significantly among rice varieties and between rice milling fractions, but the total aflatoxin content in all varieties was within the prescribed limits.

Key words: Export quality rice with milling fractions, heavy metals, aflatoxins, high power liquid chromatography.

INTRODUCTION

Rice is the leading food grain crop which has been reported to be used a human food for the last almost 5000 years (International Rice Research Institute (IRRI), 1997). The major rice growing Asian, American and African regions are in more than a hundred countries. The major rice exporting countries are the Thailand, United States, Vietnam, Pakistan and India. Pakistan falls in the region, where 90% of the world's rice is produced as the second biggest cash crop Government of Pakistan (GOP, 2013). The quality characterization of rice in relation to consumers and export is also attributed to chemical and biochemical characteristics like heavy metal contents, aflatoxin level and microbiological parameters.

These quality parameters are big threat to rice exporting countries. The Codex Alimentarius commission (CAC) has set a specific codex standards for rice (CODEX STAN 198-1995) including the general quality factors (classification of rice types and qualities), specific quality factors (moisture content, filth and other matter) and contaminants (heavy metals and pesticide residues). The main components of the European Union (EU) food legislation for rice contaminants, that is, nitrates, aflatoxins, heavy metals like lead, cadmium, mercury and 3-monchloropropane 1,2 diol. The mycotoxin contamination, particularly aflatoxin, is commonly found

Table 1. Sample codes of rice varieties with milling fractions.

S/N	Sample code	Name of sample
1	IRRI-B	Brown rice of IRRI-6 variety
2	IRRI-W	White rice of IRRI-6 variety
3	KS-B	Brown rice of KS-282 variety
4	KS-W	White rice of KS-282 variety
5	B2-B	Brown rice of Basmati 2000 variety
6	B2-W	White rice of Basmati 2000 variety
7	SB-B	Brown rice of Super Basmati variety
8	SB-W	White rice of Super Basmati variety

in locally produced agricultural crops. The aflatoxins are designated as B1, B2, G1, and G2. Aflatoxins are known to be human carcinogens based on sufficient evidence of carcinogenicity in humans (Mishra and Das, 2003). Aflatoxin B1 is the most compelling carcinogen known compound (FAO/WHO, 2004). The aflatoxin in food has been recognized as a latent hazard in humans. This health hazard may be caused by direct contamination through contaminated grains or it's by products (Kotsonis et al., 2001). The numeral death cases have been associated by the consumption of aflatoxin infected food in humans (Moss, 1996). The intake of heavy metals in food makes up a considerable amount of contamination in humans. The ingestion of heavy metals in humans may be through food chain via directly or indirectly and to some extent it is also accumulated in the human body. Lead and cadmium are among the most abundant heavy metals and are particularly toxic. Exceeding level of these toxic metals may affect the health of individuals. Cadmium has shown adverse effects on kidney's role in human body and some studies have reported it as a carcinogenic element. There are various sources of contamination of these heavy metals in food grains especially the use of contaminated water in rice fields may enhance the level of cadmium contents in rice grains. A number of cases have been reported that if rice is major food of different countries than their cadmium intake might be due to the consumption of that infected rice. The 50% cadmium intake in Indonesia has been reported due to the consumption of contaminated rice and this amount is ranged from 40 to 60% in Japan (Suzuki, 1988).

The quality characterization of rice in relation to consumers' use and export requires the assessment of the levels of potential contaminants, such as heavy metals and aflatoxin (Santini and Ritieni, 2013). Failure to meet these quality parameters can be a big threat to rice exporting countries. Pakistan is the major rice exporting country. Basmati rice produced in Pakistan is exported to more than 80 countries mainly to Gulf and European Countries. The Pakistani Basmati rice varieties have special pleasant aroma, long slender grain and with soft texture on cooking. Thus, it is of great importance to analyze the Pakistani rice varieties for their quality parameters.

The objective of this study was to determine occurrence and level of aflatoxin and heavy metals in export quality Indica rice varieties for the better understanding of world rice exporters and consumers.

MATERIALS AND METHODS

Samples

Four Rice varieties (IRRI-6, KS-282, Basmati 2000 and Super Basmati) were procured from Rice Research Institute, Kala Shah Kaku, Pakistan. The paddy of each variety was dehulled and milled by passing through stake sheller. The McGill laboratory mill (Rapsco, Inc, Brookshire, TX) was used to obtain two fractions of each rice variety, that is, brown rice and white rice. A portion of rice fractions each of white and brown rice was milled by passing through cyclone mill (Udy Corp, Fort Collins, Co) to get rice flour for further studies. The code number was assigned to each sample as described in Table 1.

Heavy metals analysis

The analyses of heavy metals were performed by following the method given in AOAC (2000). One gram sample of each fraction of rice flour was taken in a conical flask. The samples were fist digested with 10 ml HNO_3 at a temperature of 60 to 70°C for 20 min and then digested with 5 ml $HCLO_4$ at a temperature of 60 to 70°C for 20 min and subsequently raising the temperature to 195°C till the volume was near to dry or until a clear solution was obtained. For arsenic determination, the 5 ml of H_2SO_4 was also added. The digested samples were transferred to 100 ml volumetric flask and volume was made with distilled deionized water and then filtered and stored in air tight bottles for analysis of heavy metals with the help of atomic absorption spectrophotometer (Model Varian Spectra AA 250 plus) at wavelengths: 228.8 nm for cadmium, 283.3 nm for lead, 254 nm for mercury and 193.7 nm for arsenic.

Aflatoxin analysis

The level of aflatoxins B1, B2, G1 and G2 in rice flour samples was estimated by running samples through HPLC according to the procedure described by Alberts et al. (2006) with some modifications.

Sample extraction and clean-up

The extraction and clean-up of flour samples was carried out by following the procedure mentioned by Romar Labs Inc (2003) and described in Figure 1.

HPLC conditions

The levels of aflatoxin B1, B2, G1 and G2 in standards and test rice flour samples were estimated by running the extracted samples through HPLC (Perkin Elmer Series 200) equipped with an auto sampler (Perkin Elmer Series 200) using a UV detector and standard chromatogram is presented in Figure 2. The

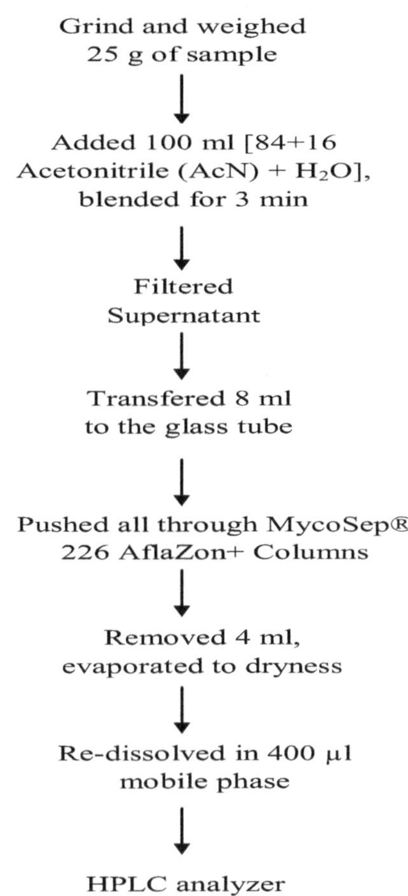

Figure 1. Sample extraction and clean-up for aflatoxin (B_1, B_2, G_1, G_2).

standard Mycotoxin mix 1(Aflatoxins) was obtained from Biopure Referenzsubstanzen GmbH (Technopark1, Tulln, Austria). A Phenomenex Luna C18 column (25 cm × 4.6 mm I.D.) was used to perform the HPLC analysis. A mobile phase of water/acetonitrile/methanol (57:17:26 v/v/v) was used for the separation of aflatoxins through the column at a flow rate of 0.5 ml/min and UV wavelength at 365 nm. The column temperature was maintained at a temperature of 30°C. The sample dilution was done 1.5 ml with water and injection volume was 100 µl.

Statistical analysis

The analysis of variance was used to analyze the data (two factor factorial) according to the method described by Steel et al. (1997). Completely randomized design (CRD) was applied on the data to assess the significance level and differences among means were compared with Duncan's multiple range test (DMRt).

RESULTS AND DISCUSSION

The flour samples of different rice varieties were analyzed for different parameters to assess their quality characteristics. The data obtained for different parameters were subjected to statistical analyses and the results are interpreted and discussed following.

Heavy metals content

The ingestion of heavy metals in humans may be through food chain via directly or indirectly and to some extent it is also accumulated in the human body. Some countries have set standards for permissible limits of heavy metals in rice. In case of rice it is very important that rice must possess heavy metals within the prescribed permissible limit. Cadmium, lead, mercury and arsenic level were estimated in different brown and white rice sample. The mean squares for cadmium, lead, mercury and arsenic concentration showed significant variation among milling fractions and rice cultivars.

The results pertaining to the cadmium content of different rice varieties and milling fractions have been presented in Table 2. The contents of cadmium ranged from 0.08 to 0.63 mg/kg among different rice varieties (when milling fractions were pooled). The cadmium content was found significantly higher in brown rice as compared to white rice. The highest cadmium level was found in rice variety IRRI-6 followed by KS-282, Basmati 2000 (B2) and Super Basmati (SB), in an ascending order. The cadmium level was found to be higher in brown rice of IRRI-6 (0.71 mg/kg) followed by white rice of IRRI-6 (0.55 mg/kg), KS-B (0.47 mg/kg), KS-W (0.41 mg/kg), B2-B (0.15 mg/kg), B2-W (0.12 mg/kg), SB-B (0.09 mg/kg). The lowest cadmium content was recorded in white rice of Super Basmati (0.07 mg/kg). The results showed that cadmium content was found significantly higher in coarse varieties as compared to Basmati varieties. In the present study, the cadmium content was found higher in brown rice samples and lower in white rice milling fractions. The higher level of cadmium in brown rice fractions may be present in the outer parts of the grain which is removed in the milling process. The interactive effect of brown rice of IRRI-6 and KS-282 varieties may be due to the uptake efficiency of element from soil to the grain which may vary depending on the type of soil. The findings of the present study are in the line with Meharg et al. (2013) who found the rice samples each of milled and brown rice had higher cadmium content of 2.2 and 3.1 mg/kg respectively. The findings of the present study were in concordance with the results observed by Kikuchi et al. (2002) and Frazzoli et al. (2007) who found the range of cadmium concentrations in different rice samples from 0.004 to 0.38 and 0.005 to 0.49 mg/kg, respectively. The cadmium concentration of different rice flour samples is much lower than the study of Chen et al. (1994) who reported that the mean level of cadmium in different samples of brown rice was (2.5 mg/kg). The variation in cadmium content may be ascribed to genetic as well as non genetic factors in different studies which have great impact on cadmium of a rice variety. Cadmium is the most abundant heavy metal and particularly toxic. The exceeding level of

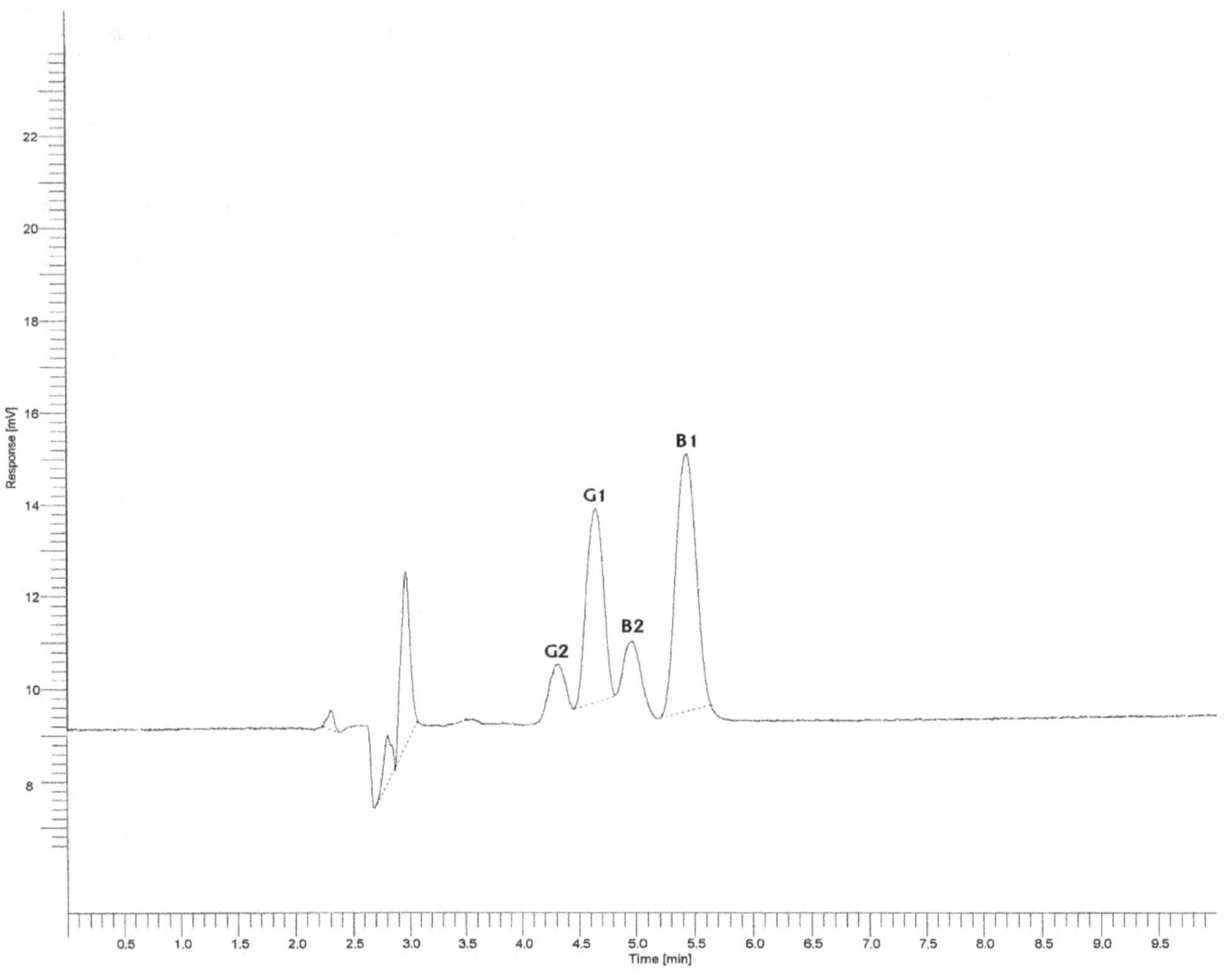

Figure 2. Representative standard chromatogram of aflatoxin (B_1, B_2, G_1 and G_2).

Table 2. Cadmium content (mg/kg) in different rice samples.

Varieties/ fractions	Irri-6	KS-282	Basmati-2000	Super Basmati	Mean
Brown	0.71[a]	0.47[c]	0.15[e]	0.09[fg]	0.35
White	0.55[b]	0.41[d]	0.12[ef]	0.07[g]	0.29
Mean	0.63	0.44	0.14	0.08	

*Means carrying same letter in a row or column are not significantly differed (P ≥ 0.05).

cadmium can affect the health. Cadmium can disturb kidney functions, and some studies indicate a cancerous effect. Research has proved that, those countries whose main food is rice, the consumption of this rice causes an intake of cadmium. 50% of cadmium intake comes from the rice consumption (Machiwa, 2010) and in Japan this amount was 40 to 60%. The Asian Governments have established critical maximum levels of heavy metals in rice to protect the health of their citizens. In Japan the maximum level of cadmium in unpolished rice is 1.0 mg/kg while in China the maximum permitted level is 0.4 mg/kg of polished rice (Chen, 2000). According to DOH/ROC (1988) the maximum permissible cadmium concentration in rice is only 0.5 mg/kg. European Union (EU) food legislation provides an overview of maximum permissible level of cadmium concentration in rice is 0.2 mg/kg. The standards for the cadmium level of different brown rice have been found within the safe limits prescribed by Japanese Standards. However, according to EU standards the cadmium content found in coarse varieties (KS-282 and IRRI-6) exceeds the prescribed limits. However, rice varieties Basmati 2000 and Super Basmati contained cadmium contents within the permissible limits prescribed by European Union

Table 3. Lead content (mg/kg) in different rice samples.

Varieties/ fractions	Irri-6	KS-282	Basmati-2000	Super Basmati	Mean
Brown	0.31	0.16	0.25	0.20	0.23[a]
White	0.23	0.13	0.19	0.14	0.17[b]
Mean	0.27[a]	0.15[d]	0.22[b]	0.17[c]	

*Means carrying same letter in a row or column are not significantly differed (P ≥ 0.05).

Table 4. Mercury content (mg/kg) in different rice samples.

Varieties/ fractions	Irri-6	KS-282	Basmati-2000	Super Basmati	Mean
Brown	0.01[d]	0.09[a]	0.03[c]	0.00[e]	0.03[a]
White	0.00[e]	0.06[b]	0.01[d]	0.00[e]	0.02[b]
Mean	0.005[c]	0.075[a]	0.020[b]	0.000[c]	

*Means carrying same letter in a row or column are not significantly differed (P ≥ 0.05).

standards.

The lead content ranged from 0.15 to 0.27 mg/kg among different rice varieties (Table 3). The brown rice yielded lead content 0.23 mg/100 g which reduced significantly to 0.17 mg/100 g in white rice milling fraction. The highest content of lead was found in rice variety IRRI-6 followed by Basmati 2000 (B2), Super Basmati (SB) and KS-282. The lead content was found significantly higher in brown rice than white rice. The lead content ranged from 0.13 to 0.31 mg/kg and found significantly higher in brown rice of IRRI-6 followed by B2-B, IRRI-W, SB-B, B2-W, KS-B and SB-W while the lowest lead concentration was found in white rice milling fraction KS-282. The results of the present study are much lower than the results of Lin (1991) and Bakhtiarian et al. (2001) who found higher lead concentrations (0.43 and 0.74 mg/kg, respectively) in brown rice of different rice varieties. These researchers indicated that many environmental factors and variable may have an effect on the absorption and storage of lead in the grains of rice. This may be the reason for variation in the results obtained in the present study. EU has set standards for maximum permissible level of lead concentration in rice and other cereals as 0.2 mg/kg. The lead concentration in rice variety IRRI-6 was 0.27 mg/kg and Basmati 2000 (0.22 mg/kg) which is higher than the prescribed limits set by EU standards. Rice variety Super Basmati (SB) and KS-282 was found within permissible limits set by EU. The lead concentration of IRRI-B, B2-B and IRRI-W was higher than EU standards and the lead contents of SB-B, SB-W, B2-W, KS-B and KS-W fall with in the limits set by EU standards. Higher concentration of lead may cause brain complications; coma and death may occur if not treated instantly (Environmental Protection Egency (EPA), 1986).

The mercury content of different rice varieties given in Table 4 indicated that the maximum level of mercury (0.09 mg/kg) was found in brown rice of KS-282 variety. The concentration of mercury ranged from 0.02 to 0.03 mg/kg between white and brown rice milling fractions, respectively. The highest mercury content was found in rice variety KS-282 followed by Basmati 2000 (0.020 mg/kg) and IRRI-6 (0.005 mg/kg) (when the fractions were combined). The highest mercury content (0.09 mg/kg) was found in brown rice of KS-282 followed by brown rice of Basmati 2000 (0.03 mg/kg) and brown rice of IRRI-6 (0.01 mg/kg). The mercury content could not be detected in white rice of IRRI-6 and white and brown rice of Super Basmati. Similarly in white rice fractions, the highest mercury content was found in white rice of KS-282 (0.06 mg/kg) followed by white rice of Basmati 2000 (0.01mg/kg) while mercury content could not be detected (ND) in white rice of IRRI-6 and Super Basmati. The content of mercury found in some rice varieties in the present study is lower than level of 0.71 µg/kg proposed by WHOM. Taiwan has also reported maximum permissible level of mercury in harvested rice to be 0.05 mg/kg In the present study the mercury content with exception of brown and white rice of KS-282 was found lower than those reported for Taiwan Standards as reported by Chen (2000).

The arsenic concentration given in Table 5 indicated that it ranged from 0.09 to 1.00 mg/kg among different rice varieties and 0.39 to 0.78 mg/kg between milling fractions, that is, white and brown rice, respectively. The arsenic content was found to be the highest in rice variety IRRI-6 followed by KS-282 (0.65 mg/kg) and Super Basmati (0.60 mg/kg). The rice variety Basmati 2000 (B2) yielded the lowest content (0.09 mg/kg) of arsenic. The arsenic level was found higher in brown rice fraction of IRRI-6 (1.30 mg/kg) followed by SB-B (0.90 mg/kg), KS-B (0.80 mg/kg) and B2-B rice (0.10 mg/kg). Higher arsenic content was found in white rice of IRRI-6 (0.70mg/kg) followed by KS-W (0.50 mg/kg), SB-W (0.30 mg/kg) and it was the lowest in rice variety Basmati 2000 (0.07 mg/kg). The results in the present study are consistent

Table 5. Arsenic content (mg/kg) in different rice samples.

Varieties/ fractions	Irri-6	KS-282	Basmati-2000	Super Basmati	Mean
Brown	1.30[a]	0.80[c]	0.10[g]	0.90[b]	0.78[a]
White	0.70[d]	0.50[e]	0.07[g]	0.30[f]	0.39[b]
Mean	1.00[a]	0.65[b]	0.09[c]	0.60[b]	

*Means carrying same letter in a row or column are not significantly differed ($P \geq 0.05$).

with the previous findings of Das et al. (2004) and Bordajandi et al. (2004) who found variation in arsenic content from 0.04 to 0.27 mg/kg with mean value of 0.17 mg/kg in brown rice samples. With respect to rice milling fractions, the results are supported by the findings of Kokot and Phuong (1999) who found that the arsenic concentration in rice samples of each white and brown rice were 0.22 and 0.29 mg/kg, respectively. The arsenic content in the samples analyzed in this study did not exceed the permissible limit set by Food hygiene concentration limit of 1.0 mg/kg as reported by Das et al. (2004). The poisoning due to arsenic can cause serious health effects including cancers, restrictive lung disease, gangrene, diabetes mellitus and hypertension (Rahman, 2002). Generally, the crops do not accrue the adequate arsenic which may cause toxicity to man. However the arsenic contaminated soil and water may raise the level of arsenic in plant tissues of edible crops (Larsen et al., 1992). The findings of these researchers pointed out that the presence of arsenic contents in rice flour sample may be due to the contaminated soil and water. However, the present study suggested the level of arsenic content in different rice varieties are within safe limits.

Aflatoxins content

Mycotoxin contamination, particularly aflatoxin contamination, is commonly found in agricultural crops such as corn, peanuts, wheat and rice. A number of mycotoxins have been investigated and their effects on human and livestock health problems are known (Hanak et al., 2002; Katiyar et al., 2000).

The aflatoxins, that is, B_1, B_2, G_1 and G_2 and totalaflatoxins detected in rice samples are given in Table 6. The samples of different rice varieties IRRI-6, KS-282 and Super Basmati (SB) were found to contain the level of aflatoxin B_1 ranging from 0.60 to 1.46 µg/kg. AFB_1 could not be detected (ND) in rice variety Basmati 2000. The highest AFB_1 content was found in rice variety IRRI-6 followed by KS-282 and Super Basmati (SB). The AFB_1 concentration ranged from 0.70 to 1.01 µg/kg between white and brown rice milling fractions, respectively (when varieties were pooled). The concentration of aflatoxin B_1 was found significantly higher in brown rice milling fraction of IRRI-6 (1.67 µg/kg) followed by KS-282 (1.63 µg/kg), Super Basmati (0.77 µg/kg) and Basmati 2000 (ND). The highest AFB_1 was found in white rice of IRRI-6 (1.24 µg/kg) followed by KS-282 (1.16 µg/kg), Super Basmati (0.43 µg/kg) and Basmati 2000 (ND).

The European Communities (EC) has set the limits of 2 µg/kg for aflatoxin B_1, 4 µg/kg for total aflatoxins in cereals, including rice and cereal products intended for direct human consumption or use as an ingredient in foodstuffs.

The amount of aflatoxin B_1 found in the present study for all the rice flour samples was within the permissible limits of 2 µg/kg for AFB_1 prescribed by the EC and Ministry of Agriculture in Turkey (Official Journal of Turkish Republic, 2002).

The AFB_2 was detected only in rice variety Super Basmati. The content of AFB_2 could not be detected in all other rice varieties IRRI-6, KS-282 and Basmati 2000 and in their different milling fractions. The highest concentration of AFB_2 was found in brown rice fraction of Super Basmati (0.95 µg/kg) and its lowest content was found in the same variety but in white milling fraction (0.66 µg/kg).

The results for the contents of AFG_1 of different rice varieties revealed that the concentration of AFG_1 ranged from ND to 2.16 and 1.21 to 1.54 µg/kg among the rice varieties and between the milling fractions of white and brown rice, respectively. The content of AFG_1 was found to be significantly the highest in Basmati 2000 (B2) followed by IRRI-6, KS-282 while it was not detected in Super Basmati (SB). When the milling fractions were compared, the AFG_1 concentration was found higher in brown rice milling fraction of Basmati 2000 (2.43 µg/kg) followed by coarse varieties IRRI-6 (1.99 µg/kg), KS-282 (1.72 µg/kg) while in white rice fractions the highest AFG_1 content was found in white rice of Basmati 2000 (1.88 µg/kg) followed by IRRI-6 (1.64 µg/kg) and KS-282 (1.33 µg/kg).

The results of the AFG_2 concentrations indicated the range from ND to 0.52 µg/kg and 0.28 to 0.44 µg/kg among the rice varieties and between milling fractions of white and brown rice, respectively. The highest AFG_2 content was found in rice variety IRRI-6 followed by Basmati 2000 (B2), KS-282 while it could not be detected in Super Basmati (SB). Among milling fractions the AFG_2 content was found higher in brown rice than white rice milling fraction. The highest content of AFG_2 (0.64 µg/kg) was found in brown rice fractions of Basmati 2000 followed by IRRI-6 and KS-282.

The brown and white rice milling fraction of Super Basmati did not contain detectable amount of aflatoxin

Table 6. Aflatoxin level (μg/kg) of different rice cultivars with their milling fractions.

Sample	AFB_1	AFB_2	AFG_1	AFG_2	Total AFs
IRRI-B	1.67 ± 0.05	ND	1.99 ± 0.08	0.61 ± 0.01	4.27 ± 0.07
IRRI-W	1.24 ± 0.08	ND	1.64 ± 0.02	0.42 ± 0.02	3.30 ± 0.09
KS-B	1.63 ± 0.05	ND	1.72 ± 0.02	0.49 ± 0.02	3.84 ± 0.06
KS-W	1.16 ± 0.04	ND	1.33 ± 0.06	0.37 ± 0.01	2.86 ± 0.11
B2-B	ND	ND	2.43 ± 0.03	0.64 ± 0.03	3.07 ± 0.03
B2-W	ND	ND	1.88 ± 0.05	0.33 ± 0.005	2.21 ± 0.02
SB-B	0.77±0.02	0.95±0.04	ND	ND	1.72±0.01
SB-W	0.43±0.01	0.66±0.006	ND	ND	1.09±0.02
Pooled mean fractions					
Brown	1.01^a	0.24^a	1.54^a	0.44^a	3.23^a
White	0.70^b	0.17^b	1.21^b	0.28^b	2.37^b
Pooled mean varieties					
IRRI-6	1.46^a	0.00^b	1.82^b	0.52^a	3.79^a
KS-282	1.39^a	0.00^b	1.53^c	0.43^b	3.35^b
Basmati 2000	0.00^c	0.00^b	2.16^a	0.49^a	2.64^c
SuperBasmati	0.60^b	0.81^a	0.00^d	0.00^c	1.41^d

Means carrying same letter in a column are not significantly differed at P ≥ 0.05; AFB_1 = Aflatoxin B_1, AFB_2 = Aflatoxin B2, AFG_1 = Aflatoxin G_1, AFG_2 = Aflatoxin G_2; Total AFs = Total aflatoxin, ND = Not Detected.

G_2. Total concentration of aflatoxin of different rice varieties and milling fractions (white and brown) have been presented in Table 6 which ranged from 1.41 to 3.79 and 2.37 to 3.23 μg/kg, respectively. The total aflatoxin content was found higher in rice variety IRRI-6followed by KS-282, Basmati 2000 (B2) and its lowest content was found in Super Basmati (SB). Among milling fractions the total aflatoxins content was found significantly higher in brown rice than white rice and total aflatoxin content ranged from 1.09 to 4.27 μg/kg in different milling fractions of rice varieties. The highest content of total aflatoxin was found in IRRI-B followed by KS-B, IRRI-W, B2-B, KS-W, B2-W, SB-B and the lowest aflatoxin content was observed in white milling fraction SB-W.

European Union (EU) food legislation in regulation (EC) No. 466/2001 and in the subsequent amendments the maximum levels for certain contaminants in food stuffs are set in order to protect public health. EU food legislation provides an overview of maximum permissible level of AFB_1 concentration and total aflatoxins in cereals including rice is 2 and 4 μg/kg, respectively.

The aflatoxin contents of all the rice varieties along with their milling fractions have been found lower than those previously described by Saleemullah et al. (2006) who estimated the aflatoxin content (mean value 17.7 μg/kg) of different rice flour samples. All the rice samples of brown and white rice did not exceed the maximum limits permitted by EU standards except the brown rice milling fraction of IRRI-6 which exceeded to these limit. The individual aflatoxin and total aflatoxins content in the rice samples analyzed in this study have been also found under the permissible limits recommended by the United States of Food and Drug Administration (USFDA) and FAO which adopted the maximum allowed levels of 10 and 50 μg/kg for all cereals, respectively (Saleemullah et al., 2006).

The findings of the present study are in close agreement to the findings found earlier by Noorlidah et al. (1998) who showed the positive report of aflatoxin G_1 and G_2 concentration (mean, 3.69 μg/kg) in different rice samples and such results are also consistent with previous findings of Food Standard Agency (2002) who reported the aflatoxin contents range from 0.2 to 1.8 μg/kg in Basmati rice, long grain rice, brown and white rice and also observed that the highest level of aflatoxin was found in brown rice which support to the present findings.

Toteja et al. (2006) observed the presence of aflatoxin B_1 at levels of 5 μg/kg in 38.5% of total number of rice samples and 17% of the total samples presence of aflatoxin B_1 above the Indian regulatory limit of 30 μg/kg.

Yoshihashi et al. (2004) observed that the incidence percent of positive samples of aflatoxin in polished and brown rice were 94 and 100%, respectively and the aflatoxin levels decreased as the rice progressed through various milling stages. The results of the present study regarding aflatoxin content in rice flour samples are also well in agreement with the findings of Toteja et al. (2006) and Yoshihashi et al. (2004). The variation between milling fractions of different rice varieties may be due to the presence of bran portion in brown rice which may increase the level of aflatoxin as reported by Castells et al. (2007) who found aflatoxin contents in all rice milling fractions but higher contamination levels was detected in bran portion as compared to white rice. The results of this study demonstrated that the concentrations of aflatoxins in all the tested rice samples are almost under the permissible levels. However, the mycotoxin contamination should be monitored routinely for food safety.

Conclusions

Though the presence of aflatoxin and heavy metals content were observed but mostly these contents were under the safe limits. It is also important to establish the permanent controlling and monitoring program from the production until consumption of cereals in order to minimize the contamination problem of AFs. On the other hand, the training programs on this problem should be developed especially for farmers and agriculturists. Using the optimum techniques for harvesting, handling and storage and selection of proper time for harvesting reduce or eliminate this problem for foods and prevent the threat to human health and the risk of great economic loss. Government must fix a legislative limit for major threats like aflatoxin, heavy metals in various foods and their products. Further work should be carried on aflatoxin and heavy metals in various foods/feeds to check health problems for humans and animals.

REFERENCES

Alberts JF, Engelbrecht Y, Styn PS, Holzapfel WH, Van-Zyl WH (2006). Biological degradation of aflatoxin B1 by Rhodococcus erythropolis cultures. Inter. J. Food Microbial. 109:121-126.

AOAC (2000). Official Methods of Analysis, 17th edition. Association of the Official Analytical Chemists, Washington D.C, USA.

Bakhtiarian A, Gholipour M, Ghazi-Khanssari M (2001). Lead and Cadmium Content of Korbal Rice in Northern Iran. Iranian J. Publ. Health 30:129-132.

Bordajandi IR, Gomez G, Abad E, Rivera J, Fernandez-Baston MDM, Blasco J, Gonzalez MG (2004). Survey of persistent Organochlorine Contaminants (PCBs, PCDD/Fs, and PAHs), Heavy Metals (Cu, Cd, Zn, Pb and Hg), and Arsenic in food samples from Huelva (Spain): Levels and Health implications. J. Agric. Food Chem. 52:992-1001.

Castells M, Ramos AJ, Sanchis V, Marin S (2007). Distribution of total aflatoxins in milled fractions of hulled rice. J. Agric. Food Chem. 55(7):2760-2764.

Chen ZS (2000). Relationship between heavy metal on concentrations in soils of Taiwan and uptake by crops. Technical bulletin, Dept of Agricultural chemistry, National Taiwan University Taipei 106. Taiwan, ROC.

Chen ZS, Lo SL, Wu HC (1994). Summary analysis and assessment of rural soils contaminated with Cd in Taoyuan. Project report of Scientific Technology Advisor Group (STAG), executive Yuan. Taipei, Taiwan. (In Chinese, with English abstract and Tables). (Unpub. Mimeograph).

Das HK, Mitra AK, Sengupta PK, Hossain, A, Islam F, Rabbani GH(2004). Arsenic concentrations in rice, vegetables and fish in Bangladesh: a preliminary study. Environ. Intr. 30: 383-387.

DOH/ROC (1988). The Critical Concentration of Cd in Diet Rice for Health. Dept. of Health (DOH), Executive Yuan, ROC.

FAO/WHO (2004). Aflatoxin Contamination in Foods and Feeds in the Philippines. FAO/WHO Regional Conference on Food Safety for Asia and Pacific Seremban, Malaysia. Agenda Item 9, Document 13.

Food Standard Agency (2002). Mycotoxins in rice 'low or non-detectable'. [online] Available at. http://www.food.gov.uk/news/newsarchive/2002/apr/mycotoxinsinrice. (Verified 14 Oct. 2006).

Frazzoli CD, Marilena, Amato, S Caroli (2007). Arsenic and other potentially toxic trace elements in rice. The determination of chemical elements in food: Applications for atomic and mass spectrometery. Jhon Wiley and sons, Inc. 12:383-397.

GOP (Government of Pakistan) (2013). Economic survey 2012-13. Govt. of Pakistan, Economic advisor wing, Financial division, Islamabad, Pakistan.

Hanak, E., P. Boutrif, P. Fabre, and M. Pinn (Scientific Editors) (2002). Food safety management in developing countries. Proceedings of the international workshop, CIRAD-FAO, 11-13 December 2000, Montpellier, France.

International Rice Research Institute (IRRI) (1997). Almanac, 2nd ed. Manila, Philippines: Int. Rice. Res. Inst, Los banos, Laguna, Philippines.

Katiyar S, Thaukar V, Gupta RC, Sarin SK, Das BC (2000). P_{53} tumor suppressor gene mutations in carcinoma patients in India. Cancer. 88:1565-1573.

Kikuchi Y, Nomiyama, T, Kumagai N, Uemura T, Omae K (2002). Cadmium concentration in current Japanese foods and beverages. J. Occup. Health. 44:240-247.

Kokot S, Phung TD (1999). Elemental content of Vietnamere rice. Analyst. 124:561-569.

Kotsonis FM, Burbock GA, Flamm WG (2001). Food Toxicology. In C.D. Klasses (Ed), Casarett and doull's Toxicology:The basic science of poisons. New York: McGraw-Hill. pp. 1049-1088.

Larsen EH, Moseholm L, Nielsen NM (1992). Atomic deposition of trace elements around point sources and health risk assessment: II. Uptake of arsenic chromium by vegetables grown near a wood preservation factory. Sci. Total Environ. 126:263-275.

Lin HT (1991). A study on the establishment of heavy metal tolerance in soil through the heavy metal concentration of crop. Unpub. M.Sc. Thesis. Research Institute of Soil Science, National chung Hsing University, taichung, Taiwan.

Machiwa JF (2010). Heavy metal levels in paddy soils and rice (*Oryza sativa*) from wetlands of lake victoria basin, tanzania. Tanz. J. Sci. (36):451-459.

Meharg AA, Norton G, Deacon C, Williams P, Adomako EE, Price A, Zhu Y, Li G, Zhao F, McGrath S, Villada A, Sommella A, Mangala P, De Silva CS, Brammer H, Dasgupta T, Islam MR (2013). Variation in Rice Cadmium Related to Human Exposure. *Environ. Sci. Technol.*47 (10):5234-5241.

Mishra HN, Das C (2003). A review on biological control and metabolism of aflatoxin. Crit. Rev. Food Sci. 43:245-264.

Moss MO (1996). Mycotoxin Fungi in microbial food poisoning. Second Edition. Edited by Eley, A.R. Published by Chapman and Hall. Nitrosamines in food by HS-SPME-GC-TEA. Food Chemistry 91:173-179.

Noorlidah A, Nawawi A, Othman I (1998). Survey of fungal counts and natural occurrence of aflatoxins in Malaysian starch-based foods. Mycopathologia. 143(1): 53-58.

Official Journal of Turkish Republic (2002). September 23, No. 24885, P. 32.

Rahman M (2002). Arsenic and contamination of drinking waterin Bangladesh: a public health perspective. J. Health Popul. Nutr. 20:193-197.

Romar Labs Inc. (2003). Clean-up for aflatoxin (B_1, B_2, G_1, G_2) and Zearalenone. 1301 Stylemaster Drive Union, MO 63084-1156, USA.

Saleemullah, Iqbal, A, Iqtidar AK, Shah H (2006). Aflatoxin contents of stored and artificially inoculated cereals and nuts. Food chem. 98:699-703.

Santini A, Ritieni A (2013). Aflatoxins - Recent Advances and Future Prospects (edit. M. Razzaghi-Abyaneh). Intech publishers. Unit 405,

Office Block, Hotel Equatorial Shanghai No.65, Yan An Road (West), Shanghai, 200040, China.

Steel RGD, Torrie JH, Dickey DA (1997). Principles and Procedures of Statistics. A Bio-Metrical Approach. 3rd Ed. McGraw Hill Book Co., Inc. New York USA.

Suzuki S (1988). Estimation of daily intake of cadmium from foods and drinks from faucets at three of kampongs of java Island. In; Suzuki, S(Ed); Health Ecology in Indonesia. Goosey. Cotokyo. pp. 65-73.

Toteja GS, Mukherjee A, Diwakar S, Singh P, Saxena BN, Sinha KK, Sinha AK, Kumar N, Nagaraja KV, Bai G, Prasad CA, Vanchinathan S, Roy R, Parker S (2006). Aflatoxin B1 contamination in wheat grain samples collected from differenr geographical regions of India: A multicenter study. J. Food Protect. 69:1463-1467.

U.S. Environmental Protection Egency (EPA) (1986). Lead effects on cardiovascular function, early development and stature: An addendum to EPA air quality. Criteria for Lead. In Air Quality Criteria for Lead, Vol. 1. Environmental Criteria and Assessment office, Research Trianglen Park. NC. EPA-600/8-83/028aF. Available For NTIS, Springfield, VA; PB87-142378. pp. A1-67.

Yoshihashi T, Nguyem T, Kabaki N (2004). Area Dependency of 2-Acetyl-1-Pyrroline Content in an Aromatic Rice Variety, Khao Dawk Mali 105. J. ARQ. 38:105-109.

Effect of different water regimes and organic manures on quality parameters of noni (*Morinda citrifolia*)

S. Muthu Kumar and V. Ponnuswami

Horticultural College and Research Institute, Tamil Nadu Agricultural University, Periyakulam, Theni District-625604, Tamil Nadu, India.

An experiment was conducted in *Morinda citrifolia* to find out the suitable water regime along with organic manure schedule for obtaining superior quality fruits. The experiment was carried out in split plot design with irrigation regimes on main plot (four levels) and organic manures on sub plot (eight levels) with two replications. Among the interaction, M_2S_4 (100% crop water requirement through drip irrigation + 50% farmyard manure + 50% vermicompost) recorded the highest scores for pulp recovery (47.12%), juice recovery (34.32%), TSS (11.26° brix), ascorbic acid (197.82 mg/100 g), titrable acidity (0.392%), fruit firmness (3.26 kg/cm^2), total carbohydrate (584.39 mg/100 g), total phenol (374.26 mg/100 g), total carotenoids (0.177 mg/100 g), protein (412.95 mg/100 g) and total flavonoids (128.53 mg/100 g) content of noni fruits.

Key words: *Morinda citrifolia,* drip irrigation, farmyard manure (FYM), vermicompost, quality, carotenoids, flavonoids, protein.

INTRODUCTION

Medicinal plants are nature's priceless gift to human beings. The demand for plant based raw materials for pharmaceuticals is increasing tremendously. Today's health care systems rely largely on plant material. Most of the world's population depends on traditional medicine to meet daily health requirements, especially within the developing countries, where plants are the main source of medicine (Meena et al., 2009).

Over the past few years as natural products have become increasingly popular, the field of natural herbal remedies have flourished. One upcoming botanical name, the fruit of *Morinda citrifolia* very popularly known as NONI belongs to the family Rubiaceae.

Noni is the biggest pharmaceutical unit in the universe because it has more than 160 nutraceuticals, vitamins, minerals, micro and macro nutrients that help the body in various ways from cellular level to organ level (Rethinam and Sivaraman, 2007). The fruit juice is in high demand in alternative medicine for different kinds of illnesses such as arthritis, diabetes, high blood pressure, muscle aches, menstrual difficulties, headaches, heart disease, AIDS, cancers, gastric ulcers, sprains, mental depression, senility, poor digestion, atherosclerosis, blood vessel problems and drug addiction (Wang et al., 2002).

The purpose of this medicinal herb will be fulfilled only if it is free from residual effects due to chemical farming. Otherwise the herb will become toxic than of medicinal value. Moreover, the medicinal plants have several active biochemical ingredients, which may get altered and deteriorated quality wise, when grown with the use of inorganic fertilizers and pesticides.

Continuous and unscrupulous use of fertilizers, pesticides and fungicides without the incorporation of organic manure cause environmental degradation especially, in the soil thereby affecting its fertility on long term basis. For maintaining optimum productivity of the land and building up of soil fertility, the addition of organic manures to crops has been suggested as one among the recommendation. Large scale cultivation under organic conditions is gaining momentum to produce toxic free medicinal plant products (Padmanabhan, 2003).

Organic agriculture is gaining importance and acceptance throughout the world with annual growth of 20 to 25% (Vanilarasu, 2011).

The availability of irrigation water is dwindling day-by-day. Adoption of conventional methods of irrigation to crops leads to an acute scarcity of water and results in reduced production and productivity of crops. Therefore, it becomes imperative to go for alternate water saving methods by more crop and income for every drop of water. Drip irrigation can be considered as an efficient irrigation system, since it causes wetting of the soil only and maintain optimum moisture content in the root zone. It also offers several water management advantages like timely application of water and water supply. Micro irrigation provides many unique agronomic, water and energy conservation benefits that address many of the challenges facing irrigated agriculture, now and in the future (Selvarani, 2009).

The water demand for ecological farming is far less and the crops grown using organic supplements and biological inputs are hardier and more drought tolerant than the ones grown with chemical inputs (Baby, 2012).

Hence, the study was undertaken to find out the best organic manure schedule along with irrigation system for production of noni fruits with supreme quality.

MATERIALS AND METHODS

This study was conducted at Horticultural College and Research Institute, TNAU, Periyakulam, Tamil Nadu, India which is situated at 77°E longitude, 10°N latitude and at an altitude of 300 m above mean sea level.

Methodology

The design as well as method used is as follows:

1. Statistical design: Split plot design
2. Factors: 2
3. Replications: 2
4. Spacing: 3.6 m × 3.6 m

Treatment details

The following are the treatment details for the main plot (irrigation):

M_1: 75% WRc (Crop water requirement through drip irrigation);
M_2: 100% WRc (Crop water requirement through drip irrigation);
M_3: 125% WRc (Crop water requirement through drip irrigation);
M_4: Check basin method of irrigation (once in 5 days)

The treatment details for the subplot (Organic manures) are as follows:

S_1: 100% farmyard manure (FYM);
S_2: 100% vermicompost (VC);
S_3: 100% coir pith compost (CPC)
S_4: 50% FYM + 50% VC
S_5: 50% FYM + 50% CPC;
S_6: 50% VC + 50% CPC;
S_7: 100% recommended dose of nitrogen (RDN) through inorganic fertilizers;
S_8: Control (no manures and no fertilizers);

All organic manures were applied on equivalent weight of RDN (60 g/plant/year - on N equivalent basis). The treatments S_1 to S_6 are applied with *Azospirillum* (10 g/ plant) + phosphobacteria (10 g/ plant) + VAM (20 g/ plant).

Crop water requirement (WRc)

Crop water requirement was calculated by using the following formula:

$$WRc = P_e \times K_p \times K_c \times A \times WP \text{ L/plant/day}$$

Where, P_e = Pan evaporation in mm; K_p = Pan Co-efficient (0.75); K_c = Crop factor (0.90 for vegetative stage, 0.95 for flowering and harvesting stage); A = Area occupied by the tree (3.6 m × 3.6 m); WP = wetted percentage (40).

Observations

Quality parameters

Pulp recovery: The pulp was separated from fully ripe fruits excluding peel and seeds and weighed and expressed in percentage.

Juice recovery: The juice was extracted from fully ripe fruits using hand pulper and expressed in percentage.

Total soluble solids (TSS): The TSS content of noni fruit was determined using a "Zeiss" hand refractometer. The readings were recorded as °brix after deducting the correction factor.

Ascorbic acid: The ascorbic acid content was estimated using the procedure of A.O.A.C. (1975) and expressed as mg/100 g of fresh sample.

Titrable acidity: The titrable acidity as percentage of citric acid was estimated following the method of A.O.A.C. (1975) and expressed in percentage.

Fruit firmness: Firmness of fruits was measured using a penetrometer (LT Lurton model FG 5000 A). Readings were taken at the proximal, distal and middle portions and the mean values were expressed as kilograms per square centimeter (kg/cm^2).

Total carbohydrate: Total carbohydrate was estimated by the method suggested Somogyi (1952) and expressed in mg/100 g.

Total phenol: Total phenol content of noni fruits was estimated by the method suggested by Bray and Thorpe (1954) using Folin Ciocalteu reagent and expressed as mg/100 g.

Total carotenoids: Total carotenoids were estimated by the method of Roy (1973) and expressed in mg/100 g.

Protein content: Protein content was estimated as per the A.O.A.C

Table 1. Effect of different water regimes and organic manures on pulp recovery (%) of noni fruits.

Treatments	M_1	M_2	M_3	M_4	Mean
S_1	37.86	40.71	40.36	35.78	38.68
S_2	38.04	42.14	42.31	36.23	39.68
S_3	37.08	39.28	39.13	35.15	37.66
S_4	38.56	47.12	43.65	36.34	41.42
S_5	37.14	39.79	40.10	35.27	38.08
S_6	37.29	41.87	41.53	35.94	39.16
S_7	38.23	41.02	41.15	36.11	39.13
S_8	34.14	34.29	34.51	31.27	33.55
Mean	37.29	40.78	40.34	35.26	38.42
	M	**S**	**M at S**	**S at M**	
SE(d)	0.209	0.275	0.555	0.549	
CD at 5%	0.664	0.563	1.225	1.125	

(1975) method and expressed in mg/100 g.

Total flavonoids: Total flavonoids were estimated by the method suggested by Singh et al. (2011) and expressed in mg/100 g.

Statistical analysis

The statistical analysis of data was done by adopting the standard procedures of Panse and Sukhatme (1985). The AGRES software (version 3.01) was used for analysis of data.

RESULTS

Pulp recovery

With regard to main plots, the treatment M_2 (100% WRc through drip irrigation) exhibited the highest pulp recovery of 40.78% while the treatment M_4 (check basin method of irrigation) recorded the lowest pulp recovery of 35.26% (Table 1). Between the sub plot treatments, S_4 (50% FYM + 50% VC) registered the highest pulp recovery of 41.42%. The pulp recovery was found to be the lowest in S_8 (no manures and no fertilizers) with 33.35%. The treatment combination, M_2S_4 (100% WRc through drip irrigation + 50% FYM + 50% VC) produced the highest pulp recovery of 47.12% and this was followed by M_3S_4 (125% WRc through drip irrigation + 50% FYM + 50% VC) with 43.65%. The pulp recovery was found to be the lowest (31.27%) in M_4S_8 (check basin method of irrigation + no manures and no fertilizers).

Juice recovery

The main plot M_2 (100% WRc through drip irrigation) recorded the highest juice recovery of 28.14%. The treatment M_4 (check basin method of irrigation) registered the least score (24.02) for juice recovery percentage (Table 2). In the sub plot, the treatment S_4 (50% FYM + 50% VC) exhibited the highest juice recovery of 29.19% and this was followed by S_2 (100% VC) with 27.30%. The juice recovery of fruits was found to be the lowest in S_8 (no manure and no fertilizers) with 22.44%. In the interactions, the combination of the treatment M_2S_4 (100% WRc through drip irrigation + 50% FYM + 50% VC) exhibited the highest juice recovery of 34.32% and this was followed by M_3S_4 (125% WRc through drip irrigation + 50% FYM + 50% VC) with 31.03%. The juice recovery was found to be the lowest in M_4S_8 (check basin method of irrigation + no manure and no fertilizers) with 21.09%.

TSS

Among the main plot treatments, M_2 (100% WRc through drip irrigation) registered the highest score for TSS with 10.13° brix and this was on par with M_3 (125% WRc through drip irrigation) with TSS content of 10.06° brix (Table 3). While the treatment M_4 (check basin method of irrigation) noticed the lowest score for TSS content (8.57° brix) of noni fruits. Among the sub plot, S_4 (50% FYM + 50% VC) recorded the highest TSS (10.18° brix) content and this was followed by S_2 (100% VC) with TSS of 9.95° brix. Whereas S_8 (no manures and no fertilizers) exhibited the lowest score for TSS (7.83 $^\circ$ brix) of noni fruits.

The highest TSS content of fruits (11.26° brix) was recorded in the treatment combination M_2S_4 (100% WRc through drip irrigation + 50% FYM + 50% VC) and this was followed by M_3S_4 (125% WRc through drip irrigation + 50% FYM + 50% VC) with TSS of 10.74° brix. The TSS content was found to be the lowest (7.21° brix) in the treatment M_4S_8 (check basin method of irrigation + no

Table 2. Effect of different water regimes and organic manures on juice recovery (%) of noni fruits.

Treatments	M_1	M_2	M_3	M_4	Mean
S_1	26.02	27.84	27.55	24.22	26.41
S_2	26.11	29.03	29.24	24.80	27.30
S_3	25.42	26.93	26.71	23.82	25.72
S_4	26.50	34.32	31.03	24.89	29.19
S_5	25.67	27.14	27.32	24.16	26.07
S_6	25.88	28.75	28.52	24.51	26.92
S_7	26.24	28.22	28.34	24.66	26.87
S_8	22.73	22.86	23.06	21.09	22.44
Mean	25.57	28.14	27.72	24.02	26.36
	M	**S**	**M at S**	**S at M**	
SE(d)	0.144	0.189	0.381	0.377	
CD at 5%	0.458	0.386	0.842	0.773	

Table 3. Effect of different water regimes and organic manures on TSS (° brix) of noni fruits.

Treatments	M_1	M_2	M_3	M_4	Mean
S_1	9.55	10.23	10.30	8.70	9.70
S_2	9.71	10.53	10.64	8.90	9.95
S_3	9.42	9.92	9.88	8.59	9.45
S_4	9.74	11.26	10.74	8.96	10.18
S_5	9.49	10.16	10.07	8.63	9.59
S_6	9.63	10.48	10.42	8.76	9.82
S_7	9.59	10.34	10.37	8.83	9.78
S_8	7.96	8.13	8.02	7.21	7.83
Mean	9.39	10.13	10.06	8.57	9.54
	M	**S**	**M at S**	**S at M**	
SE(d)	0.051	0.069	0.138	0.137	
CD at 5%	0.162	0.140	0.303	0.280	

manures and no fertilizers).

Ascorbic acid

Among the irrigation regimes, M_2 (100% WRc through drip irrigation) recorded the highest ascorbic acid content (165.80 mg/100 g) and this was on par with M_3 (125% WRc through drip irrigation) with 164.80 mg/100 g of ascorbic acid (Table 4). Whereas the lowest ascorbic acid content (126.99 mg/100 g) was noticed from treatment comprising check basin method of irrigation (M_4).

Regarding the sub plots, application of 50% FYM + 50% VC (S_4) exhibited the superior scores for ascorbic acid content (165.93 mg/100 g) and this was followed by S_2 (100% VC) with 160.97 mg/100g. While the treatment S_8 (no manures and no fertilizers) showed very poor performance for ascorbic acid content with 119.76 mg/100 g.

Among the interactions, the treatment combination comprising 100% WRc through drip irrigation + 50% FYM + 50% VC (M_2S_4) registered the highest ascorbic acid content (197.82 mg/100 g) and this was followed by M_3S_4 (125% WRc through drip irrigation + 50% FYM + 50% VC) with ascorbic acid content of 191.35 mg/100 g. Whereas the ascorbic acid content was found to be the lowest (116.08 mg/100 g) in the treatment combination M_4S_8 (check basin method of irrigation + no manures and no fertilizers).

Titrable acidity

Concerning the main plot, M_2 (100% WRc through drip irrigation) produced the fruits with the highest titrable acidity (0.352%) and this was on par with M_3 (125% WRc through drip irrigation) with titrable acidity of 0.350% (Table 5). The titrable acidity was found to be the lowest (0.292%) in M_4 (check basin method of irrigation).

Table 4. Effect of different water regimes and organic manures on ascorbic acid (mg/100 g) content of noni fruits.

Treatments	M_1	M_2	M_3	M_4	Mean
S_1	136.21	166.85	162.59	127.69	148.34
S_2	142.06	184.74	186.52	130.56	160.97
S_3	135.83	151.28	149.55	124.10	140.19
S_4	142.57	197.82	191.35	131.97	165.93
S_5	136.10	157.77	155.02	124.92	143.45
S_6	138.53	174.27	181.65	129.22	155.92
S_7	144.63	172.56	170.39	131.38	154.74
S_8	120.47	121.14	121.35	116.08	119.76
Mean	137.05	165.80	164.80	126.99	148.66
	M	**S**	**M at S**	**S at M**	
SE(d)	0.855	1.066	2.169	2.132	
CD at 5%	2.720	2.183	4.826	4.367	

Table 5. Effect of different water regimes and organic manures on titrable acidity (%) of noni fruits.

Treatments	M_1	M_2	M_3	M_4	Mean
S_1	0.320	0.356	0.353	0.292	0.330
S_2	0.328	0.371	0.375	0.302	0.344
S_3	0.318	0.343	0.340	0.285	0.322
S_4	0.330	0.392	0.383	0.310	0.354
S_5	0.319	0.351	0.347	0.287	0.326
S_6	0.325	0.364	0.368	0.296	0.338
S_7	0.334	0.362	0.361	0.306	0.341
S_8	0.270	0.274	0.276	0.256	0.269
Mean	0.318	0.352	0.350	0.292	0.328
	M	**S**	**M at S**	**S at M**	
SE(d)	0.002	0.002	0.005	0.005	
CD at 5%	0.006	0.005	0.011	0.010	

Regarding the manure treatments, application of 50% FYM + 50% VC (S_4) had resulted in the highest titrable acidity of 0.354%. While S_8 (no manures and no fertilizers) registered the least score for titrable acidity (0.269%).

Pertaining to interaction, the highest titrable acidity of 0.392% was recorded from the treatment combination comprising 100% WRc through drip irrigation + 50% FYM + 50% VC (M_2S_4) and this was on par with M_3S_4 (125% WRc through drip irrigation + 50% FYM + 50% VC) which recorded titrable acidity of 0.383%. The treatment combination, M_4S_8 (check basin method of irrigation + no manures and no fertilizers) recorded the least score (0.256%) for titrable acidity of fruits.

Fruit firmness

Concerning the main plot, M_2 (100% WRc through drip irrigation) registered the highest fruit firmness (2.62 kg/cm^2) and this was on par with M_3 (125% WRc through drip irrigation) with 2.59 kg/cm^2 (Table 6). While the lowest fruit firmness (1.63 kg/cm^2) was recorded by M_4 (check basin method of irrigation).

Pertaining to the sub plot treatments, S_4 (50% FYM + 50% VC) produced the fruits with high firmness of 2.63 kg/cm^2 and this was followed by S_2 (100% VC) with 2.55 kg/cm^2. While the treatment S_8 (no manures and no fertilizers) recorded the lowest fruit firmness of 1.33 kg/cm^2.

Among the interactions, M_2S_4 (100% WRc through drip irrigation + 50% FYM + 50% VC) registered the superior score for fruit firmness (3.26 kg/cm^2) and this was followed by M_3S_4 (125% WRc through drip irrigation + 50% FYM + 50% VC) with fruit firmness of 3.15 kg/cm^2. Fruit firmness was found to be the lowest (1.20 kg/cm^2) in the treatment combination M_4S_8 (check basin method of irrigation + no manures and no fertilizers).

Table 6. Effect of different water regimes and organic manures on fruit firmness (kg/cm^2).

Treatments	M_1	M_2	M_3	M_4	Mean
S_1	2.14	2.80	2.71	1.71	2.34
S_2	2.26	3.03	3.10	1.80	2.55
S_3	1.97	2.43	2.37	1.55	2.08
S_4	2.29	3.26	3.15	1.83	2.63
S_5	2.06	2.52	2.48	1.58	2.16
S_6	2.20	2.84	2.92	1.73	2.42
S_7	2.08	2.67	2.59	1.65	2.25
S_8	1.32	1.42	1.37	1.20	1.33
Mean	2.04	2.62	2.59	1.63	2.22
	M	**S**	**M at S**	**S at M**	
SE(d)	0.013	0.016	0.033	0.033	
CD at 5%	0.040	0.034	0.073	0.067	

Table 7. Effect of different water regimes and organic manures on total carbohydrate content (mg/100 g) of noni fruits.

Treatments	M_1	M_2	M_3	M_4	Mean
S_1	543.27	562.78	563.29	530.68	550.01
S_2	549.48	575.44	577.32	533.86	559.03
S_3	540.38	557.08	556.37	529.15	545.75
S_4	553.61	584.39	581.26	537.25	564.13
S_5	541.26	560.72	557.84	529.84	547.42
S_6	545.44	571.52	569.74	533.15	554.96
S_7	552.84	567.20	567.81	536.49	556.09
S_8	509.21	512.69	512.10	503.64	509.41
Mean	541.94	561.48	560.72	529.26	548.35
	M	**S**	**M at S**	**S at M**	
SE(d)	1.461	1.958	3.943	3.915	
CD at 5%	4.649	4.010	8.691	8.020	

Total carbohydrate

Among the irrigation regimes, application of 100% WRc through drip irrigation (M_2) had resulted in the utmost total carbohydrate content (561.48 mg/100 g) while lower amount of total carbohydrate (529.26 mg/100 g) was found to be with check basin method of irrigation (M_4). Regarding the sub plots, application of 50% FYM + 50% VC (S_4) exhibited the superior scores for total carbohydrate content (564.13 mg/100 g) and this was followed by S_2 (100% VC) with 559.03 mg/100 g. While the treatment S_8 (no manures and no fertilizers) showed very poor performance for total carbohydrate content with 509.41 mg/100 g (Table 7). Among the interactions, the treatment combination comprising 100% WRc through drip irrigation + 50% FYM + 50% VC (M_2S_4) registered the highest total carbohydrate content of 584.39 mg/100 g and this was on par with M_3S_4 (125% WRc through drip irrigation + 50% FYM + 50% VC) with 581.26 mg/100 g. The total carbohydrate content was found to be the lowest (503.64 mg/100 g) in the treatment combination M_4S_8 (check basin method of irrigation + no manures and no fertilizers).

Total phenol

Among the main plot treatments provision of 100% WRc through drip irrigation (M_2) was found to have profound influence on the total phenol content of noni fruits (Table 8). The highest total phenol content (316.92 mg/100 g) of fruits was observed from the treatment M_2 (100% WRc through drip irrigation) and this was on par with M_3 (125% WRc through drip irrigation) with total phenol content of 315.82 mg/100 g. The treatment M_4 (check basin method of irrigation) recorded the lowest total phenol content of229.38 mg/100 g.

Between the sub plots, application of 50% FYM + 50%

Table 8. Effect of different water regimes and organic manures on total phenol content (mg /100 g) of noni fruits.

Treatments	M_1	M_2	M_3	M_4	Mean
S_1	275.42	323.51	318.64	230.49	287.02
S_2	280.96	351.08	354.28	240.57	306.72
S_3	272.14	302.35	308.68	226.06	277.31
S_4	284.12	374.26	358.50	246.35	315.81
S_5	270.02	315.27	317.89	231.87	283.76
S_6	279.05	342.56	340.02	237.68	299.83
S_7	285.56	331.46	334.83	248.52	300.09
S_8	187.06	194.85	193.74	173.52	187.29
Mean	266.79	316.92	315.82	229.38	282.23
	M	**S**	**M at S**	**S at M**	
SE(d)	1.539	2.052	4.136	4.104	
CD at 5%	4.899	4.203	9.122	8.406	

VC (S_4) registered the highest scores for total phenol content (315.81 mg/100 g) and this was followed by S_2 (100% VC) with 306.72 mg/100 g of total phenol content, while the lowest total phenol content (187.29 mg/100 g) was recorded from the treatment S_8 (no manures and no fertilizers).

The experimental plots receiving 100% WRc through drip irrigation + 50% FYM + 50% VC (M_2S_4) produced the fruits with high total phenol content (374.26 mg/100 g) and this was followed by M_3S_4 (125% WRc through drip irrigation + 50% FYM + 50% VC) with total phenol content of 358.50 mg/100g. While the lowest total phenol content (173.52 mg/100 g) was recorded from the treatment combination M_4S_8 (check basin method of irrigation + no manures and no fertilizers).

Total carotenoid

Increased carotenoid content of fruits were registered in the treatment M_2 (100% WRc through drip irrigation) with 0.160 mg/100 g as compared to the treatment M_4 (check basin method of irrigation) with 0.128 mg/100 g (Table 9). In the sub plots, S_4 (50% FYM + 50% VC) exhibited the highest score of 0.161 mg/100g carotenoid content and the treatment S_8 (no manures and no fertilizers) with the lowest carotenoid content of 0.108 mg/100 g. Among the interactions, the treatment M_2S_4 (100% WRc through drip irrigation + 50% FYM + 50% VC) showed the highest carotenoid content (0.177 mg/100 g) and decreased contents were seen in the treatment from M_4S_8 (check basin method of irrigation + no manures and no fertilizers) with 0.104 mg/100 g.

Protein content

Concerning the main plot, M_2 (100% WRc through drip irrigation) registered the highest protein content of 393.13 mg/100 g and this was on par with M_3 (125% WRc through drip irrigation) with 392.54 mg/100 g (Table 10). While the lowest protein content (364.03 mg/100 g) was recorded by M_4 (check basin method of irrigation). Application of 50% FYM + 50% VC (S_4) produced the fruits with higher protein content (394.29 mg/100 g) and this was followed by 100% VC (S_2) with protein content of 389.91 mg/100 g. The treatment comprising no manures and no fertilizers (S_8) registered the fruits with decreased protein content of 352.40 mg/100 g. Among the interaction, the highest protein content of 412.95 mg/100 g was recorded from the treatment combination comprising 100% WRc through drip irrigation + 50% FYM + 50% VC (M_2S_4) and this was on par with M_3S_4 (125% WRc through drip irrigation + 50% FYM + 50% VC) with protein content of 410.79 mg/100 g. The treatment combination, M_4S_8 (check basin method of irrigation + no manures and no fertilizers) recorded the least score (347.62 mg/100 g) for protein content of fruits.

Total flavonoids

The highest total flavonoid content of 119.52 mg/100 g was recorded by the treatment M_2 (100% WRc through drip irrigation) in main plot and the treatment M_4 (check basin method of irrigation) registered the lowest score of 103.76 mg/100 g (Table 11). Among the sub plot treatments application of 50% FYM + 50% VC (S_4) recorded the highest values for total flavonoid content (119.54 mg/100 g) which is on par with 100% VC (S_2) with 118.37 mg/100 g. While the treatment no manures and no fertilizers (S_8) recorded the lowest total flavonoid content of 95.69 mg/100 g. Interaction effects of the treatment M_2S_4 (100% WRc through drip irrigation + 50% FYM + 50% VC) exhibited the highest total flavonoid content (128.53 mg/100 g) which is on par with M_3S_4

Table 9. Effect of different water regimes and organic manures on total carotenoid content (mg/100 g) of noni fruits.

Treatments	M_1	M_2	M_3	M_4	Mean
S_1	0.145	0.164	0.162	0.131	0.151
S_2	0.153	0.171	0.172	0.134	0.158
S_3	0.141	0.157	0.156	0.125	0.145
S_4	0.154	0.177	0.174	0.138	0.161
S_5	0.144	0.160	0.159	0.127	0.148
S_6	0.148	0.170	0.168	0.132	0.155
S_7	0.149	0.166	0.167	0.135	0.154
S_8	0.108	0.111	0.110	0.104	0.108
Mean	0.143	0.160	0.159	0.128	0.147
	M	**S**	**M at S**	**S at M**	
SE(d)	0.0004	0.0005	0.0011	0.0011	
CD at 5%	0.0013	0.0011	0.0024	0.0022	

Table 10. Effect of different water regimes and organic manures on protein (mg/100 g) content of noni fruits.

Treatments	M_1	M_2	M_3	M_4	Mean
S_1	374.65	389.55	391.20	364.62	380.01
S_2	381.24	406.38	404.79	367.24	389.91
S_3	373.29	388.73	386.98	362.68	377.92
S_4	384.26	412.95	410.79	369.14	394.29
S_5	375.29	391.56	394.24	364.44	381.38
S_6	378.66	402.53	399.75	366.75	386.92
S_7	381.63	397.53	398.26	369.76	386.80
S_8	351.88	355.80	354.29	347.62	352.40
Mean	375.11	393.13	392.54	364.03	381.20
	M	**S**	**M at S**	**Sat M**	
SE(d)	1.018	1.361	2.743	2.722	
CD at 5%	3.241	2.788	6.047	5.576	

Table 11. Effect of different water regimes and organic manures on total flavonoid (mg/100 g) content of noni fruits.

Treatments	M_1	M_2	M_3	M_4	Mean
S_1	111.06	119.78	119.31	104.58	113.68
S_2	114.70	126.13	126.68	105.95	118.37
S_3	111.43	118.25	117.56	102.63	112.47
S_4	115.65	128.53	127.24	106.74	119.54
S_5	112.39	118.69	119.02	105.20	113.83
S_6	114.04	124.48	123.80	105.52	116.96
S_7	115.23	122.69	123.17	107.28	117.09
S_8	95.74	97.63	97.21	92.18	95.69
Mean	111.28	119.52	119.25	103.76	113.45
	M	**S**	**M at S**	**S at M**	
SE(d)	0.608	0.813	1.638	1.625	
CD at 5%	1.936	1.665	3.611	3.330	

(125% WRc through drip irrigation + 50% FYM + 50% VC) with 127.24 mg/100 g. The lowest total flavonoid content of92.18 mg/100 g was expressed in the treatment M_4S_8 (check basin method of irrigation + no manures and no fertilizers).

DISCUSSION

The increase in pulp and juice recovery percentage with the application of 100% WRc through drip irrigation + 50% FYM + 50% VC (M_2S_4) might be due to optimum level of water and nutrient availability for fruit development. Increased soil moisture under drip irrigation might have led to effective absorption and utilization of nutrients resulting better source sink relationship which facilitates better fruit quality parameters. Reduced moisture availability in check basin method of irrigation resulting in water deficit might manifest many changes in plant anatomy such as decrease in size of cells and inter cellular spaces limiting cell division and elongation resulting in overall decrease in quality parameters as reported by Brantley and Warren (1960) and Prakash (2010).

The present study indicated that combined application 100% WRc through drip irrigation + 50% FYM + 50% VC (M_2S_4) significantly improved the TSS content of fruits. Optimum moisture supply by drip irrigation might have enhanced the enzymatic activities thus resulting in translocation and accumulation of assimilates in fruits. This was line with the findings of Prakash (2010) in mango. Similarly, this might be due to the major role played by organic manures particularly FYM. FYM application facilitates the higher availability of micronutrients which would have played an important role in plant metabolism through their involvement in various enzymatic reactions. This was in line with the findings of Nayaki (2000) in tomato and sweet potato.

Application of organic manures particularly FYM and VC contain appreciable amount of micronutrients especially ferrous. It has been found that Fe is important as an essential component of the many respiratory enzymes like catalase, cytochrome A, B and C which are involved in the respiratory process in the cell system. Enhancement of respiration in a cell system or plant system will naturally result in the conversion of reserve food materials to soluble simple components which could be utilized for either growth or maintenance. This might be the cause for the increase in TSS content of noni fruits. The higher TSS achieved in this elite treatment may be attributed to accelerated mobilization of photosynthates from the leaves by auxins produced by *Azospirillum* (Tien et al., 1989).

Ascorbic acid content and titrable acidity was the highest in the treatment combination of 100% WRc through drip irrigation + 50% FYM + 50% VC (M_2S_4). The availability of adequate soil moisture with the required plant nutrients might have resulted in more uptake of plant by more profused rooting system in the plant which resulted in balanced nutrition and thereby increased the quality of fruits. These were conformity with findings of Kadam and Mayar (1992), Bafna et al. (1993) and Balasubramanian (2008).

Another possible reason for higher level of ascorbic acid content and titrable acidity might be due interaction effects of water regimes and organic manures which showed positive effect on ascorbic acid content (Jeeva, 1997). The increase in ascorbic acid content may be due to the optimum availability of N due to organic manures application (Randhawa and Bhail, 1976).

The increased level of ascorbic acid content may also be due to the action of micronutrients and growth hormones particularly gibberellins produced by the rhizosphere *Azospirillum,* phosphobacteria and *Azotobacter* and their activity of number of enzymes. Tien et al. (1989) also was of the opinion that gibberellins could either augment the biosynthesis of ascorbic acid or block the oxidation of synthesized ascorbic acid by ascorbic acid oxidase.

Fruit firmness was significantly higher in fruits harvested from plants that received 100% WRc through drip irrigation + 50% FYM + 50% VC (M_2S_4) than those from check basin method of irrigation + no manures and no fertilizers (M_4S_8). This increased fruit firmness might be due to the optimum level of potassium availability through FYM and VC. The K related increase in fruit firmness was associated with increased 'tissue pressure potential'. Tissue pressure potential was in turn, positively correlated with TSS and total carbohydrate content of fruits (Prakash, 2010). The increased total phenol content in 100% WRc through drip irrigation + 50% FYM + 50% VC (M_2S_4) may due to presence of phenyl alanine in FYM which is the precursor for several phenolic substances would also have contributed to the increase in the total phenol content.

Marked increase in total carbohydrates in noni fruits with 100% WRc through drip irrigation + 50% FYM + 50% VC (M_2S_4) may be due to more accumulation of photosynthates because of optimum N and K nutrients through organic manures and optimum moisture status through drip irrigation. It would have helped in accumulation of photosynthates through better availability of nutrients and water during cropping period ultimately favouring the increase in TSS and carbohydrate content in noni fruits. Similarly, the humic substances in FYM might have influenced the carbohydrate metabolism of plants and promoted the accumulation of more carbohydrates in the fruits. The earlier findings of Prakash (2010) are in validation with the present study.

In the present study, the protein content was significantly increased by combined application of 100% WRc through drip irrigation + 50% FYM + 50% VC (M_2S_4). FYM, rhizosphere *Azospirillum*, phosphobacteria and *Azotobacter* physiologically influence the activity of number of enzymes which leads to increased cell metabolisms, enzymatic activity which in turn changes the biochemical composition of the produce (Okon, 1985). Rhizosphere *Azospirillum* and phosphobacteria due to the physiological influence on the activity of number of enzymes altered the proteins to a desired level. Hence, the protein content was increased. The enhanced absorption of nitrogen and its direct participation in protein synthesis might also be the reason as postulated by Subbiah and Ramanathan (1982).

Total flavonoid content was found to be the highest in M_2S_4 (100% WRc through drip irrigation + 50% FYM + 50% VC) as against the lowest in M_4S_8 (check basin method of irrigation + no manures and no fertilizers). This may due to optimum soil moisture and nutrient status in the elite treatment.

In case of check basin method of irrigation, the interval between the two successive irrigations was higher due to which the available soil moisture content varied from the field capacity (at the time of irrigation) to stress condition (just before consecutive irrigation). These two extremes of moisture availability cause poor physiological activity of the crop, ultimately reflecting on the quality of the crop as already reported by many earlier workers *viz.*, Selvaraj (1997), Chakraborty et al. (1998) and Vijayselvaraj (2007). Similarly nutrient starvation prevailing in M_4S_8 (check basin method of irrigation + no manures and no fertilizers) may affect the physiological process which inturn affect the fruit quality parameters.

The treatment combination M_2S_4 (100% WRc through drip irrigation + 50% FYM + 50% VC) exhibited superior performance for all the quality parameters studied. Hence, it is recommended for production of best quality noni fruits with higher bioavailability of nutrients.

REFERENCES

A.O.A.C (1975). Official methods of analysis. 12th edition. Washington DC, USA, Association of Official Analytical Chemists.

Baby PT (2012). Influence of ultra high density planting and organics in enhancing leaf production and protein content in annual moringa (*Moringa oleifera* Lam.) cv.PKM-1. MSc Thesis, Tamil Nadu Agricultural University, Coimbatore, Tamil Nadu, India.

Bafna AM, Daftardar SY, Khade KK, Patel PV, Dhotre RS (1993). Utilization of nitrogen and water by tomato under drip irrigation system. J. Water Manage. 1:1-5.

Balasubramanian P (2008). Comparative analysis of growth, physiology, nutritional and production changes of tomato (*Lycopersicum esculentum* Mill.) under drip, fertigation and conventional systems. PhD Thesis, Tamil Nadu Agricultural University, Coimbatore, Tamil Nadu, India.

Brantley BB, Warren GF (1960). Effect of nitrogen on flowering, fruiting and quality of the watermelon. J. Am. Soc. Hort. Sci. 75:644-649.

Bray HG, Thorpe MU (1954). Analysis of phenolic compounds of interest in metabolism. Meth. Biochem. Anal. 9:27-52.

Chakraborty D, Singh AK, Kumar A, Uppal KS, Khanna M (1998). Effect of fertigation on nitrogen dynamics in broccoli. Proceedings of seminar on micro Irrigation and sprinkler irrigation systems held at New Delhi. pp. 185-189.

Jeeva S (1997). Studies on certain aspects of growth, maturity and yield of okra cv. Arka Anamika under the influence of organic nutrition and growth regulators. PhD Thesis, Tamil Nadu Agricultural University, Coimbatore, Tamil Nadu, India.

Kadam JR, Mayar SS (1992). Effect of different irrigation methods on yield contributing parameters of tomato. Maharashtra. J. Hort. 6:79-85.

Meena AK, Bansal P, Kumar S (2009). Plants herbal wealth as a potential source of ayurvedic drugs. Asian J. Traditional. Med. 4(4):152-170.

Nayaki DA (2000). Studies on *in situ* vegetable crop residue management with organic and inorganic combinations for sustainability. PhD Thesis, Tamil Nadu Agricultural University, Coimbatore, Tamil Nadu, India.

Okon Y (1985). *Azospirillum* as potential inoculants for agriculture. TIBTECH. 3:223-228.

Panse VG, Sukhatme PV (1985). Statistical methods for agricultural workers, Indian Council of Agricultural Research, New Delhi.

Padmanabhan K (2003). Effect of organic manures on growth, root yield and alkaloid content of ashwagandha (*Withania somnifera* (L.) Dunal) 'Jawahar'. MSc Theis, Tamil Nadu Agricultural University, Coimbatore, Tamil Nadu, India.

Prakash K (2010). Studies on influence of drip irrigation regimes and fertigation levels on mango var. Alphonso under ultra high density planting. MSc Theis, Tamil Nadu Agricultural University, Coimbatore, Tamil Nadu, India.

Randhawa GS, Bhail AS (1976). Growth yield and quality of cauliflower influenced by N, P and boron. Indian J. Hort. 33:83-91.

Rethinam P, Sivaraman K (2007). Noni (*Morinda citrifolia* L) the miracle fruit - A holistic review. Int. J. Noni Res. 2(1-2):4-37.

Roy SK (1973). A simple and rapid method of estimation of total carotenoid pigment in mango. J. Food Sci. Tech. 10(1):45.

Selvaraj PK (1997). Optimization of irrigation scheduling and nitrogen fertigation for maximizing the water use efficiency of turmeric in drip irrigation. PhD Thesis, Tamil Nadu Agricultural University, Coimbatore, Tamil Nadu, India.

Selvarani A (2009). Drip fertigation studies in maize (*Zea mays*) - bhendi (*Abelmoschus esculentus*) cropping sequence. PhD Thesis, Tamil Nadu Agricultural University, Coimbatore, Tamil Nadu, India.

Singh DR, Minj D, Mathew J, Salim KM, Kumari C, Varughese A (2011). Morphological and biochemical characterization of ecotypes of *Morinda citrifolia* ecotypes. Proceedings of sixth national symposium on Noni - A panacea for wellness held at Chennai, India. pp. 121-132.

Somogyi M (1952). Notes on sugar determinations. J. Biol. Chem. 195:19-23.

Subbiah K, Ramanathan KM (1982). Influence of N and K_2O on the crude protein, carotene, ascorbic acid and chlorophyll contents of amaranthus. South Indian Hort. 30(2):82-86.

Tien TM, Gaskens MH, Hubbell DH (1989). Plant growth substances produced by *Azospirillum brasilense* and their effect on growth of pearl millet (*Pennisetum americanum* L.). Appl. Environ. Microbiol. 37:1016-1024.

Vanilarasu K (2011). Standardization of organic nutrient schedule in banana cv. Grand Naine. MSc Thesis, Tamil Nadu Agricultural University, Coimbatore, Tamil Nadu, India.

Vijayselvaraj KS (2007). Standardization of irrigation and fertigation techniques in jasmine (*Jasminum grandiflorum*) var. CO 2. MSc Thesis, Tamil Nadu Agricultural University, Coimbatore, Tamil Nadu, India.

Wang MY, West B, Jensen CJ, Nowicki D, Su C, Palu AK, Anderson G (2002). *Morinda citrifolia* (Noni): a literature review and recent advances in Noni research. Acta Pharmacologica Sinica. 23:1127–1141.

16

Cellulolytic bacterial biodiversity in long-term manure experimental sites

Niharendu Saha[1*], Stephan Wirth[2] and Andreas Ulrich[2]

[1]AICRP on Soil Test Crop Response Correlation, Directorate of Research,
Bidhan Chandra Krishi Viswavidyalaya Kalayni-741235, Nadia, West Bengal, India.
[2]Centre for Agriculture and Land use Research (ZALF), Institute of Microbial Ecology, MÜncheberg, Germany.

To study the Influence of long-term application of manure on cellulolytic bacterial diversity, 54 efficient cellulolytic bacterial cultures were isolated from two long-term manure experimental field sites of Berlin-Dahlem (established in 1923) and BadLauchstädt (established in 1902). The sequence divergence of highly conserved region of 16S rDNA was exploited by restriction analysis of PCR-amplified 16S rDNA using restriction enzyme *ScrFI* to assay the evolutionary relatedness of isolates. Restriction analysis identified 10 genetically diverse pattern groups comprising five groups each of bacterial and actinomycetal domain. Irrespective of manuring, a dominant pattern group (H_2) was identified, containing 31.48% of total isolates. On the other hand, two site-specific pattern groups highly specific for the brown soil (H_2, 9.25% of total isolates) and for black soil (J_2, 11.11% of total isolates), respectively were identified. In general, the composition of cellulolytic isolates in two sites displayed differences with respect manure application and soil properties. Manure strongly influenced the abundance of cellulolytic bacterial diversity in brown soil. The terminal restriction fragment length polymorphism (T-RFLP) data revealed a distinct relationship of total bacterial diversity with long-term manure application. This influence is more prominent in nutrient poor brown soil. Based on 16S rDNA sequence analysis, isolates of the dominant as well as the specific pattern groups could be assigned to the genus *Streptomyces* comprising species of diverse phylogenetic affiliation. Furthermore, sequencing of 16S rDNA of isolates of five bacterial pattern groups revealed a high phylogenetic diversity among these isolates, including *Streptococcus, Paenibacillus, Bacillus, Bacillus megaterium* and *Bacillus pumilus.*

Key words: Cellulolytic bacteria, 16SrDNA gene, restriction analysis, restriction fragment length polymorphism (T-RFLP), terminal restriction fragment length polymorphism (T-RFLP), phylogenetic diversity.

INTRODUCTION

The process of photosynthesis is the main route for the acquisition of energy in plant biomass of which cellulose is the major component. Release of energy and return of bio-sequestered carbon to soil environment are primarily the concern of cellulose utilizing special group of microorganisms called cellulolytic organisms. The carbon cycle is closely associated with the activities of cellulose-utilizing microorganisms in soil. Thus, microbial cellulose utilization is responsible for one of the largest material flows in the biosphere and is of interest in relation to analysis of carbon flux at both local and global scales. Plant biomass is the only foreseeable sustainable source of fuels and materials available to humanity (Lynd et al., 2002). In this context, cellulosic materials are particularly attractive because of their relatively low cost and plentiful supply. The central technological impediment to more widespread utilization of this important resource for soil enrichment is the general absence of appropriate technology for overcoming the recalcitrance of cellulosic biomass (Halliwell, 1965) and a little knowledge about the

*Corresponding author. E-mail: nihar_bckv@rediffmail.com.

participating bacterial communities. A promising strategy to overcome this impediment involves the utilization of native cellulolytic microbes producing wide array of cellulase enzymes. Notwithstanding its importance in various contexts, fundamental understanding of microbial cellulose utilization is in many respects rudimentary. Thus, interest in cellulose decomposition and in cellulolytic microorganisms has been stimulated by the need of greater understanding of this important process in nature (Coughlan and Mayer, 1992; Eriksson et al., 1990). In this context, cellulolytic bacteria play the pivotal role in the transformation of cellulosic residues in the soil ecosystem (Mullings and Parish, 1984; Szegi, 1988). The group of aerobic cellulolytic bacteria is rather diverse and affiliated to various taxonomic groups (Eriksson et al., 1990). There is a general interest in studying the diversity of indigenous bacteria capable of degrading cellulosic materials because of their inherent capacity to elaborate cellulase enzyme complex for sequential degradation of pure crystaline cellulose, more so, cellulose derivatives of higher resistance, hugely available in wastes of farm and industrial origin. Efforts have been made to isolate potential degraders and characterize bacterial communities as well as their response to the substrate cellulose, more so to identify the genes involved in the degradation process. A range of approaches is available for assessing the composition, diversity of soil microbial communities, including 16S rDNA analysis, assays of substrate utilization profiles (Ulrich and Wirth, 1999; Ulrich et al., 2008). While examining the distribution of cellulolytic species across taxonomic groups, it deems useful to consider microbial taxonomy based on phylogeny, rather than a set of arbitrary morphological or biochemical characteristics as used in classical taxonomy. Current views of the evolutionary relatedness of organisms are based largely on phylogenetic trees constructed from measurements of sequence divergence among chronometric macromolecules, particularly small-subunit - rRNAs (16S rRNA of procaryotes and 18S rRNA of eukaryotes Olsen and Woese, 1993; Woese, 2000). An inspection of these trees reveals that the ability to digest cellulose is widely distributed among many genera in the domain *Eubacteria* and in the fungal groups within the domain *Eucarya*, although no cellulolytic members of domain *Archaea* have yet been identified. Within the eubacteria there is considerable concentration of cellulolytic capabilities among the predominantly aerobic order *Actinomycetales* (phylum *Actinobacteria*) and the anaerobic order *Clostridiales* (phylum *Firmicutes*) (Lynd et al., 2002).

Most studies of aerobic cellulolytic bacteria involve techniques based on pour-plate serial dilution, which is inexpensive and suboptimal, necessarily selective since an unknown part of the indigenous microbial population is considered to be non-cultured (Olson and Bakken, 1987). Moreover, a range of different cellulosic substrates is used to isolate and detect cellulolytic bacteria and their activities but not all are equally utilized by all cellulolytic bacteria lacking of wide array of cellulase system (Beguin and Aubert, 1994). Thus, knowledge of the diversity and community level physiological profile of cellulolytic bacteria remains obscure. Despite methodological constrains, evidence for distinct differences in community composition could recently be demonstrated by application of ribosomal RNA sequence analysis McCraig et al. (2001) offering new perspectives to the traditional phenotypic classification system. Existing scientific literatures show huge bacterial diversity under the influence of long term manure application in field scale (Stellwag et al., 1995; Teather and Wood, 1982; Pourcher et al., 2001; Sekiguchi et al., 2007; Ulrich et al., 2008). But evidences regarding the long term effect of farm yard manure on bacterial diversity in different soil types are sparse. Thus, the aim of the study was to explore the effect of long-term manure application on abundance of phylogenetically diverse culturable cellulolytic bacteria in two different agricultural soils.

MATERIALS AND METHODS

Study site and soil sampling

Over hundred year old long-term manure experimental sites were selected for the isolation and characterization of cellulolytic bacterial communities. Nutrient-poor Orthic Luvisol (Brown soil) and Haplic Chernozem (Black soil) were studied at long–term field experimental sites, located in Berlin-Dahlem and Bad Lauchstadt (Table 1), respectively. At both sites, winter wheat was grown in three replicated plots with or without farmyard (FYM) treatments and were sampled in April, 2003 at full growth stage of wheat. Ten samples were collected form plough layer of each plots using soil corer and pooled as composite sample, finally sieved (<5 mm) and stored at 4°C until processing.

Isolation of soil cellulolytic bacteria

The culturable soil cellulolytic bacteria were isolated using soil serial dilution and pour plate technique. Soil samples (1 g dry weight basis) were suspended in 99 ml sterile distilled water by magnetic stirring in order to dislodging bacterial cells for clay particles. Replica aliquots (1 ml) were poured and dispersed by swirling with Ken Knight's Agar Medium prepared form K_2HPO_4-1.0 g, $NaNO_3$-0.1 g, KCl-0.1 g, $MgSO_4$, $7H_2O$-0.1 g, Cellulose powder-10 g, Agar agar-15.0 g, and Distilled water 1 L. Cycloheximide (50 mg^{-1}) was added to suppress fungal growth. pH was adjusted at 7.0 prior to autoclaving. Congo red (0.2% w/v) was used as an indicator for the detection of cellulolytic bacteria as described by Teather and Wood (1982). In dilution plate, detection of cited bacteria was carried out after 3 and 5 days of incubation at 30°C. After an incubation of 2 to 5 weeks at 20°C, cellulolytic isolates were transferred to Tryptic Soya Agar, pH 7.0 (Difco).

DNA isolation and PCR amplification

For template preparation, DNA was isolated form bacteria grown in

Table 1. Site properties of the experimental plots.

Long-term land use experiment at Berlin-Dahlem (DIII)	
Location	Berlin, Deutschland; 52°28′N, 13°18′E
Established	1923
Climate	Semi-continental
Mean annual rainfall	547 mm
Mean annual temperature	9.2°C
Soil classification	Brown soil/Orthic Luvisol
Soil type	Silty sand
Crop rotation	Fodder beet- winter wheat –potato- winter wheat
Manure	30 ton ha^{-1} every second year (since1939)
Static, long-term fertilizer experiment at Bad Lauchstädt	
Location	Saxony-Anhalt, Deutschland, 51°24′N, 11°53′E
Established	1902
Climate	Semi-continental
Mean annual rainfall	483 mm
Mean annual temperature	8.7°C
Soil classification	Blach soil/ Haplic Chernozem
Soil type	Loan
Crop rotation	Sugar beet –spring barley-potato- winter wheat
Manure	30 ton ha^{-1} in every year (since 1902)

tryptic soy broth (Difco) for 1 to 4 days at 28°C. Cells were harvested (0.5 to 1.5 ml) by centrifugation, washed with 0.3% NaCl, resuspended in 50 µl 0.3% NaCl, and ground using mortar and pestle. Finally, 5 to 10 µl of the crushed cells were added to 20 µl of 25 mM NaOH/0.25% SDS and heated for 15 min at 95°C. Aliquots (0.2 µl) of the resulting lysate were directly used for PCR without further purification. The primers fDI and 926r (Ulrich and Wirth, 1999; Ulrich et al., 2008) used in this study are homologous to the consensus sequence of the 16S rDNA genes and are capable of amplifying almost the complete 16S rDNA of most eubacteria. Routinely, a 50 µl reaction mixture containing 1 µl of template DNA, 5 µl of 10 × reaction buffer (Applied BiosystemWeiterstadt, Germany), 3.5 µl of $MgCl_2$ 0.5 µl of deoxynucleoside triphosphate, 0.25 µl of each primer, 0.5 µl Taq-polymerase (PE Biosystem) and 39.5 µl water was used. The amplifications were performed in a GeneAmp PCR system 2400 (Perkin-Elmer Corporation, Norwaik, CT) with the following protocol: initial denaturation at 95°C for 2 min; 25 cycles of 30 s at 94°C, 40 s at 54°C, 1.3 min at 72°C, a single final extension at 72°C for 8 min and a final soak at 4°C (Ulrich and Muller, 1998). After the reaction, aliquots (3 µl) were mixed with 5V stopper and the PCR products were examined in a 1% agarose gel.

RFLP analysis

PCR product (2 to 8 µl) were digested with *Scr*FI (New England Biolabs, Beverly, MA) and separated in 2.5% Metaphor agarose gels (FMC Bioproducts, Rockland, ME). The DNA molecular weight markers V and VI, respectively, were used as a size standard (Boehringer Mannheim, Germany). The gels were stained with ethidium bromide and finally documented using video camera image system (EasyImage Plus, Herolab, Wiesloch, Germany). Isolates with identical restriction patterns were designated as a single 16S rDNA ribotype group.

Sequencing

Representative isolate of these pattern groups were used for 16S rDNA sequence determination. PCR products of 16S rDNA were purified by Qiaquick PCR Purification Kit (Quiagen, Santa Clarita, CA) according to the manufacturer's instructions. A cycle sequencing protocol was applied for sequencing both complementary strands using primers 346r, 399f and fdl with ABI Prism 310 Sequencer (Applied Biosystems) using a BigDye Terminator Cycle Sequencing Kit. The nearly complete 16S rDNA sequences were compared to sequences available from the Ribosomal Database Project and EMBL/Gene Bank database.

Analysis for phylogenetic tree

To understand the phylogenetic relationships among the pattern groups, a phylogenetic tree was constructed. The similarity values of isolate ribotype were based on a pairwise comparison of sequences and pairwise distance (DNADIST). For phylogenetic analysis, the DNA sequences were aligned using the W algorithm (program version Clustal X ,1.83) and trees were constructed using the Neighbour-Joining and Maximum-likelihood algorithms (PHYLIP Programme package 3.57) (Felsenstein, 1993). The tree topologies were evaluated by bootstrap analysis of the Neighbour-Joining tree using original dataset and 1000 bootstrap data sets.

Analysis of bacterial communities

Total DNA was extracted from 0.25 g of composite soil using a Fast DNA Spin Kit for Soil (Q BIOgene, Carlsbad, CA) following manufacturer's direction. Moreover, a wash procedure with 5.5 M guanidine thiocyanate was performed twice after the extracted DNA to completely remove humic acids which interferes during

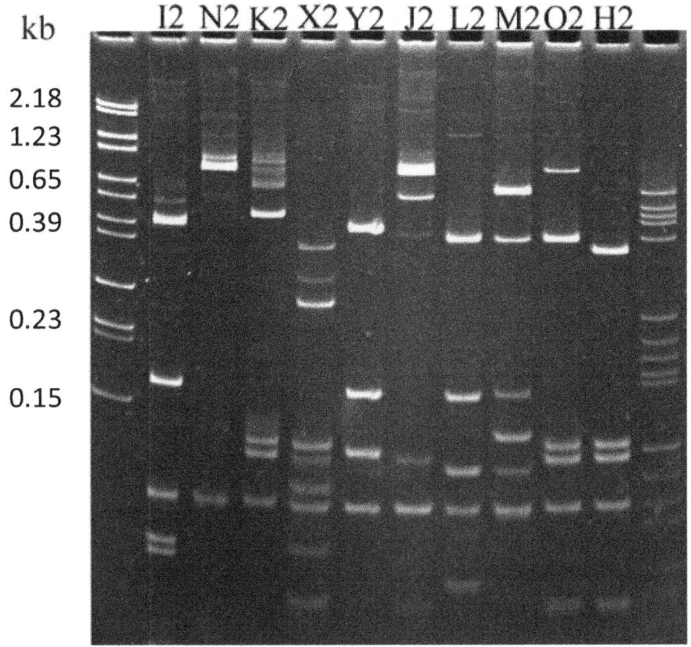

Figure 1. Restriction patterns of anplified 16S rDNA digested with endonuclease ScrFI. Lanes I2 to H2 corresponds to pattern groups as shown in Table 3. First and last lanes represent molecular weight markers V and VI, respectively (Boehringer Mannheim, Germany).

amplification in PCR. T-RFLP analysis was performed as described by Ulrich and Becker (2006). The 16S rRNA gene was amplified using eubacterial primers 8f labeled with 6-FAM and 926r. The polymerase chain reaction (PCR) products were digested with HhaI and subsequently separated with GeneScan 1000 Rox (Applied Biosystems, Foster City, CA) as an internal size standard on an ABI 310 DNA sequencer (Applied Biosystems) using POP6 polymer. Terminal fragments were evaluated by the GeneScan Analytical Software version 3.1.2. T-RFLP profiles were standardized in a similar way as suggested by Dunbar et al. (2001). After the standardization of the profiles, only peaks above a threshold of 50 fluorescence units were considered. Cluster analysis of the T-RFLP profiles was performed using the GelCompar II software v. 2.5 (Applied Maths, Saint-Martens-Latem, Belgium). The ABI files were converted into the Gel Compar curve format (Ulrich and Becker, 2006).. The optimization procedure of the peak tolerance resulted in a value of 0.01%. To consider both the presence and the relative abundance of the terminal restriction fragments (TRFs), densitometric curves of whole profiles were analyzed using Pearson's correlation coefficients. The resulting similarity matrix was the basis for clustering by the Ward algorithm (Ward, 1963). Densitometric curves in the range of 30 to 900 bp were used for the calculation.

RESULTS

Biodiversity of cellulolytic bacterial isolates

A great phenotypical variability among the isolates was observed when grown on SEA medium. Isolates were selected on the basis of phenotypic characteristic, subcultured and purified following standard microbiological protocol. The rDNA of the isolates were amplified using primer pair fdl and 926r, resulting in characteristic single band of about 1400 bp. Restriction analysis of PCR products of the 54 isolates using restriction enzyme *Scr*FI revealed various master patterns as demonstrated in Figure 1. DNA fragment smaller than 70 bp were not properly resolved by electrophoresis and therefore were not used for the comparison of the patterns. Ultimately, five to eight restriction fragments per pattern were used as tools to differentiate the isolates. Using the restriction enzyme *Scr*FI, 54 isolates were classified into 10 genetically diverse pattern groups.

Five isolates out of the 10 RFLP groups were assigned to the *Actinomycetales* branch, and rest to the non-actinomycetales as supported by high bootstrap values (Table 3). At all sampling positions whether manured or unmanured, a dominant pattern group (H_2) was identified, containing 31.48% of total isolates.

The restriction pattern of master group H_2 and O_2 are very similar in band pattern. Only two bands is different (Figure 1). Sequence data of the representative member of those groups showed 99.1 and 99.3% similarity with reference sequence of *Streptomyces avermitlis* and *Streptomyces* sp., respectively. Two site-specific pattern groups could be identified, representing a part of the total population, which was highly specific for the brown soil treated with FYM (H_2, 9.25% of total isolates) and for black soil treated with no FYM (J_2, 11.11% of total isolates), respectively. In general, the composition of cellulolytic isolates in two sites displayed differences with respect manure application and soil properties. Cellulolytic bacterial diversity in brown soil had been influenced by manuring while the same was not noticed in black soil treated with manure. But the composition of cellulolytic 16S rDNA RFLP groups was impacted by manure application with highest evidence in the brown soil. Based on 16S rDNA sequence analysis, isolates of the dominant as well as the specific pattern groups could be assigned to the genus *Streptomyces*. Furthermore, sequencing of 16S rDNA of isolates of five pattern groups of L_2, X_2, K_2, N_2 and I_2 revealed a high phylogenetic diversity among these isolates, representing sequence similarity of *Streptococcus* (98.8% similarity), *Paenibacillus* (98.0% similarity), *Bacillus* (99.1% similarity), *Bacillus megaterium* (98.0% similarity) and *Bacillus pumilus* (98.7% similarity), respectively. The genotypic group K_2 has a phylogenetic relationship with an unknown soil bacterium with a close relationship with bacilli. The master group Y_2 is phylogenetically very distant as compared to other groups and classified as Gram-negative genus of *Fateuria* (Figure 2). Phylogenetic study reveals that the cellulose decomposing ability is widely distributed among many genera including

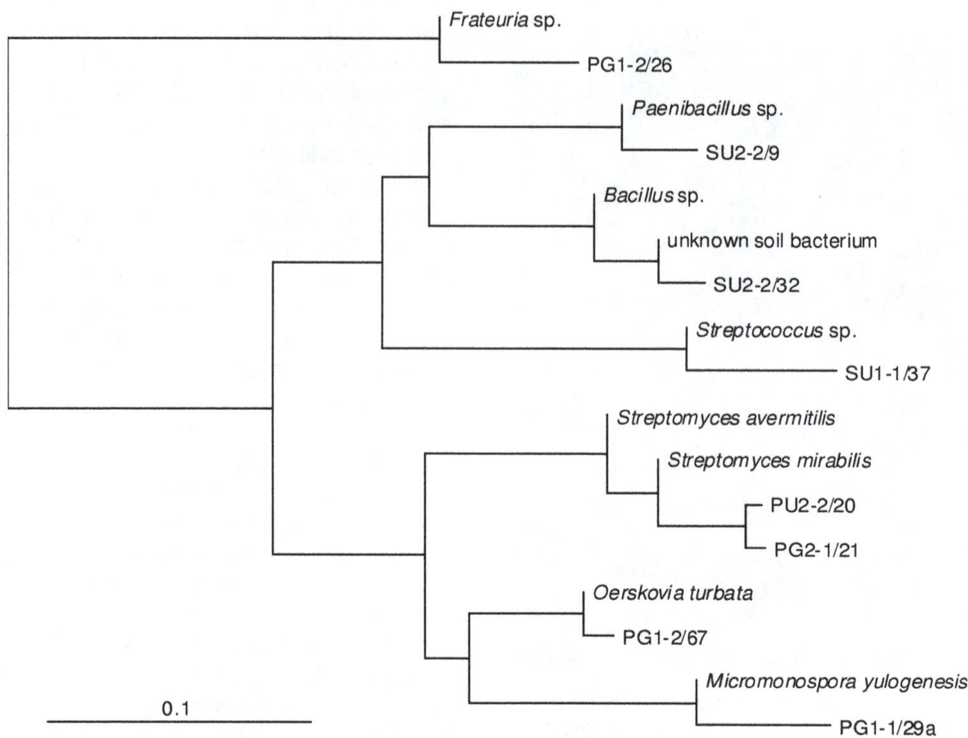

Figure 2. Phylogenetic tree showing the relationship of the cellulolytic isolates and related reference species. The tree constructed using neighbour-joining method is based on nearly complete 16S rDNA sequences. The tree was rerooted by outgroup of E.coli. The bar, 0.1 indicates the relative sequence divergence. SG - Schwarzerde stallmistgedüngt, SU - Schwarzerde Kontrolle, PU - Parabraunerde Kontrolle, PG - Parabraunerde stallmistgedüngt mean black soil manured,.black soil unmanured, brown soil unmanured and brown soil manured, respectively.

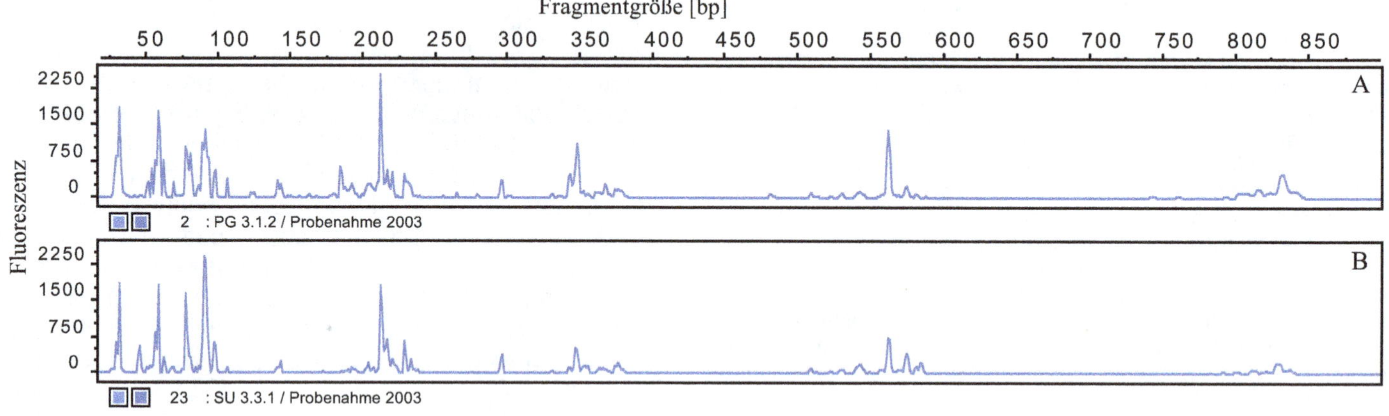

Figure 3. T-RFLP pattern of bacterial community under brown soil (A) and black soil (B).

the domain *Eubacteria* comprising aerobic order *Actinomycetales* (phylum *Actinobacteria*) largely comprises of *Streptomyces* species of diverse phylogenetic affiliation; and the aerobic genera of *Streptococcus, Paenibacillus* and *Bacillus.* The phylogenetic relationship between cellulolytic bacterial isolates and reference strains is demonstrated in Figure 2.

Terminal restriction fragment length polymorphism (T-RFLP) analysis of 16S rDNA for charcterization of total bacterial community of long-term manure experimental sites

The huge number of T-RFs in the individual T-RFLP profiles (Figure 3) of soil of experimental sites representing a greater abundance of bacterial groups.

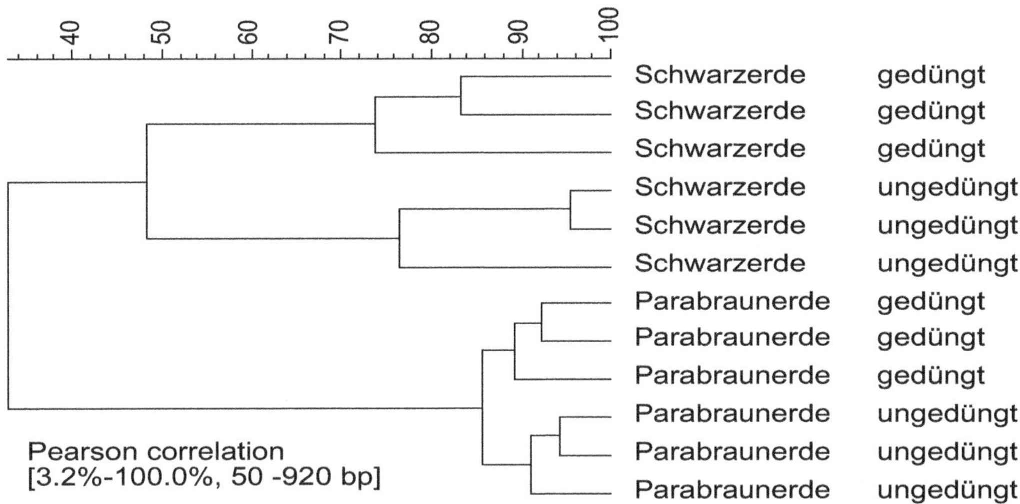

Figure 4. Dendogram showing the similarity between the T-RFLP profiles (community structure) derived from the manured (gedungt) and control (ungedungt) plots of the black soil (Schwarzerde) and brown soil (Parsbraunerde).

After recognizing the predominent T-RFs of the complete T-RFLP Profile, a hierarchical cluster analysis was performed to identify relatively homogeneous group of isolates based on T-RFLP profile characteristics, using an algorithm (complete linkage). Distance or similarity measures are generated by the proximities procedure following WARD-clustering. Cluster analysis of the T-RFs grouped the community and yielded a different basic dendogram topology representing different bacterial clusters. In agreement with the results from the cluster analyses, the combined dataset revealed very tight clustering of the soil bacterial community and a very clear separation among the community (Figure 4).

The primary branch differentiates the community structures of the two soils investigated. The secondary branch represents the T-RFLP profiles obtained from both manured and unmanured plots of experimental sites. In the tertiary branches, all profiles obtained form manured brown soils/and or black soils are distinguished from the control unmanured plots of respective soil (Figure 4).

DISCUSSION

Bibliography antecedent reveals that most of the studies so far conducted on cellulolytic bacteria in agroeco-system were based mostly on the enumeration of colony forming units on agar plate using serial dilution and pour plate technique where the community composition was rarely analyzed (Ulrich et al., 2008).

Moreover, phylogenetic studies of cellulolytic bacterial communities published so far have focused on microbial communities associated with terrestrial and marine herbivores (Stahl et al., 1988; Stellwag et al., 1995) or other highly specific environment (Marri et al., 1997; Rainey et al., 1993). This present study deals with cellulolytic bacterial diversity in two long-term manure experimental sites of different soil and environmental condition with a view to study the soil cellulolytic bacterial biodiversity along gradients of fertility where trends of diversity of bacteria can be explored (Giller et al., 1997). The experimental site in Berlin-Dhalem is nutrient poor sandy soil while in BadLauchstädt is highly fertile loamy soil. Thus, a broad range of different soil ecological situation is present as charcterised by soil properties (Tables 1 and 2). Therefore, distinct soil ecological conditions would expect to harbour diverse cellulolytic bacterial communities.

Single methodology is not sufficient to resolve the community composition in soil. So, in this study a polyphasic approach comprising of culture dependent and culture independent methodlogy were adopted. Culture dependent methods encopassing screening of cellulolytic bacteria and a subsequent classification using RFLP of amplified 16S rDNA was applied, revealing a simple community of culturable soil cellulolytic bacteria.Whereas, culture independent methodlogy consists of total soil DNA analysis and characterizing through T-RFLP profiling. Culture dependent methods provide insufficient information about soil cellulolytic bacterial community composition since a major part of the soil residing cellulolytic bacteria is considered to be nonculturable due to the lack of proper designing of media with appropriate composition. Thus, phylogenetic diversity on the basis of database obtained from RFLP profiling are considered to be representative for a culturable subset of the potentially cellulolytic bacterial community. Despite these methodological constraints, evidence for differences in community composition could be demonstrated.

RFLP profiling of 54 bacterial isolates using single

Table 2. Characteristic of experimental soil.

Parameter	Berlin-Dahlem (Orthic- Luvisol)		Bad Lauchstädt (Haplic- Chernozem)	
	Manured	Control	Manured	Control
C_{org} (%)	0.86 ± 0.16*	0.62 ± 0.03	2.27 ± 0.05*	1.68 ± 0.01
N_t (%)	0.07 ± 0.01	0.05 ± 0.00	0.19 ± 0.00	0.13 ± 0.00
CEC (cmol kg^{-1})	8.8 ± 0.34	9.6 ± 1.17	39.1 ± 3.81*	29.8 ± 1.26
pH	6.0 ± 0.11	5.6 ± 0.16	6.7 ± 0.06	6.3 ± 0.29
Texture				
Clay (%)	2.9		21.0	
Silt (%)	25.0		67.8	
Sand (%)	72.1		11.2	

Table 3. 16S rDNA genotype groups identified and their relative proportion at the four experimental plots.

Pattern group	Brown soil manured	Brown soil unmanured	Black soil manured	Black soil unmanured	Total isolate	Percent of similarity to the closest relative	Putative assignment
H_2	5 (9.25)	4 (7.40)	4 (7.40)	4 (7.40)	17 (31.48)	99	*Streptomyces avermitilis*
O_2	1 (1.85)	1 (1.85)	2 (3.70)	1 (1.85)	5 (9.25)	99.3	*Streptomyces*
M_2	0	1 (1.85)	2 (3.70)	0	3 (5.55)	98.5	*Streptomyces*
L_2	1 (1.85)	1 (1.85)	0	3 (5.55)	5 (9.25)	98.4	*Streptomyces lavendulae*
J_2	2 (3.70)	0	0	6 (11.11)	8 (14.81)	98.8	*Streptococcus*
Y_2	1 (1.85)	0	0	2 (3.70)	3 (5.55)	98	*Streptomyces cyaneus*
X_2	2 (3.70)	0	1 (1.85)	1 (1.85)	2 (3.70)	98.3	*Paenibacillus*
K_2	2 (3.70)	0	0	2 (3.70)	4 (7.40)	99.1	*Bacillus*
N_2	1 (1.85)	0	2 (3.70)	1 (1.85)	5 (9.25)	98.7	*Bacillus megaterium*
I_2	0	0	0	1 (1.85)	2 (3.70)	98	*Bacillus pumilus*

Pattern groups are based on the restriction analysis by the enzyme ScrFI.

endonuclease yielded in altogether 10 16S rDNA genotypic master groups. As single endonuclease digestion is not sufficient to resolve the bands of amplified gene, most of the pattern groups probably include more than one species. This corroborates the findings of earlier workers (Ulrich and Wirth, 1999; Ulrich and Becker, 2006; Ulrich and Muller, 1998). After sequencing of 16S rDNA, isolates of genotypic group H_2 were classified to genus *Streptomyces*. Thus, it may be concluded that group H_2 represents several *Streptomyces* species. Although there exists differences in community composition, the community structure of that functional group was rather simple. Only a single predominant RFLP group (H_2) that comprised 31.48% of the total isolates were identified. This is the line with the observation of Akasaka et al. (2003). They also found the predominance of *Cellulomonas* species during the degradation cellulose rich paddy straw. This phenomenon is further supported by the report of many scientists (Ulrich and Wirth, 1999; Lynd et al., 2002; Lynd et al., 1999; Ulrich et al., 2008) who observed site specific predominance of a single group of bacteria.

In order to evaluate the phylogenetic range of the cellulolytic bacterial isolates, single representatives of another five genotype groups were identified by 16S rDNA sequencing. Besides the predominance of *Streptomyces* species, the sequence data revealed non actinomyetes bacterial species of *Streptococcus* (98.8% similarity), *Paenibacillus* (98.0% similarity), *Bacillus* (99.1% similarity), *Bacillus megaterium* (98.0% similarity) and *Bacillus pumilus* (98.7% similarity), indicating a diversity of species within the Gram-positive cellulolytic soil bacteria. DNA sequencing in combination with RFLP analysis revealed instead of pattern group H_2, isolates of pattern groups O_2, M_2, L_2 and Y_2 also belong to the *Actinomycetales.* Thus, the results indicate the predominance of the *Actinomycetales* within the culturable cellulolytic bacteria. This result correspond to findings of Errikson et al. (1990) and Lynd et al. (2002) who discussed the striking role of actinomycetes in lignocellulosic materials decomposition in soil.

Although, the pattern of distibution of RFLP groups of culturable cellulolytic bacteria exhibited some similarities among the plots both manured and unmanured, for example, predominant occurance of the pattern group, H_2 several characteristic differences among the distribution of other RFLP groups were found. RFLP group X_2 was present in black soil under both the manured and unmanured plots but was distinctly absent in brown soils under similar treatments. Thus, it may be concluded that such diversity of cellulolytic bacteria is related to the site properties. Similarly, the existence of greater number of RFLP groups in black soil as compared to that of brown soil might be related to higher organic carbon content and higher fertility status as well as better texture of soil. Results, thus, substaintiated the earlier findings of numerous authors that diversity of microorganisms largely depends on the soil properties, particularly soil texture (Ulrich and Becker, 2006).

A distinct site-specific bacterial community structure was indentified on the basis of T-RFLP analysis of 16S rDNA gene fragments. The community was further differentiated on application of manure. Thus, long-term manure application was proven to cause discribible impacts on the community structure. In general, site properties displayed stronger effects on the community structure than the long-term application of manure. This is the line with the observation of Girvan et al. (2003) who demonstrated soil type to be the primary determinant of community structure in arable soils under three different farm mangement practices. Soil parent material and texture were particularly proven to have a decisive influence on the composition of the bacterial communities (Ulrich and Becker, 2006). Other factors such as fertilizers, were also considered as potent drivers of soil microbia change and community structure (Marschner et al., 2003; O'Donnell et al., 2001). Long-term manure application in the present experiment caused a differentiation in the community structure of total bacterial community. This finding, thus, corroborates the earlier reports of Hartmann et al. (2006) that application of farmyard manure not only suppressed disease producing pathogens but also influenced on the bacterial community structure. This is the line with observation of Sekiguchi et al. (2007). This is further supported by Ulrich et al. (2008) who recorded a shift in bacterial communities in two contrasting soils under long-term manure application. This present experiment demonstrated the strong impact of both soil type and long-term manure application on the total bacterial community. This result has further supported by the unique distribution of RFLP genotypic groups of cellulolytic bacteria in two experimental sites as well as under the influence of manure application.

Contrasting soil property with a strong fertility gradient and long-term manure application as cellulose source is an idle experimental site to study the diversity of cellulolytic bacteria. In this present investigation application of molecular techniques followed by phylogenetic analysis of sequence data adopted though differentiated the community structure of culturable cellulolytic bacteria, this analytical approach of using a single endonuclease could not explore the full range of diversity of this functional group. A finer resolution using a further combination of restriction enzymes is deemed necessary to revealed the diversity at a finer scale. Soil is a heterogenous system where the microbial load is subject to spatiotemporal variation. So, single assessment technique is not sufficient to work out the whole range of diversity. A polyphasic approach consisting of molecular approach coupled with functional attributes as well as structural diversity will be more effective to study the cellulolytic bacterial diversity in a rational way.

ACKNOWLEDGEMENT

The author (NS) gratefully acknowledges the financial assistance offered by German Govt. (InWEnt-International, Capacity Building) to carry out the programme in collaboration with ZALF (Center for Agriculture and Land Use Research, Institute for Primary Production, Muenceberg, Germany) during 2002-2003.

REFERENCES

Akasaka H, Izawa T, Ueki K, Ueki A (2003). Phylogeny of numerically abundant culturable anaerobic bacteria associated with degradation of rice plant residue in Japanese paddy field soil. FEMS Microbial. Ecol. 43:149-161.

Beguin P, Aubert LP (1994). The biological degradation of cellulose. FEMS Microbiol. Rev. 13:25-58.

Coughlan MP, Mayer F (1992). The cellulose-decomposing bacteria and their enzyme systems. In: Balows, A Truper, HG Dworkin, M Harder, W Schleifer KH (eds.) The Procaryotes, 2nd edition. Springer verlag, New York. pp. 460-515.

Dunbar J, Tick nor LO, Kuske CR (2001). Phylogenetic specificity and reproducibility and new method for analysis of terminal restriction fragment profiles of 16S rRNA genes from bacterial communities. Appl. Environ. Microbiol. 67:190-197.

Eriksson KEL, Blanchette RA, Ander P (1990). Microbial and enzymatic degradation of wood and wood components, Ch. 2.6. Cellulose degradation by bacteria, Springer Series in Wood Science (Timell TE, ed.). Springer, Berlin.

Felsenstein J (1993). PHYLIP (Phylogeny interference pachage), 3.5c edu.Seattle, University of Washington.

Giller KE., Beare MH, Lavelle R, Izac AMN, Swift JJ (1997). Agricultural intensification, soil biodiversity and agroecosystem fallow. Appl. Soil Ecol. **6**:3-16.

Girvan MS, Bullimore J, Pretty JN, Osborn AM, Ball AS (2003). Soil type is the primary determinant of the composition of the total andactive bacterial communities in arable soils. Appl. Environ. Microbiol. 69:1800–1809.

Halliwell G (1965). Hydrolysis of fibrous cotton and reprecipitated cellulose by cellulolytic enzymes from soil microorganisms. Biochem. J. 95:270-281.

Hartmann M, Fliessbach A, Oberholzer HR, Widmer F (2006). Ranking the magnitude of crop and farming systems on soil microbial biomass and genetic structure of bacterial communities. FEMS Microbiol. Ecol. 57:378–388

Lynd LR, Weimer PJ, van Zyl WH and Pretorius IS. (2002).Microbial cellulose utilization: fundamentals and biotechnology. Microbiol .Mol. Biol. Rev. 66:506–577

Lynd LR, Wyman CE, Gerngross TU (1999). Biocommodity engineering. Biotechnol. Prog. 15:777–793.

McCraig AE, Grayston SJ, Prosser JI, Glover LA (2001). Impact of cultivation on characterization of species composition of soil. International J. Food Microbiol. 4:315-325.

Marri I, Barboni E., Irdani T, Perito B, Mastromei G (1997). Restriction enzyme and DNA hybridization analysis of cellulolytic Streptomuces isolates of different origin. Can. J. Microbiol. 43:395-399.

Marschner P, Kandeler E, and Marschner, B (2003). Structure and function of the soil microbial community in a long-term fertilizer experiment. Soil Biol. Biochem. 35:453–461

Mullings R, Parish JH (1984). Mesophilic aerobic Gram negative cellulose degrading bacteria from aquatic habitats and soils. J. Appl. Bacteriol. 57:455-468.

O'Donnell AG, Seasman M, Macrae A, Waite I, Davies JT (2001).Plants and fertilisers as drivers of change in microbial community structure and function in soils. Plant Soil 232:135–145

Olsen GJ,Woese CR (1993). Ribosomal RNA: a key to phylogeny. FASEB J. 7:113-123.

Olson RA, Bakken LR (1987).Variability of soil bacteria: optimization of plate counting technique and comparison between total counts and plate counts within different size groups. Microb. Ecol. 13:59-74.

Pourcher AM, Sutta L, Hebe L, Mogueder G, Bollet C, Simoneau P, Gardan L (2001). Enumeration and characterization of cellulolytic bacteria from refuse of landfill. FEMS Microbiol. Ecol. 34:229-241.

Rainey FA, Jansen PH, Morgan HW, Stuchebrandt E (1993). A biphasic approach to the determination of the phenotypic and genotypic diversity of some anaerobic, cellulolytic, thermophilic, rod-shaped bacteria. Antonie von Leeuwenhoek.64:341-355.

Sekiguchi H, Kushida A, Takanaka S (2007). Effects of cattle manure and green manure on the microbial community structure in upland soil determined by denaturing gradient gel electrophoresis. Microb. Environ. 22;327-335.

Stahl DA, Plesher B, Mansfield HB, Montgomery L (1988). Use of phylogenetically based hybridization probes for studies of ruminal microbial ecology. Appl. Environ. Microbiol. 54:1079-1084.

Stellwag EJ, Smith TD, Luczkovich B (1995). Characterization and ecology of carboxymethylcellulose-producing anaerobic bacterial communities associated with the intestinal tract of the pinfish, *Lagodon romboides*. Appl. Environ. Microbiol. 61:813-816.

Szegi J (1988). Cellulose decomposition and soil fertility. Academiai Kiado, Budapest. p. 186.

Teather RM, Wood PJ (1982). Use of Congo red-polysaccharide interactions in cnumeration and characterization of cellulolytic bacteria fron bovine rumen. Appl. Environ. Microbiol. 43:777-780.

Ulrich A, Wirth S (1999). Phylogenetic diversity and population densities of culturable cellulolytic soil bacteria across an agricultural encatchment. Microb. Ecol. 37:238–247

Ulrich A, Becker R (2006).Soil parent material is a key determinant of the bacterial community structure in arable soils. FEMS Microbiol. Ecol. 56:430–443

Ulrich A, Muller T (1998). Heterogeneity of plant-associated streptococci as characterized by phenotypic features and restriction analysis of PCR amplified 16S rDNA. J. Appl. Microbiol. 84:293-303.

Ulrich A, Klimke G, Wirth S (2008). Diversity and Activity of Cellulose-Decomposing Bacteria, Isolated from a Sandy and a Loamy Soil after Long-Term Manure Application. Microbiol. Ecol.55:512-522.

Ward JH (1963). Hierarchical grouping to optimize an objective function. J. Am. Stat. Ass. 58:236.

Woese CR (2000). Interpreting the universal phylogenetic tree. Proc. Natl. Acad. Sci. USA 97:8392–8396.

17

Innovations in the soyabean innovation system in Benue State, Nigeria

Daudu S.[1] and Madukwe M. C.[2]

[1]Department of Agricultural Extension and Communication, University of Agriculture, Makurdi, Benue State, Nigeria.
[2]Department of Agricultural Extension, University of Nigeria, Nsukka, Enugu State, Nigeria.

The study determined the factors hindering innovation in the soyabean innovation system in Benue State, Nigeria. A total of 100 respondents were selected using simple random sampling technique. Data were analyzed by use of frequency, percentage and mean statistic. The result of the study indicated that a number of innovations were introduced and classified into produce/process innovations, the basic research innovation and the marketing/managing strategy innovations. The study also revealed that among the product/processing innovation introduced in the soyabean innovation system, soya milk (x = 2.78) was the most preferred innovation, while the least preferred innovation was Soya garri (x = 1.55). The study also revealed that ten factors were identified as major factors hindering innovation in the soyabean innovation system. These were lack of new knowledge (x = 2.32), lack of research policy (x = 2.26), lack of capital (x = 2.78), lack of markets (x = 2.34), lack of competent human capital (x = 2.11), lack of research and extension (x = 2.26), lack of patency (x = 2.12), lack of regulation in the innovation system (x = 2.32), lack of functional processing industrial system (x = 2.74), lack of stakeholder interaction (x = 2.19). It was recommended that the introduced innovation should be nurtured and patented by regulatory bodies while the preferred products innovation given high acceptability rating be guaranteed funding by commercial banks and such finance institutions to commercialize these products to stimulate further innovations. Finally, the factors isolated as hindering innovation in the soyabean innovation should be ameliorated through policy options and conscious intervention measures.

Key words: Innovation system, factors, critical actors, soyabean.

INTRODUCTION

Benue State is the largest producer of soyabean (*Glycine max*) in Nigeria (BNARDA, 2000). In 1985 Benue State was declared a special soyabean producing area by the Federal Government in order to concentrate efforts where comparative advantage is greatest in line with its ecological specialization policy. The importance of soyabean as a high protein, primary input in vegetable oil, dairy and feed industries is not in doubt. The International Institute for Tropical Agriculture (IITA) has enhanced its protein content to 40% level. This makes the crop a suitable substitute to animal protein (Ayoola, 2001).

In agriculture an innovation can be a practical technique or different way of doing things like when a farmer makes a new arrangement about how land should be used with a neighbor. Basically, innovation is universal and occurs in every environment depending on the complexity and commercial value. The basic rule is that it usually arises as a result of interaction (LEISA, 2000). Three broad categories of innovation in the Nigeria agriculture innovation systems are (a) innovation in product technology (b) innovation in genetic engineering and bio-control and (c) innovation in process technology (Ayoola, 2001). With respect to product innovation, following the Structural Adjustment Programme there was lack of foreign exchange to import animal protein. To alleviate the problem soyabean utilization programmes were mounted and many recipes were prepared to reduce malnutrition among the low income people. The products locally produced include soymilk, soybread, soy-garri, soysnacks (cakes), soy-cooking oil, fried soyabean flour (Ayoola, 2001; Agbulu and Wombo, 2005; Lawal et al., 2005). According to Lawal et al. (2005), soyabean milk has the highest acceptability while soyabean cake was least used.

With respect to soyabean innovation system, aside from the product innovation the genetic engineering innovation needs to be highlighted. In the genetic innovation category the production of soyabean is based on traditional farming system using varieties such as the pod shattering Malaysian varieties. This was introduced by the early Christian missionaries. Current improved varieties include those produced by IITA (TGX and TGM varieties) and Institute of Agricultural Research (IAR), Ahmadu Bello University, Zaria (Samsoy varieties). Improvement in yield rose from mere 200 kg per ha to above 1.5 tons per ha (Ayoola, 2001). About 70 improved varieties of soyabean have been developed for use in the Nigerian farming system (Conroy, 2003). The dominant varieties disseminated by the Benue State extension system are: Early/Median Maturing (100 to 120 days): Sam-soy 2 (M216), Samsoy 1 (M79); TGX 536 – 026, TGX 539 – 5E, TGX 923 - 2E, TGX 292 – 172C, M98 and M351, TGM 579. The late maturing (> 120 days) TGX 306 – 036C, TGX 725 – 011D, TGM 244 (BNARDA, 2000).

The promotion of joint private enterprises and joint venture investment involving government to increase the value added utility of soyabean resulted in the establishment of the Taraku Mill. The Taraku soyabean mill has the Capacity to process 72,000 metric tons of soyabean into about 10,000 metric tons of refined soyabean oil, 61,000 tonnes of soyameal and 900 metric tonnes of soap per annum (Ayoola, 2001). However the mill, producer of "Golden Soya" cooking oil, operates at 30% installed capacity despite the fact that it is located in Benue State where soyabean production is concentrated. According to Ayoola (2001) various strategies have been implemented to accelerate the production of soyabean in Nigeria. These include technology inducement, awareness campaign, promotion of ecological specialization and traditional support services such as extension and input services.

Special significance to the innovation system is the interactive activities of stakeholders in the soyabean industry. The National Soyabean Association (NSA) provides a forum for interaction of individuals, enterprises and agencies involved in the soyabean industry. At the forum, research scientists derive their research priorities and problems directly from farmers, ADPS, processors, marketers, and consumers. To enhance academic interaction newsletters, proceedings of conferences/ meetings and the Tropical Oil Seed Journal are published, and these support interaction with international community on soyabean matters. These activities enhanced net working among actors in the soyabean industry (Ayoola, 2001).

Despite these initiatives soyabean innovation system in Benue State has not recorded any marked adequacy of quantum of soyabean innovations and partnership. The soyabean mill at Taraku produces below capacity. Furthermore, importation of soyabean, the primary raw materials is done from Brazil and neighbouring West African countries unabated (Abagu, 2004). Ayoola (2001) suggested the following measures to mitigate this situation: (1) institutional integration where the relevant agencies and corporate bodies come together in the process of generating commercial research findings and technological breakthrough and (2) monitor the appropriateness of the existing and newly emerging innovations.

The (Technical Centre for Agricultural and Rural Cooperation /United Nations University-Institute for New Technologies/Koninklijk Instituut Voor de tropens, CTA/UNU – UNITECH/KIT, 2005) noted that national markets in the developing world are highly fragmented and national research and development institutions are poorly linked to the production centres. As a result centers do not sufficiently and adequately supply science, technology and innovation to facilitate meeting the challenges of competing effectively in the global market. Consequently, strengthening innovation system is extremely important to improve the interface between scientists, the dissemination and utilization subsystem, policy makers and decision makers as relate to science, technology and innovation policy formulation. This calls for a study of the innovation system of soyabean in Benue State in order to suggest policy options on the types of innovation introduced in the soyabean innovation system the preference level of innovation introduced and the factors that inhibit innovation among the critical actors in the soyabean innovation system.

Purpose and objective

The overall purpose of the study was to assess the soyabean innovation system in Benue State. The specific objectives of the study were to; (i) identify the innovations introduced in the soyabean innovation system; (ii) determine the preference level of the innovations introduced in the soyabean innovation system, and (iii) determine the factors that hinder innovation among the critical actors in the soyabean innovation system.

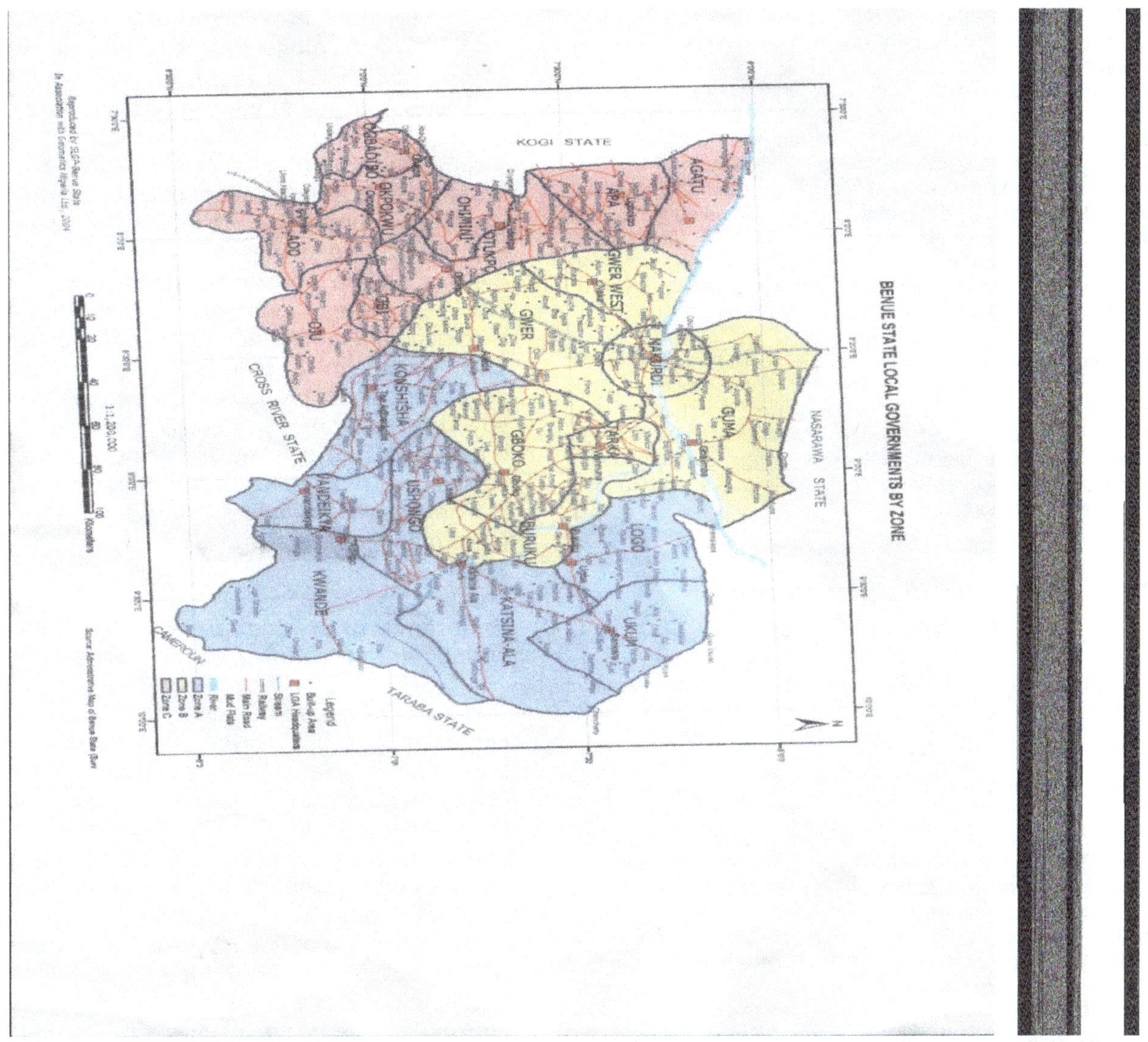

Figure 1. Map of each State.

METHODOLOGY

Benue State derives its name from River Benue, the second largest River in Nigeria. The State is located in the central agro-ecological region of Nigeria. According to the 2006 census the population of the State is 4,219,244 (NPC, 2006). It occupies a total land mass of 3,0855 square kilometres (Benue State Government, 2002). Majority of the people are subsistent arable farmers while the riverine inhabitants engage in fishing as their primary occupation. Benue State is the largest producer of soyabean, beniseed, cassava and yam in Nigeria (Benue State Government, 2002). Subsistence and rainfed traditional farm practices are the dominant mode of farming. Benue State is comprised of 23 administrative local government areas. The state is divided into three agricultural zones. Northern zone comprising of 7 LGAs with Gboko as headquarters. Eastern Zone has 7 LGAs with headquarters at Adikpo while the Central zone is made up of 9 LGAs with headquarters at Otukpo. The Northern and Eastern zones are inhabited by Tiv, while the central zone is inhabited by Idoma and Igede. Soyabean is mainly produced in the Northern and Eastern Zones (BNARDA, 2000).

The population of this study was stakeholders (Critical actors) in the soyabean industry. Specifically it consisted of sample drawn from two local government areas from each of the northern and eastern agricultural zones of Benue State where soyabean is mostly produced (Figure 1).

Selection of samples in each of the LGAS was based on the segmentation of actors in the soyabean innovation system into components. According to Paterson et al. (2003) and CTA/UNU-INTECH/KIT (2005) the component of the agricultural science technology innovation (ASTI) system can be classified into five namely: Demand enterprise, diffusion research and infrastructure.

(i) Demand component: Actors includes consumers of food and food producers in rural and urban areas; consumers of industrial raw materials, international commodity market,
(ii) Enterprise components: Farmers; input suppliers (seeds, agro-chemicals, animal feed), service suppliers (advice, credit, insurance, machinery rentals etc); commodity traders, transporters; agricultural processing industries; farmer and trade organization representing business interest,
(iii) Diffusion components: Extension service (public/private), non

governmental organizations (NGOS) and community based organizations (CBOs), farmer and trade organizations, input and service suppliers,
(iv) Research component: National and international agricultural research organizations; universities and other institution of higher learning, private research foundations; private companies and NGOS with own research facilities,
(v) Infrastructure component: Policy making process and agencies; banking and financial system, transport and marketing system, information and communication infrastructure; professional networks, including farmer and trade organizations, regulatory agencies (Sanitary and phyto-sanitary regulations etc), standard setting bodies.

Five respondents who were identified as critical actors in the soyabean innovation system were selected by purposive sampling methods from each of the 5 components from each of the four sampled LGAs: Gboko, Gwer, Katsina-Ala and Makurdi. The four LGAs were selected by simple random sampling method, a total of 100 respondents were selected through purposive and random sampling method for the study.

Instrument for data collection

Data were collected through the use of interview schedule and questionnaire. Both instruments were validated to reflect all the objective of the study. These were administered to respondents in each LGA.

Measurement of variables

Objective 1 was achieved by listing the innovation introduced; objective 2: The preference level of innovations were measured by asking the respondents to indicate the perceived preference level of each of the listed innovations on a 3-point Likert type scale ranging from highly preferred = 3, moderately preferred = 2 and not preferred = 1. These values were added to get a value of 6 which was later divided by 3 to get a mean score of 2.0. The respondents mean was obtained on each of the items. Any mean (X) score ≥2.0 was regarded as preferred while any mean (X) score less than 2.0 were regarded as not preferred. Objective 3: The factors hindering innovation were measured by asking the respondents to indicate their perceived level of importance of each of the listed factors on a 3-point Likert type scale ranging from very important = 3; important = 2 and not important = 1. These values were added to get a value of 6 which was later divided by 3 to get a mean score of 2.0. The respondents mean was obtained on each of the items. Again, any mean (X) score ≥ 2.0 was regarded as important; while any mean (X) score less than 2.0 was regarded as not important. Specifically objective 1 was analyzed by use of frequency and percentage, objectives 2 and 3 were analyzed by use of mean statistic.

RESULTS AND DISCUSSION

Innovations introduced in the soyabean innovation system

Data in Table 1 indicate efforts of the soyabean innovation system in Benue State in introducing innovations. The results show that 24 innovations have been introduced in the soyabean innovation system in Benue State. The two most mentioned innovations in each categories were soyabean milk (70%), soyabean oil (64%); in the product/processing innovation category, successful trial of improved high yielding, disease resistant, non pod shattering hybrid seeds (TGX - 1448-2E) (11%); appropriate fertilizer type and rate recommendation (9%) were innovations introduced in the basic research innovation category; while compulsory soyabean producers cooperative meeting (10%), appropriate marketing strategies by by-passing middle men to enable soyabe an farmers earn more income (10%) were innovations introduced in the marketing/management strategy innovation category. The overwhelming preference for soyabean milk and soyabean oil is probably because soyabean milk and soyabean oil are traditionally used as compliment to breast milk and cooking oil for adults respectively. This implies that many innovations have been introduced and that adoption of these innovations will enhance the performance of the soyabean innovation system.

Preference of product/processing innovations in the soyabean innovation system

Results of performance of products/processing innovations in the soybeans innovation systems are presented in Table 2. Respondents in the demand components preferred soymilk (x = 2.80) and soy cooking oil (x = 2.80), this was followed by soy bread (x = 2.15) and fried soyabean flour (x = 2.15). This is probably because these products serve as small businesses in the innovation system. Similarly, respondents in the enterprise component preferred soy cooking oil (x = 3.40), soymilk (x = 2.60), fried soyabean flour (x = 2.40) and soy bread (x = 2.10). The higher preference for soy cooking oil than soya milk by respondents in the enterprise components is probably because soy cooking oil has a longer shelf value than soya milk which is perishable. Thus soy cooking oil can be kept and sold gradually.

The respondents in the diffusion component preferred only soymilk (x = 25), soy cooking oil (x = 2.75) and fried soyabean flour (x = 2.05). This is probably because these products serve as source of protein and small businesses in the innovation system. The commercialization of these innovations will boost commercial activities and enhance the soyabean innovation system. Respondents in research component showed prepared to soymilk (x = 2.90), soy cooking oil (x = 2.90), fried soyabean flour (x = 2.25) and soybread (x = 2.05). These products are useful sources of business. The higher preference of soy milk by the research component is probably because the product is commonly utilized by researchers to provide nutrients as a replacement to animal protein and has been promoted by the UNICEF to replace or compliment breast milk in Benue State recently.

Respondents in the infrastructure components preferred soy cooking oil (X = 2.75) soy milk (x = 2.70),

Table 1. Distribution of respondents by innovation introduced in the soyabean innovation system in Benue State. (n=100).

S/N	Product/processing F innovation	%	Basic research innovation F	%	Marketing/management F strategy innovation	%
1	Soyabean milk 70	70.00	(1) Successful trial of 11 improved high yielding, disease resistant, non pod shattering hybrid seeds (Tgx – 14 48 –2E)	11.00	(1) Compulsory 20 soyabean producers cooperative meetings	20.00
2	Soyabean oil 64	64.00	(2) Appropriate fertilizer 9 type and rate recommendation for soyabean production	9.00	(2) Appropriate marketing strategy by by-passing 10 middlemen to enable soyabean farmers earn more income	10.00
3	Fried soya 30 bean flour *(mumu)*	30.00	(3) Appropriate soyabean weeding regimes	6.00		
4	Soyabean 11 Akara/ cake	11.00	(4) Best storage methods	5.00		
5	Soyabean 11 Pomade	11.00	(5) Appropriate chemical weed control	5.00		
6	Soyabean 10 Garri	10.00	(6) Appropriate prototype threshing machines	5.00		
7	Soyabean 10 condiment *(nune)*	10.00	(7) Appropriate soyabean plant spacing (40 × 10 cm) and time of planting	5.00		
8	Soyabean 9 livestock feeds	9.00	(8) Appropriate threshing methods	5.00		
9	Soyabean 4 meal	4.00				
10	Soyabean 4 Soup	4.00				
11	Soyabean flour 4	4.00				
12	Soyabean Moimoi 2/budding	2.00				
13	Soyabean 2 Potash	2.00				

F = Frequency, * Multiple Responses, Source: Field Survey, 2006.

and fried soyabean flour (x = 2.05). The implication of this finding is that the patenting and commercialization of these innovations shall enhance the soyabean innovation system. The findings of this study show that the two most consistently preferred soyabean product innovations among respondents in the five components of the soyabean innovation system in Benue State were soymilk and soya cooking oil. This confirms the findings of Lawal et al. (2005) that soyabean milk has the highest acceptability rating in Benue State. This implies that processing soya beans into *Akpukpa* and soy milk were most profitable in the Benue State soyabean innovation system. Therefore, the appropriate packaging and commercialization of these soyabean products may probably enhance the innovative performance of the soyabean innovation system in Benue state (Table 2).

Factors that hinder innovation among critical actors in the soybean innovation system

The serious factors hindering innovation as perceived by actors in the demand component were: Lack of knowledge (x=2.40), lack of research

Table 2. Mean soyabean product preference.

S/N	Innovation	Mean (X) preference score of components					
		Demand	Enterprise	Diffusion	Research	Infrastructure	Mean (X)
1	Soymilk	2.80*	2.60*	2.8*	2.90*	2.70*	2.76*
2	Soybread	2.15*	2.10*	1.8	2.05*	1.95	2.01*
3	Soygarri	1.35	1.65	1.45	1.65	1.80	1.58
4	Soy cooking oil	2.80*	2.40*	2.75*	2.90*	2.75*	2.72*
5	Fried soyabean flour	2.15*	2.40*	2.05*	2.25*	2.05*	2.18*

*, Preferred product/processing innovation. Source: Field survey, 2006.

policy (x = 2.15), lack of capital (x = 2.70), lack of markets (x = 2.35), lack of competent human capital (x = 2.25), lack of research and extension (x = 2.15), lack of patency (x = 2.0), lack of regulation in the innovation system (x = 2.15), lack of functional processing industries (x = 2.86) and lack of adequate stakeholder interaction (x = 2.00) (Table 3). The lack of capital was the most serious problem hindering innovation among the critical actors. Lack of finance could deter actors from engaging in new areas to innovate. The implication of this finding is that the listed factors were probably responsible for the few innovations introduced in the soyabean innovation system in Benue State. Therefore, to enhance innovation among the critical actors in the soyabean innovation system in Benue State, the identified serious factors must be mitigated.

According to the respondents in the Enterprise component (Table 3) indicated that the serious factors hindering innovation in the soyabean innovation system in Benue State were: lack of new knowledge (x = 2.35), lack of research policy (x = 2.50), lack of capital (x = 2.55), lack of markets (x = 2.05), lack of competent human capital (x = 2.05), lack of research and extension (x = 2.30), lack of networking (x = 2.15), lack of patency (x = 2.30), lack of regulation in the innovation system (x = 2.40), and lack of functional processing industries (x – 2.65). Lack of finance was again identified as the serious hindering factor to innovation. This is probably because commercial banks tend to give loans based on collateral which the small scale innovators cannot guarantee. Therefore policy option to make credit facility available to small scale processors should be considered by stakeholders interested in promoting innovations in the sayabean innovation system in Benue State.

The factors hindering innovation as perceived by actors in the diffusion component were: lack of research policy (x = 2.15), lack of capital (x = 3.10), lack of markets (x = 2.40), lack of research and extension (x = 2.00), lack of regulation in the innovation system (x = 2.10), lack of functional processing industries (x = 2.65), lack of adequate stakeholder interaction (x = 2.10).

The factors in this study may have been responsible for the low performance of the soyabean innovation in Benue State. The improved innovative performance of the system policy options must be put in place to alleviate the identified factors hindering innovations as identified by actors in the diffusion component. The factors identified by actors in the research component were: lack of new knowledge (x = 2.70), lack of research policy (x = 2.45), lack of capital (x = 3.10), lack of markets (x = 2.40), lack of competent human capital (x = 2.15), lack of research and extension (x = 2.60), lack of patency (x = 2.35), lack of regulation in the innovation system (x = 2.55), lack of functional processing industries (x = 2.90) and lack of adequate stakeholder interaction (x = 2.55). These factors are probably responsible for the few innovations introduced in the soyabean innovation system. Policy options should therefore be put in place to mitigate the factors hindering innovation for effective performance of the innovation system.

Respondents in the infrastructure component identified the following factors as responsible for hindering innovation in the soyabean innovation system in Benue State: lack of new knowledge (x = 2.25), lack of research policy (x = 2.45), lack of capital (x = 2.75), lack of markets (x = 2.53), lack of competent human capital (x = 2.25), lack of net working (x = 2.00), lack of regulation in the innovation system (x = 2.40), lack of functional processing industries (x = 2.70), and lack of adequate stakeholder interaction (x = 2.00). These factors were probably responsible for the low interaction among actors in the infrastructure component of soyabean innovation in Benue State. To improve the innovative performance of the infrastructure component and hence the soyabean innovation systems policy options to ameliorate these hindering factors must be given adequate consideration.

The factors identified in each component were probably responsible for the low total interaction score (TIS) by the five components and hence few innovations introduced in the innovation system. The factors hindering the soyabean innovation system in Benue State are similar to those identified by John and Jacobsson (2002) in their studies on innovation in the German, Dutch and Swedish wind turbine industries. It is probable that factors that hinder innovation in industries are similar to those that hinder innovation in agricultural innovation systems. Policy options should be considered by policy makers and implementers to alleviate these serious factors in

Table 3. Mean factors hindering innovation among critical actors in the soyabean innovation system in Benue State. (n=100).

S/N	Factor	Mean (X) score of components					
		Demand	Enterprise	Diffusion	Research	Infrastructure	Mean (X)
1	Lack of new knowledge	2.40*	2.35*	1.90	2.70*	2.25*	2.32*
2	Lack of research policy	2.15*	2.50*	2.05*	2.15*	2.45*	2.26*
3	Lack of capital	2.70*	2.55*	2.80*	3.10*	2.75*	2.78*
4	Lack of markets	2.35*	2.05*	2.40*	2.40*	2.53*	2.34*
5	Lack of competent human capital	2.25*	2.05*	1.85	2.15*	2.25*	2.11*
6	Lack of research and extension	2.15*	2.30*	2.00*	2.60*	2.25*	2.26*
7	Lack of networking	1.85	2.15*	1.8	1.85	2.00*	1.93
8	Lack of patency	2.20*	2.30*	1.95	2.35*	1.80	2.12*
9	Lack of regulation in the innovation system	2.15*	2.40*	2.10*	2.55*	2.40*	2.32*
10	Lack of functional processing industries	2.80*	2.65*	2.65*	2.90*	2.70*	2.74*
11	Lack of adequate stakeholder interaction	2.00*	1.85	2.10*	2.55*	2.47*	2.19*

Source: Field survey, 2006,*serious factor.

order to enhance innovative performance of the soyabean innovation system in Benue State.

Conclusion

As a result of the major findings of the study the following conclusions were drawn:

(1) The major innovation introduced in the soyabean innovation system were soyabean milk and soyabean oil in the product/processing innovation category, the successful trial of improved high yielding, disease resistant, non pod shattering hybrid seeds (TGX – 1448 – 2E) and appropriate fertilizer type and rate recommendation in the basic research innovation category and the compulsory soyabean producers cooperative membership and the marketing strategy to by-pass middlemen in the sale of soyabean seeds.
(2) Among the product/processing innovation introduced in the soyabean innovation system, soymilk was the most preferred innovation and was followed by soy cooking oil; while soya garri was least preferred.
(3) Ten factors were implicated as major factors hindering innovation in the soyabean innovation system. The major factors were lack of new knowledge, lack of research policy, lack of capital, lack of markets, lack of competent human capital, lack of research and extension, lack of patency, lack of regulation in the innovation system, lack of adequate stakeholders interaction and functional processing industries.

RECOMMENDATIONS

(1) Policy makers and implementers should mitigate the ten factors highlighted as major factors hindering innovation in the soyabean innovation system. Specifically, by encouraging the creation of new knowledge and research through legislation of private sector funding of research and training of competent manpower, legislation of funding arrangement that will make aggressive private capital available from banks and interested bodies for establishment of soyabean processing industries. Also provision of tax holidays and infrastructure to encourage the investment in this regard. Also policies to generally improve markets, infrastructures, market information, information and communication technology (ICT) infrastructure and private sector funding of soyabean activities should be put in place and implemented. Also, exportation of soyabean should be encouraged by legislation to ensure that banking facilities are made available for investors interested in soyabean exportation. Above all, farmers who are capable of cultivating soyabean on a large scale and traditional soyabean farmers are encouraged through provision of inputs and extension services and insurance cover.
(2) The innovation introduced in the system

should be nurtured and patented by the regulatory bodies. This will encourage further innovation in the soyabean innovation system.

(3) The four product innovation, given high acceptability ratings by respondents shall be guaranteed funding by commercial banks to commercialize them to stimulate further innovation in the system.

REFERENCES

Abagu JI (2004). Address by the Honourable commissioner for commerce and Industries at Makurdi International Trade Fair.

Agbulu ON, Wombo AB (2005). Indigenous Vs modern approaches to processing storage and marketing of soyabean for Hanger and poverty reduction in Benue State. In: C.P.O Obinne (ed): Readings on indigenous Processing, Storage and Marketing for Poverty Reduction in Nigeria: pp. 33-44.

Ayoola GB (2001). Essays on the Agricultural Economy: A book of Readings on Agricultural Policy and Administration in Nigeria. Vol. 1 Ibadan, Nigeria, TMA Publishers.

Benue State Agricultural and Rural Development Authority (BNARDA) (2000). Agricultural Production recommendation for Benue State. Ext. Bull. P. 3.

CTA/UNU-INTECH/KIT (2005). African Sub-Regional Workshop, West and Central Africa: Analysing their agricultural science, technology and innovation system (ASTI) held at Abia State University, Nigeria. 18th-12th August, 2005. The Technical Centre for Agriculture and Rural Cooperation/United Nations University-Institute for New Technologies/Koninklijk Instituut voor de tropens (CTA/UNU-INTECH/KIT).

Conroy C (2003). New Directions for Nigeria's Basic Agricultural Services: A Discussion Paper. (1):63 National Stakeholders Workshop on Basic Agricultural Services held in Abuja, 18th - 19th November, 2003.

Konin Klijk Instituut Vor de tropen (KIT) (2005). Agricultural Innovation Systems, Multi-stakeholder processes and client-oriented delivery of agri-services. Practical Experiences of the Royal Tropical Institute with facilitating rural innovation processes in Africa Development Policy and Practices, Amsterdam, February 2005.

Lawal WL, Chisonum MO, Ater PI (2005). Gross Margin analysis of Indigenous Processing of Soybeans in Benue State: In: C.P.O Obinne (ed): Readings on indigenous processing, storage and marketing for poverty reduction in Nigeria. pp. 123-131.

Low External Input for Sustainable Agriculture (LEISA) (2000). Unleashing the creativity of farmers. LEISA Mag. 2(2):1-6.

Correlation analysis of some agronomic traits for biomass improvement in sorghum (*Sorghum Bicolor* L. Moench) genotypes in North-Western Nigeria

Lawali Abubakar[1] and T. S. Bubuche[2]

[1]Department of Crop Science, Faculty of Agriculture, Usmanu Danfodiyo University, P. M. B. 2346, Sokoto State, Nigeria.
[2]Department of Agricultural Education, College of Education, P. M. B. 1012, Argungu, Kebbi State, Nigeria.

Ten Sorghum (*Sorghum bicolor* L. Moench) land races of North Western, Nigerian origin were evaluated during 2010 at Usmanu Danfodiyo University, teaching and research farm, Sokoto, Sokoto State and during the 2011 rainy season at Bubuche in Augie Local Government Area, Kebbi State in North-Western Nigeria. The objective of the study was to determine the correlation between the traits of sorghum that have direct bearing with biomass improvement. The procedure outlined in the IBGR/ICRISAT sorghum descriptor was used to measure each trait. Results obtained revealed that, leaf number (LN) has significant positive correlation with plant height (r=0.275), leaf area index (LAI) (r=0.308), and straw weight (STRW) (r=0.433) but negative significant correlation with flag leaf area (FLA) (r= -0.401). Leaf length (LL) had significant positive correlation with flag leaf length (FLL) (r= 0.299) and straw weight (STRW) (r=0.516). Plant height (PH) recorded significant positive correlation with leaf number (LN) (r=0.275) and STRW (r=0.360) but negatively correlated with FLL (r=0.118) and total grain yield (TGY) (r=0.102). LAI has significant positive with LN (r=0.308), PH (r=0.274) and negatively correlated with hundred grain weight (HUNDGWT) (r= -0.158) and TGY (r= -0.190). FLA had significant positive correlation with only FLL (r=0.266) and significantly negatively correlated with LN (r= -0.401), and STRW (r= -0.540). STRW has the significant and positive correlation with LN (r=0.433), with LL (r=0.516) and PH (r=0.360) but significantly and positively correlated with only FLA (r= -0.540). Selection for straw weight can therefore be carried out simultaneously with plant height, leaf length and leaf number.

Key words: Correlation, agronomic, traits, biomass, sorghum.

INTRODUCTION

Sorghum (*Sorghum bicolor* L. Moench) is one of the most important food crops of the world (Mahajan et al., 2011), and provide bulk of raw materials for the livestock and many agro-allied industries in the world (Dogget, 1970). It is drought tolerant which allows farmers to use one third less water than similar crops in its cultivation (Kumar et al., 2012). Sorghum is indigenous to Africa, and many of today's varieties originated on that continent. Sorghum was also grown in India before recorded history and in Assyria as early as 700 BC. The crop reached China during the thirteenth century and the Western Hemisphere much later (Undersander et al., 2013).

Sorghums in general can be classified into two types: Forage types (mainly for forage or animal feed) and grain types (mainly for human consumption). The forage sorghums are further grouped into four types: (a) hybrid forage sorghum, (b) sudangrass, (c) sorghum x sudan hybrids (also known as sudan hybrids), and (d) sweet sorghum. The latter is used mainly for molasses but more recently for biofuel production as well (Newmann et al., 2010).

The estimated world production of grain sorghum in 2011 was 4,198,010 tonnes, with an average yield of 15, 274 kg/ha produced over 35,482,800 ha (FAOSTAT, 2011). In Africa 20,780,959 tonnes were produced at an average yield of 10,623 kg/ha across 19,561,929 ha during the same period. Nigeria was ranked second in sorghum production worldwide after India in 2011 (FAOSTAT, 2011), with production of 6,897,060 tonnes, harvested from 4, 891,150 ha at with an average yield of 14,101 kg/ha.

Yield is a complex character, which depends upon many independent contributing characters. Knowledge on type of association between yield and its components themselves greatly help in evaluating the contribution of different components towards yield, information on the nature of association between yield and its components help in simultaneous selection for many characters associated with yield improvements (Kumar et al., 2012). To determine relationships, correlation analyses are used such that the values of two characters are analyzed on a paired basis, results of which may be either positive or negative. When there is positive association of major yieldcharacters component breeding would be very effective but when these characters are negatively associated, it would be difficult to exercise simultaneous selection for them in developing a variety (Nemati et al., 2009).

Grain yield and fodder production in sorghum are complex characters controlled by many genes, therefore their improvement will lead to increased overall productivity in sorghum production (Sadia et al., 2001). To exploit the potentiality of sorghum therefore several crop improvement programmes have been undertaken (Mahajan et al., 2011). There is therefore the need to study the relationship of sorghum characters for all breeding objectives, as this will facilitate selection.

The objective of the study was therefore to determine the correlation between the traits of Sorghum land races of North Western Nigerian origin that have direct bearing with biomass improvement, with a view of using the result as a basis for selection in conventional breeding programmes.

MATERIALS AND METHODS

Ten local sorghum varieties were evaluated during 2010 the Usmanu Danfodiyo University, teaching and research farm, Sokoto, Sokoto State and during the 2011 rainy season at Bubuche in Augie Local Government Area of Kebbi State all in North-Western Nigeria. Sokoto is located in the Sudan Savanna agro-ecological zone of Nigeria on latitude 13° 01 N; longitude 5° 15 E and altitude of about 350 m above sea level (ASL). Mean annual rainfall is about 752 mm. The minimum and maximum temperatures are 26 and 35°, respectively, and relative humidity of 23 to 41%. The area is characterized by long dry season with cool air during Hammattan (November – February), dry air during hot season from March – May followed by a short rainy season, (Bello, 2006) and Bubuche is located in Augie local government Kebbi State on latitude 13° 05 N; longitude 4° 12 E and altitude of 345 m above the sea level, temperature ranges from 27 to 34° and relative humidity of 24 to 44% with mean annual rainfall of 6700 to 7600 mm (Kebbi State, 2009). The texture of the soil was loamy sand and the soil is deep, loose and well drained, chemical analysis shows that the soil is slightly acidic, low to medium in organic carbon, low total nitrogen, low exchangeable cat ions low in cat ions exchange capacity (CEC), very low available P and K Ca and Mg contents and low bulk density (Table 1).

Planting materials

The materials used in this study consisted of ten indigenous grain sorghum genotypes representing the widely grown Sorghum types in North-Western Nigeria (Table 2) which were collected by National Center for Genetic Resources and Biotechnology (NACGRAB), Moor plantation, Ibadan, Nigeria.

The experiment was laid out in a Randomized Complete Block Design (RCBD) in three replications. Each pot size was 6 × 3 m, 75 cm as inter row spacing and intra-row spacing of 30 cm and a total of 240 plants per plot after thinning. Before sowing, seeds were treated with Apron-plus 3 g/kg seed against soil fungi and insects. Sowing was on 10th of June, 2010 and 2011 respectively. Five seeds were sown in each hole. Seedlings were thinned to three plants per hole after three weeks from sowing. Hand hoeing weeding was done thrice; the first one at two weeks after sowing and the subsequent weeding were carried out at three weeks interval each.

Data collection

Data were collected on days to 50% flowering (DF), plant height (PH), leaf length (LL), leaf number (LN), leaf area index (LAI), flag leaf area (FLA), flag leaf length (FLL), total grain yield (TGY), 100-grain weight (HGW) and total biomass weight (STRAW) were recorded at both locations and during both seasons. The data was collected according to standard procedures described in the IBPGR/ICRISAT (1993) Sorghum descriptor. LAI was calculated on the basis of the length and width of the third top leaf multiplied by the coefficient of 0.71 (Krishnamurthy et al., 1974).

Statistical analysis

Data were subjected to analysis of variance using SAS ver. 9.1 (SAS, 2004) to estimate variance for all traits. All factors (accession, block, and environment) were treated as random variables. The Pearson correlation coefficient was calculated for every pair of traits using the PROC CORR procedure.

Correlation

The correlation estimate was carried out using the formulae described by (Wright, 1921). Phenotypic correlation:

$$rp_{xy} = \frac{\sigma p_{xy}}{\sqrt{(\sigma^2 p_x)(\sigma^2 p_y)}}$$

Table 1. Physical and chemical properties of the soil at two experimental sites (Sokoto, 2010; Bubuche, 2011) cropping seasons.

S/N	Properties	Sokoto (2010)	Bubuche (2011)
	Physical properties		
1	Sand (g kg^{-1})	890	517
2	Silt (g kg^{-1})	45	377
3	Clay (g kg^{-1})	43	103
4	Textural class	Sand	Sandy loam
	Chemical properties		
4	Soil pH (H_2O) 1:2	5.90	5.80
5	Soil pH ($CaCl_2$) 1:2	5.40	5.30
6	Organic carbon (g kg^{-1})	2.40	1.76
7	Total Nitrogen (g kg^{-1})	0.40	0.66
8	Available P (mg kg^{-1})	0.80	0.76
9	Exchangeable K (Cmol kg^{-1})	0.44	0.58
10	C.E.C. (Cmol kg^{-1})	5.50	11.80
11	Na (Cmol kg^{-1})	0.43	0.50
12	Ca (Cmol kg^{-1})	0.35	0.31
13	Mg (Cmol kg^{-1})	208	0.51

Source: Agric. Chemical Lab. UDUS.

Table 2. Sorghum landraces used in the study.

S/N	Name	Area collected	Grain colour	Major use
1	Zago.Ex-BATSARI	Katsina State	Brown	Food
2	NG/SA/07/005	Niger State.	White	Food
3	NG/SA/07/125	Zamfara State	White.	Food
4	NGB/06/001	Kaduna State	White	Food
5	NG/SA/DEC/07/0049	Niger State	White	Food
6	NG/SA/DEC/07/0108	Niger State	White	Food
7	NG/SA/DEC/07/0213	Kaduna State	White	Food
8	NG/SA/DEC/070123	Kano State	White	Food
9	NG/SA/DEC/07/0036	Niger State	White	Food
10	EX-ARGUNGU (Kaura)	Kebbi State	Red	Food

Where, rp_{xy} = Phenotypic correlation coefficients between traits x and y; σp_{xy} = phenotypic variance of traits x and y; σp_x = Phenotypic variance of trait x; σp_y = Phenotypic variance of trait y.

Genotypic correlation

$$rg_{xy} = \frac{\sigma g_{xy}}{\sqrt{(\sigma^2 g_x)(\sigma^2 g_y)}}$$

Where, rg_{xy} = Genotypic correlation coefficients between traits x and y; σg_{xy} = genotypic covariance of traits x and y; σg_x = genotypic variance of traits x and y; σg_y = genotypic variance of trait y.

Environmental correlation

$$\text{Environmental correlation } (r_e) = \frac{S^2{}_e X.Y}{\sqrt{(S^2_e X)(S^2_e Y)}}$$

Where, $S^2{}_e$ X.Y =environmental correlation between traits x and y; $S^2{}_e$ X = environmental variance of the traits x; $S^2{}_e$ Y = environnemental variance of traits y.

RESULTS AND DISCUSSION

Knowledge of the relationship among yield components is essential for the formulation of breeding programmes aimed at achieving the desired combinations of various components of yield. The estimates of correlation

Table 3. Performance of sorghum varieties evaluated at Usmanu Danfodiyo University Sokoto teaching and research farm during 2010 rainy season..

Genotypes	LN (cm)	LL (cm)	PH (cm)	LAI (cm2)	FLA (cm^2)	FLL (cm)	STRW (kg/ha)	100-SWT (g)	TAY (kg/ha)
Zago Ex-Batsari	14.357^a	76.793^{ab}	124.84^b	0.8667^{ab}	41.433^a	32.980^{ab}	5.203^{bc}	4.533^b	381.9^b
NG/SA/07/005	14.183^a	75.533^{ab}	184.60^a	2 0333^a	45.810^a	29.237^b	7.693^{ab}	5.867^{ab}	353.^{ab}
NG/SA/07/125	13.130^a	80.180^a	194.55^a	0.7667^c	35.970^a	29.370^b	7.207^{ab}	9.767^{ab}	1154.10^{ab}
NGB/06/001	14.267^a	75.733^a	183.17^a	1.9000^{ab}	46.003^a	35.673^{ab}	8.473^a	12.667^{ab}	528.5^{ab}
NG/SA/DEC/07/0049	14.690^a	79.267^a	185.38^a	1.2000^{abc}	38.370^a	32.233^{ab}	7.493^{ab}	12.067^{ab}	1629.2^a
NG/SA/DEC/07/0108	14.707^a	74.700^{ab}	190.02^a	0 7667^c	41.267^a	31.087^{ab}	7.170^{ab}	14.863^a	400.2^{ab}
NG/SA/DEC/07/0213	12.767^{ab}	78.133^{ab}	171.27^{ab}	1.6833^{abc}	38.727^a	34.943^{ab}	7.000^{ab}	8.213^{ab}	419.3^{ab}
NG/SA/DEC/07/0123	13.433^a	71.833^b	173.94^{ab}	1.6000^{abc}	46.447^a	34.883^{ab}	4.233^c	8.513^{ab}	280.2^b
NG/SA/DEC/07/ 0036	14.400^a	75.500^{ab}	181.90^a	1.2000^{abc}	39.367^a	33.363^{ab}	7.890^a	13.370^{ab}	314.7^b
EX-Argungu (Kaura)	10.633^b	72.733^b	175.93^{ab}	1.2333^{abc}	44.867^a	38.633^a	3.357^c	10.288^{ab}	387.3^b

Mean with the same letter(s) in a column are not significantly different at 5% level of significance according to DMRT (Duncan's multiple range tests). LN, leaf number; LL, leaf length in cm; PH, plant height in cm; LAI, leaf area index cm^2; FLA, flag leaf area in cm^2; FLL, flag leaf length in cm; STRW, straw weight in kg; 100-SWT, 100-seed weight in g; TGY, total grain yield in kg/ha .

coefficients among different characters indicate the extent and direction of association. The correlation co-efficients provide a reliable measure of association among the characters and help to differentiate vital associations useful in breeding from those of the non-vital ones (Falconer, 1981).

Results of the evaluation at Sokoto during the 2010 rainy season indicated that there was significant ($P \leq 0.05$) differences between the varieties with respect to leaf number/plant, leaf length, plant height, leaf area index, flag leaf length, straw weight, 100 seed weight and total grain yield (Table 3). There was however, non significant difference between the varieties in terms of flag leaf area. Evaluation of the sorghum varieties in Bubuche during 2011 rainy season showed significant ($P \leq 0.05$) difference between the varieties in terms of leaf number/plant, leaf length, plant height, flag leaf area, flag leaf length and 100 seed weight. However, with non significance difference in leaf area index, straw weight and total grain yield (Table 4). Evaluation of the sorghum varieties in Bubuche during 2011 rainy season showed significant ($P \leq 0.05$) difference between the varieties in terms of leaf number/plant, leaf length, plant height, flag leaf area, flag leaf length and 100 seed weight. However, with non significance difference in leaf area index, straw weight and total grain yield (Table 4).

The combined analysis of the results across season and locations indicated significant difference ($P \leq 0.05$) between the varieties in all the characters studied (Table 5). The result is not surprising as Maarouf and Moataz (2009) had reported variation among sorghum genotypes developed for forage production.

Correlation analysis of the combined results revealed that leaf number (LN) has significant positive correlation with plant height (r=0.275), leaf area index (LAI) (r=0.308), and highly significant positive correlation with straw weight (STRW) (r=0.433) but negative significant correlation with flag leaf area (FLA) (r= -0.401) (Table 6). This suggests selection for leaf number, plant height, leaf area index and straw weight can be carried out simultaneously, with however, an inverse selection pattern between leaf number and flag leaf area. Tesso et al. (2011) reported that, leaf number, leaf length and leaf area index had high significant and positive correlation with straws weight, plant height had high negative correlation with days to 50% flowering and 100-grain weight.

The study also revealed that straws weight have highly significant and positive correlation with plant height (0.360), leaf length (0.516), indicating that selection for the traits can be carried out simultaneously. Also Kumar et al. (2012) reported that days to 50% flowering showed positive significant correlation with stover yield per plant, and plant height showed positive significant association with 100- seed weight and stover yield per plant.

Leaf length (LL) had significant positive correlation with only flag leaf length (FLL) (r= 0.299) and highly significant positive correlation with straw weight (STRW) (r=0.516). El Naim et

Table 4. Mean performance of Sorghum genotypes evaluated at Bubuche Augie local government, Kebbi State during 2011 rainy season.

Genotypes	LN (cm)	LL (cm)	PH (cm)	LAI (cm2)	FLA (cm^2)	FLL (cm)	STRW (kg/ha)	100-SWT (g)	TAY (kg/ha)
Zago Ex-Batsari	10.33^{ab}	62.567^{abc}	237.60^{ca}	0.7333^{b}	125.13^{ab}	31.067^{abcd}	3.480^{a}	11.417^{abc}	1751^{a}
NG/SA/07/005	13.433^{a}	47.600^{c}	142.51^{bcde}	0.9333^{b}	88.43^{b}	21.967^{d}	3.117^{a}	9.630^{abc}	362^{a}
NG/SA/07/125	$8.567.^{ab}$	66.067^{ab}	61.73f	0.9333^{b}	207.43^{a}	40.533^{a}	3.073^{a}	16.177^{a}	3374^{a}
NGB/06/001	9.833^{ab}	70.300^{ab}	199.53^{ab}	1.2333^{b}	142.67^{ab}	33.667^{abc}	3.080^{a}	14.560^{ab}	750^{a}
NG/SA/DEC/07/0049	7.900^{b}	79.133^{a}	94.38^{e}f	0.3333^{b}	191.30^{ab}	39.000^{ab}	4.070^{a}	7.370^{bc}	603^{a}
NG/SA/DEC/07/0108	11.833^{ab}	69.667^{ab}	161.88^{bcd}	2.3000^{a}	223.20^{a}	39.000^{ab}	2.630^{a}	8.590^{abc}	425^{a}
NG/SA/DEC/07/0213	12.533^{ab}	68.767^{ab}	186.58^{abc}	1.1333^{b}	173.60^{ab}	33.733^{abc}	5.297^{a}	4.417^{c}	370^{a}
NG/SA/DEC/07/0123	9.000^{ab}	61.100^{bc}	133.14^{cde}	0.7333^{b}	123.90^{ab}	30.867^{abcd}	2.355^{a}	10.333^{abc}	336^{a}
NG/SA/DEC/07/0036	11.467^{ab}	60.400^{bc}	133.51^{de}f	0.9667^{b}	142.13^{ab}	28000^{cd}	2.963^{a}	9.037^{abc}	401^{a}
EX-Argungu (Kaura)	11.800^{ab}	52.667^{bc}	162.94^{bcd}	1.2333^{b}	124.60^{ab}	30.500^{cbd}	2.263^{a}	8.297^{abc}	1395^{a}

Mean with the same letter(s) in a column are not significantly different at 5% level of significance according to DMRT; LN, leaf number; LL, leaf length in cm; PH, plant height in cm; LAI, leaf area index cm^2; FLA, flag leaf area in cm^2; FLL, flag leaf length in cm; STRW, straw weight in kg; 100-SWT, 100-seed weight in g; TGY, total grain yield in kg/ha.

Table 5. Combined mean performance of Sorghum genotypes evaluated across the locations during 2010 and 2011 rainy seasons.

Genotypes	LN (cm)	LL (cm)	PH (cm)	LAI (cm2)	FLA (cm^2)	FLL (cm)	STRW (kg/ha)	100-SWT (g)	TAY (kg/ha)
Zago Ex-Batsari	12.345^{ab}	9.680^{bc}	181.22^{ab}	0.8000^{cd}	83.28^{ab}	32.023^{a}	4.3417^{abc}	7.975^{ab}	106.66^{ab}
NG/SA/07/005	13.808^{a}	61.567^{c}	163.56^{abcd}	1.483^{abc}	67.12^{b}	25.602^{b}	5.4050^{a}	7.748^{ab}	358.10^{b}
NG/SA/07/125	10.848^{b}	73.123^{ab}	128.14^{d}	0.8500^{abc}	121.70^{ab}	34.952^{a}	5.1400^{ab}	12.972^{a}	2264.0^{a}
NGB/06/001	12.050^{ab}	73.017^{ab}	191.35^{a}	1.5667e	94.34^{ab}	34.670^{a}	5.7767^{a}	13.613^{a}	639.4^{b}
NG/SA/DEC/07/0049	11.295^{ab}	79.200^{c}	139.88^{cd}	0.7667^{d}	114.84^{ab}	35.983^{a}	5.7817^{a}	9.718^{ab}	1116.2^{ab}
NG/SA/DEC/07/0108	13.270^{ab}	72.183^{ab}	175.95^{abc}	1.533^{ab}	132.23^{a}	35.043^{a}	4.9000^{ab}	11.707^{ab}	472.7^{b}
NG/SA/DEC/07/0213	12.650^{ab}	73.440^{ab}	17.92^{abc}	1.3833^{abcd}	106.16^{ab}	34.338^{a}	6.1483^{a}	6.315^{b}	394.6^{b}
NG/SA/DEC/07/0123	11.217^{ab}	66.467^{ab}	153.54^{abcd}	1.667^{abcd}	85.17^{ab}	32.875^{a}	3.2933^{bc}	9.423^{ab}	808.3^{b}
NG/SA/DEC/07/0036	12.933^{ab}	67.950^{bc}	147.71^{bcd}	1.0833^{abcd}	90.75^{ab}	31.082^{ab}	5.4267^{a}	11.203^{ab}	357.7^{b}
EX-Argungu (Kaura)	10.848^{b}	62.700^{bc}	169.44^{abc}	1.233^{abcd}	84.78^{ab}	34.567^{a}	2.800^{c}	9.290^{ab}	891.3^{ab}

Mean with the same letter(s) in a column are not significantly different at 5% level of significance according to DMRT. LN, leaf number; LL, leaf length in cm; PH, plant height in cm; LAI, leaf area index cm^2; FLA, flag leaf area in cm^2; FLL, flag leaf length in cm; STRW, straw weight in kg; 100-SWT, 100-seed weight in g; TGY, total grain yield in kg/ha.

al. (2012) reported that Head weight (g) had highly significant and positive correlation with hay weight, plant height, number of head per plot and highly significant negative correlation with days to 50% flowering, 100 grain weight. They reported that hay weight had highly significant positive correlation with plant height, yield weight, number of head per plot, it also had highly significant negative correlation with 100 grain weight. Plant height was also reported by the authors to have highly significant positive correlation with yield /ha. Plant height (PH) was significantly positively correlated with leaf area index (LAI) (r=0.274) and highly significant positive correlation with straw weight (STRW) (r=0.360). Flag leaf area (FLA)

Table 6. Correlation relationship between characters of Sorghum evaluated at Sokoto (Sokoto State) and Bubuche (Kebbi State).

Parameter		LN	LL	PH	LAI	FLA	FLL	STRAW	HUNDGWT	TGY
LN	Pearson correlation	1	0.250	0.275*	0.308*	-0.401**	-0.219	0.433**	-0.051	-0.122
	Sig. (2-tailed)		0.054	0.033	0.017	0.002	0.093	0.001	0.699	0.351
	N		60	60	60	60	60	60	60	60
LL	Pearson correlation	0.250	1	0.196	0.233	-0.127	0.299*	0.516**	0.110	0.055
	Sig. (2-tailed)	0.054		0.133	0.074	0.333	0.020	0.000	0.401	0.675
	N	60	60	60	60	60	60	60	60	60
PH	Pearson correlation	0.275*	0.196	1	0.274*	-0.233	-0.118	0.360**	0.027	-0.102
	Sig. (2-tailed)	0.033	0.133		0.034	0.073	0.368	0.005	0.837	0.439
	N	60	60	60	60	60	60	60	60	60
LAI	Pearson correlation	0.308*	0.233	0.274*	1	0.081	0.234	0.159	-0.158	-0.190
	Sig. (2-tailed)	0.017	0.074	0.034		0.541	0.072	0.226	0.227	0.147
	N	60	60	60	60	60	60	60	60	60
FLA	Pearson correlation	-0.401**	-0.127	-0.233	0.081	1	0.266*	-0.540**	0.021	0.076
	Sig. (2-tailed)	0.002	0.333	0.073	0.541		0.040	0.000	0.876	0.566
	N	60	60	60	60	60	60	60	60	60
FLL	Pearson correlation	-0.219	0.299*	-0.118	0.234	0.266*	1	-0.035	-0.017	0.019
	Sig. (2-tailed)	0.093	0.020	0.368	0.072	0.040		0.791	0.897	0.887
	N	60	60	60	60	60	60	60	60	60
STRAW	Pearson correlation	0.433**	0.516**	0.360**	0.159	-0.540**	-0.035	1	0.022	-0.018
	Sig. (2-tailed)	0.001	0.000	0.005	0.226	0.000	0.791		0.868	0.893
	N	60	60	60	60	60	60	60	60	60
HUNDGWT	Pearson correlation	-0.051	0.110	0.027	-0.158	0.021	-0.017	0.022	1	0.154
	Sig. (2-tailed)	0.699	0.401	0.837	0.227	0.876	0.897	0.868		0.241
	N	60	60	60	60	60	60	60	60	60
TGY	Pearson correlation	-0.122	0.055	-0.102	-0.190	0.076	0.019	-0.018	0.154	1
	Sig. (2-tailed)	0.351	0.675	0.439	0.147	0.566	0.887	0.893	0.241	
	N	60	60	60	60	60	60	60	60	60

*. Correlation is significant at 0.05, ** Correlation is significant at 0.01.

recorded significant positive correlation with flag leaf length (FLL) (r=0.266) and negative significant correlation with straw weight STRAW (r = -0.54) (Table 5). Straw weight can therefore be selected along with leaf length, flag leaf length,flag leaf area, while the length of the flag leaf has an inverse relation with straw weight. Khaliq et al. (2008) reported that flag leaf area was positively correlated with flag leaf length and play a vital role in drought tolerance. Breeding for biomass in sorghum can be carried out based on the character association observed most especially when breeding for livestock.

Conclusions

Results of this study confirmed that several traits are directly or indirectly associated with biomass yield which is important in breeding for biomass yield improvement in sorghum. The study concludes that selection for straw weight is also selection for traits such as plant height, leaf length and leaf number as they add to the final straw weight which is very important in selection for biomass yield improvement.

ACKNOWLEDGEMENT

We wish to acknowledge the National Center for Genetic Resource and Biotechnology (NACGRAB), Moor Plantation, Ibadan, Nigeria for supplying the collected varieties for this research.

REFERENCES

Bello MS (2006). Effect of spacing and Potassium on growth and yield of sweet Potato (*Ipomoea batatas* (L.) LAM) in the Sudan savanna of Nigeria. Unpublished M sc. thesis Submitted to the Post graduate School of UsmanuDanfodiyo University Sokoto. P. 85.

Dogget H (1970). Sorghum. Longmans, London, P. 403

El Naim AM, Ibrahim MI, Abdel Rahman1 ME, Ibrahim EA (2012). Evaluation of Some Local Sorghum (*Sorghum bicolor* L. Moench) Genotypes in Rain-Fed. Int. J. Plant Res. 2(1):15-20.

Falconer DS (1981). *Introduction to Quantitative Genetics.* Second edition. Longman, New York.

FAOSTAT Data (2011). Food and Agriculture Organization of the United Nations.http://wwwfaostat.org.

IBPGR and ICRISAT (1993). *Descriptors for sorghum* [*Sorghum bicolor* (L.) Moench]. International Board for Plant Genetic Resources. Rome, Italy.–ICRISAT, Patancheru, India.

Kebbi State (2009). *Kebbi State Diary,* Nigeria. P. 218.

Khaliq I, Irshad A, Ahsan M (2008). Awns and flag leaf contribution towards grain yield in spring wheat (*Triticum aestivum* L.). Cer. Res. Commun. 36:65–76.

Krishnamurthy K, Jagannath MK, Rajashekara BG (1974). Estimation of leaf area in grain sorghum from single leaf measurements. Agron. J. 66:544– 545.

Kumar NV, Reddy CVCM, Reddy PVRM (2012). Study on Character Association in Rabi Sorghum (*Sorghum bicolor* L. Moench). Plant Arch. 12(2):1049–1051.

Maarouf IM, Moataz AM (2009). Evaluation of New Develpoed Sweet Sorghum (*Sorghum bicolor*) Genotypes for some forage Attributes. American-Eurasian J. Agric. Environ. Sci. 6(4):434-440.

Mahajan RC, Wadikar PB, Pole S P, Dhuppe MV (2011). Variability, Correlation and Path Analysis Studies in Sorghum. Res. J. Agric. Sci. 2(1):101-103.

Nemati A, Sedghi M, Sharifi RS, Seiedi MN (2009). Investigation of correlation between traits and path analysis of Corn (*Zea mays* L.) grain yield at the climate of Ardabil region (Northwest Iran). Not. Bot. Hort. Agrobot. Cluj). 37(1):194-198.

Newman Y, Erickson J, Vermerris W, Wright D (2010). Forage Sorghum (*Sorghum bicolor*): Overview and Management. University of Florida IFAS Extension. http://edis.ifas.ufl.edu

Tesso T, Tirfessa A, Mohammed H (2011). Association between morphological traits and yield components in the durra sorghums of Ethiopia. Hereditas 148(3):98-109.

Sadia A, Asghar A, Qamar I A, Arshad M, Salim S (2001). Correlation of economically important traits in sorghum varieties. Department of Biologycal Science, University of Arid agriculture. Pak. J. Biol. Sci. 1(5):330-331.

SAS (2004). Statistical Analysis system version 9.1 The PROC GLM procedure Surlan-momirovic G, Rankonjac, V, Pronovic S, Živanovic T (2005). Genetika i oplemenjivanje biljaka – praktikum, Beograd, pp. 231-242.

Undersander DJ, Smith LH, Kamiski AR, Kelling KA, Doll JD (2013). Sorghum-Forage.Alternative Field Crops Manual. University of Wisconsin Cooperative Extension. http://www.hort.purdue.edu/newcrop/afcm/index.html.

Wright S (1921). Systems of mating. Genetics 6:111-178.

19

Comparative study of maturity stages influenced by tomato under two different farming systems

V. Arumugam and E. Vadivel

Tamil Nadu Agricultural University Coimabtore 641003, Tamil Nadu, India.

In the present study, comparative advantages were computed to check on the existing status of cultivation practices with the precision farming practices. The study would be helpful to the tomato industries to manage their supply chain effectively by minimizing the cost and increasing the marketing efficiency and thereby enhancing their profit. Study was conducted in the Department of vegetable crops, Horticultural College and Research Institute, Tamil Nadu Agricultural University, Coimbatore, Tamil Nadu from 2007 to 2010. Field experiments were conducted in two different fields of villages at Saragapalli, Krishnagiri district to obtain the best results from harvesting of tomatoes at three different maturity stages from two different farming systems. Harvested fruits were transported from Hosur to Coimbatore. The treatment F_2M_1 (Harvesting at mature green stage of precision farming tomato) was found to be better because the lowest enzymatic activities with increased shelf life of 28.50 days.

Key words: Maturity stages, tomato, precision farming, post harvest, shelf life.

INTRODUCTION

Tomato is known for its rapid deterioration and poor shelf life due to diverse factors such as thin fruit peel, low firmness, slow rate of conversion of sugars, high pH, high rate of deterioration and high ethylene production. Tomato is being cultivated in India with the area of 826 million hectare, production of 16.53 million metric tonnes (National Horticulture Database, 2011). The shelf life is the resultant of pericarp thickness, acidity, ascorbic acid and TSS (Balasubramanian, 2008). Consequent to the introduction of precision farming system to Tamil Nadu, India, nearly 80% of the area under tomato was brought under this system and nearly 20% alone being cultivated under conventional system. Precision farming or satellite farming is a farming management concept based on observing and responding to intra-field variations.

Today, precision agriculture is about whole farm management with the goal of optimizing returns on inputs while preserving resources. In as much between 2006 to 2009 to 2009 to 2010 and tomatoes are predominantly being grown under precision system in the districts, enhanced productivity and extended harvest may lead to immediate market surplus. In order to sustain the productivity and profitability of tomato in the years to come, both the post harvest management of produce and supply chain management need to be studied together to address the issues of post harvest loss and to maximize the farmer's share of consumer price.

In this context, the present study on comparative study of maturity stages influenced by tomato growing under different production systems was taken up with the objective to study the effect of maturity stages as the elements of post harvest management of tomatoes grown under conventional and precision system of farming. Development of geomatics technology in the later part of the 20th century has aided in the adoption of site specific management systems using remote sensing, global positioning system and geographical information system. This approach is called precision farming or site specific management (Carr et al., 1991; Palmer, 1996). It is a paradigm shift from conventional management practice of soil and crop in consequence with spatial variability. It is a refinement of good whole field management, where management decisions are adjusted to suit variations in resource conditions.

MATERIALS AND METHODS

Tomato Syngenta Hybrid 516 was cultivated in different farming systems under open condition. Precision farming is getting differed from conventional farming by means of practicing advanced technologies like portray nursery, drip irrigation, fertigation, plant protection measures at appropriate time. Tomato fruits were harvested at different maturity stages of mature green (M1), turning stage (M2), and light red stage (M3). Factors considered as different production systems were conventional and precision farming. Farm practices for conventional and precision farming were followed as per Crop production guide, 2004 and Tamil Nadu Precision Farming guide. Biochemical analysis was carried out at the Department of Vegetable crops, Horticultural College and Research Institute, TNAU, Coimbatore. Coimbatore was represented as terminal market. The experimental data was analysed with factorial randomized block design as statistical design. Number of replications was 6.

The parameters considered for the experiment were physiological loss of weight (%) (PLW), shelf life (number of days) (shelf life was measured by using score chart based on color, size, shape, etc.), pectin methyl esterase enzymes activity. These parameters were estimated using the method suggested by A.O.A.C. (1975).

Physiological loss in weight

The initial weight of fresh fruit was recorded and subsequently the weights were taken on all days. The physiological loss in weight was estimated as given below and expressed as percentage:

$$\text{PLW (\%)} = \frac{\text{Initial weight of the fruit-Final weight of the fruit}}{\text{Initial weight of the fruit}} \times 100$$

Pectin methyl esterase

Pectin methyl esterase (PME) enzyme assay was done as per Ranganna (1986). The PME was assayed by the addition of 2 ml of 1.5 M NaCl to 10 ml of 1% protein solution. A few drops of Hinton's indicators were then added and it was titrated to pH 7.5 with 0.02 N NaOH. The mixture was transferred to constant temperature water bath maintained at 30°C. When the pectin solution has attained the temperature of bath, enzyme sample and water was added to adjust the volume to 20 ml. Immediately, time and volume of alkali required to maintain the pH value at the constant value was recorded and expressed the results in PME units (PME U/g of the expression of the ester hydrolysed per minute per gram of enzyme;

Statistical analysis

The data collected from the investigations were analysed by adopting the statistical procedures of Panse and Sukhatme (1985). The significance of the mean difference between the treatments was determined by computing the standard error and critical difference. Statistical analysis was done with MS Excel software.

RESULTS AND DISCUSSION

Physiological loss of weight

Tomatoes harvested at mature green stage (M_1) registered the lowest physiological loss of weight (PLW) and fruits harvested at light red stage recorded the highest PLW at 8, 16 and 24 days after storage (DAS). It ranged from 8.3 to 12.7% at 8 DAS, 17.3 to 24.7% at 16 DAS and 27.6 to 36.4% at 24 DAS (Table 1).

Significant differences were also observed among the treatment combinations of farming system and stage of harvest at 8, 16 and 24 DAS. Among the combinations, the PLW ranged from 7.53 to 13.65% at 8 DAS, 15.07 to 25.84% at 16 DAS and 24.61 to 37.12% at 24 DAS. Harvesting at mature green stage in the precision farming system (F_2M_1) resulted in significantly lower PLW at 8, 16 and 24 DAS. The highest PLW were recorded with light red stage in conventional farming system (F_1M_3) at 8, 16 and 24 DAS. F_2M_2 also recorded a lower PLW content at 8, 16 and 24 DAS. The fruit weight decreased gradually as the days to storage advanced. Moisture loss through transpiration represents the assessment of saleable weight and eventually the material becomes unusable as a result of wilting and shrinking. The loss is continuous during storage and is due to moisture loss (Balasubramanian, 2004).

In the present investigation, the PLW was highest in the fruits of light red stage maturity. This may be attributed to the fact that, the fruits harvested light red stage was in rapid growth, with a high rate of respiration accompanied by simultaneous loss of water from tissues. Differing stages of maturity can have considerable effect on water loss as the structure of outer layers change during the development of fruit. Maximum physiological loss in weight was recorded in fruits of conventional farming (34.77%) than the precision farming (30.90%) at 24 DAS. The results revealed that systems of irrigation and fertilization significantly influenced the growth characters of tomato. This might be due to the fact that frequent irrigation cum fertilization maintained at the root zone with well aerated condition and at adequate soil moisture content that did not fluctuate between wet and dry extremes (Patil and Janawade, 1999).

The PLW also increased as the period of storage advanced, irrespective of treatments. In the present study, the lowest weight loss was recorded in fruits of mature green stage from precision farming system and is continued throughout the storage period. The decrease in the fruit weight obtained during ripening would be the result of moisture loss as well as due to climacteric degradation of substrate (Salunkhe and Desai, 1984).

Table 1. Effect of maturity stages on the physiological loss of weight (%) of conventional and precision farming tomato at 8, 16 and 24 days after storage (DAS).

Treatments	8 DAS			16 DAS			24 DAS		
	F_1	F_2	Mean	F_1	F_2	Mean	F_1	F_2	Mean
M_1	9.11	7.53	8.32	19.46	15.07	17.26	30.65	24.61	27.63
M_2	12.45	9.32	10.89	24.25	20.22	22.24	36.54	32.46	34.50
M_3	13.65	11.69	12.67	25.84	23.45	24.65	37.12	35.64	36.38
Mean	11.74	9.51	10.63	23.18	19.58	21.38	34.77	30.90	32.84
	M	**F**	**M X F**	**M**	**F**	**M X F**	**M**	**F**	**M X F**
SEd	0.1069	0.0873	0.1512	0.2170	0.1772	0.3069	0.3334	0.2722	0.4715
CD (0.05)	0.2202	0.1798	0.3114	0.4470	0.3650	0.6321	0.6867	0.5607	0.9711

M1, Mature green; M2, breaker; M3, light red; F1, conventioanal farming; F2, precision farming.

Table 2. Effect of maturity stages on Shelf life (No. of days) of conventional and precision farming tomato.

Treatments	F_1	F_2	Mean
M_1	16.00	28.50	22.25
M_2	13.00	24.50	18.75
M_3	9.50	20.50	15.00
Mean	12.83	24.50	28
	M	**F**	**M X F**
SEd	0.2092	0.1709	0.2959
CD (0.05)	0.4310	0.3519	0.6095

M1, Mature green; M2, breaker; M3, light red; F1, conventioanal farming; F2, precision farming.

This is in line with the reports of Rajadurai (2008) and Priyadharshini (2009) in tomato.

Shelf life

When the effects of the farming systems were compared, precision farming system (F_2) registered significantly the highest number of days (24.50) followed by conventional system (F_1) with 12.8 days (Table 2 and Figure 1). Significant differences were observed due to the farming system in maturity stages. F_2M_1 registered significantly the most number of days (28.50) for shelf life and it was followed by F_2M_2 (Figure 1). The lowest number of days (9.50) for shelf was recorded in F_1M_3. In the present investigation, fruits harvested at mature green stage from precision farming recorded the highest shelf life (28.50 days). Sufficient supply of nutrients might have increased the production of Indole Acetic Acid (IAA) which consequently would have shown stimulatory action, in terms of cell elongation and thus resulting in increased plant growth. Pafli (1965) suggested that nitrogen, being the chief constituent of chlorophyll, protein and amino acids, is accumulated in the shoot through increased supply of nitrogen to the plants at appropriate time. Thus, fertigation deserves as an important aspect that contribute to increased plant growth. Fruits from well grown plants with well nutrient supplement give the highest shelf life (Balasubramanian, 2008).

Higher availability of nutrients and moisture in the root zone might have induced more root growth; hence, higher root volume was obtained which helps to get higher shelf life. Similar results were reported by Besford (1979) and Pandey et al. (1996). This result was in agreement with the findings of Puttaraju and Reddy (1997) and Kapse and Kalrodia (1997) in mango, Sreemathy (2004) in tomato, Rajadurai (2008) in tomato and Priyadharshini (2009) in tomato.

Enzyme activity influenced by maturity stage at harvest

Pectin methyl esterase

The pectin methyl esterase enzyme activity of fruits was influenced by farming systems and maturity stages (Table 3). When the effects of the farming systems were

Figure 1. Effect of different maturity stages on two different farming systems of tomato (a) All treatments (b) Comparison of control with best treatment.

Table 3. Effect of maturity stages on Pectin methyl esterase activity (meq/g/hr) of conventional and precision farming tomato.

Treatments	F_1	F_2	Mean
M_1	1.38	0.92	1.15
M_2	1.56	1.32	1.44
M_3	1.78	1.62	1.70
Mean	1.57	1.29	1.43
	M	**F**	**M X F**
SEd	0.0178	0.0145	0.02517
CD (0.05)	0.0367	0.0299	0.0518

M1, Mature green; M2, breaker; M3, light red; F1, conventioanal farming; F2, precision farming.

compared, precision farming system (F_2) registered significantly the lowest activity followed by the conventional farming system (F_1) for pectin methyl esterase enzyme activity (Figure 1). Among the maturity stages, harvest at mature green stage (M_1) recorded significantly the least value of pectin methyl esterase activity followed by breaker stage (M_2). The highest enzyme activity was recorded with light red stage. Significant differences were observed due to the farming system in maturity stages. F_2M_1 registered significantly the lowest enzyme activity and it was followed by F_2M_2. The highest value of pectin methyl esterase enzyme activity was recorded in F_1M_3.

Maturity is the most important determinant of storage life and final fruit quality. Maturity stages are determined for vegetables for enhanced shelf life. Harvesting crops at proper maturity allows handlers to begin their work with the best possible quality produce. Produce harvested too early lack flavor and may not ripen properly, while produce harvested too late may be fibrous or over ripe.

Generally, fruits become sweeter, more colourful and softer as they mature. A few fruits are usually harvested mature but unripe so that they withstand post harvest handling system when shipped to long distances. Most of currently used maturity evaluation is based on a compromise between those methods that would ensure the best eating quality and those that provide flexibility in transportation and marketing. Optimum harvesting stage depends upon type of fruit or vegetable and their final use. Tomatoes are harvested at firm ripe stage when it needs for processing whereas it is not suited for transport purpose.

Definition of maturity as stage of development giving minimum acceptable quality to ultimate consumer implies measurable points in commodity's development, and need for techniques to measure maturity. Maturity index for a commodity is a measurement or measurements that can be sued to determine whether a particular commodity is mature. These stages are important to trade regulation, marketing strategy and for the efficient use of labor and resources.

The major causes for spoilage of tomato are harvesting at immature stage, physical damage during harvest, fungal infections, improper storage and internal breakdown. Tomato is a climacteric fruit and hence it is highly perishable after harvest due to high metabolic rate and ethylene production. A fruit can have a long storage life only if the metabolic rate is low, with consequent minimal loss of quality. So efforts to increase the shelf life of tomato should necessarily focus on decreasing the metabolic rate and reducing the synthesis of ethylene in harvested produce (Rao and Rao, 1979).

Postharvest metabolic changes are of particular importance because the fruits are harvested at unripe and inedible stage and the quality of fruits ultimately depends upon the postharvest handling and storage methods. A proper understanding of the morphological, physiological and biochemical changes that occur in the fruit during ripening is essential for the development of good storage techniques (Sudheer and Indira, 2007).

Conclusion

The treatment F_2M_1 (Harvesting at mature green stage of precision farming tomato) was found to be better because the lowest PLW 7.53% on 8 DAS with 28.50 days of shelf life. The highest firmness with reduced spoilage was found. Lowest enzyme activity of pectin methyl esterase was found in the best treatment.

ACKNOWLEDGEMENT

Authors express their thanks to TNAU Research Assistantship, Dr. L. Pugalendhi, Dr. D. Malathi, Dr. T. N. Balamohan, Mr. R. Venkatachalam and Dr. K. Mahendran.

REFERENCES

AOAC (1975). Official methods of analysis. Association of Agricultural chemists, 9th Ed, Washington D.C. 2004.

Balasubramanian, P (2004). Studies on the effect of pre and postharvest treatments on dehydration and storage of bhendi var. Arka Anamika. M.Sc., (Hort.) Thesis. Tamil Nadu Agriculture University, Coimbatore, P. 3.

Balasubramanian P (2008). Comparative analysis of growth, physiology, nutritional and production changes of tomato (*Lycopersicon esculentum* Mill.) under drip, fertigation and conventional systems. Ph.D. (Hort.) Thesis. Tamil Nadu Agriculture University, Coimbatore, India.

Besford RT (1979). Uptake and distribution of phosphorus in tomato plants. Plant Soil 51: 331-340.

Carr PM, Carlson GR, Jacobsen JS, Nielsen GA and Skogley EO (1991). "Farming by Soils, Not Fields: A Strategy for Increasing Fertilizer Profitability." J. Prod. Agric. 4:57–61

Indian Horticulture Database (2011). (Ed) Bijay Kumar. http://nhb.gov.in/area-pro/database-2011.pdf accessed on 10.07.2013

Kapse BM, Kalrodia JS (1997). Studies on hydro cooling in Kesar mango (*Mangifera indica*). Acta. Hort. 455:707-717.

Palmer RJ (1996). In Proc. Site-Specific Management for Agric. Syst.,Minneapolis, MN, ASA-CSSA-SSSA, Madison, WI, 27–30 March, pp. 613–618.

Pandey RP, Solanki, PN, Saraf RK, Parihar, MS (1996). Effect of nitrogen and phosphorus on growth and yield of tomato varieties. Punjab Veg. Growers 31:1-5.

Panse VG, Sukhatme PV (1985). Statistical methods for Agricultural workers. Indian Council for Agricultural Research, New Delhi.

Patil VS, anawade AD (1999). Soil water plant atmosphere relationships. In : Proc. Advances in microirrigation and fertigation. June 21-30. Dharwad. Karnataka. pp. 19-32.

Priyadharshini G (2009). Pre and Post harvest handling, storage, packaging and value addition in Hybrid Tomato (*Solanum lycopersicon* Mill.) M.Sc. (Hort.) Thesis. Tamil Nadu Agriculture University, Coimbatore, India.

Puttaraju TB, Reddy TV (1997). Effect of pre cooling in the quality of mango (cv. Mallika). J. Food Sci. Technol. Quoted in post harvest News Info. 8:2088.

Rajadurai,V (2008). Standardization of pre and post harvest protocols for hybrid tomato (*Solanum lycopersicon Mill.*.) under open condition. M.Sc. (Hort.) Thesis. Tamil Nadu Agriculture University, Coimbatore, India.

Ranganna S (1986), Handbook of Analysis of quality control for fruit and vegetable products. 2nd Edition. Tata McGrow Hill Publication, New Delhi,

Rao DM, Rao MR (1979). Post harvest changes in banana cv. Robusta. Indian J. Hort. 36:387-393.

Salunkhe KK, Desai BE (1984). In: Post harvest Biotechnology of fruits Vol.1. Boca Raton, FL: CRC press, London, p. 168.

Sreemathy B (2004). Pre and post harvest treatments to enhance shelf life in tomato (*Lycopersicon esculentum* Mill). M.Sc. (Hort.) Thesis. Tamil Nadu Agricultural University, Coimbatore.

Sudheer, Indira (2007). Post harvest technology of horticultural crops. Ed. K. V. Peter. New India Publishing Agency, New Delhi.

Changes in agro-biodiversity as a result of sugarcane farming in mumias division, western Kenya

Nelly Nambande Masayi[1] and Godfrey Wafula Netondo [2]

[1]School of Environment and Earth Sciences, Maseno University, P. O. Box 333- 40105 Maseno, Kenya.
[2]Department of Botany and Horticulture, Faculty of Science, Maseno University, P. O. Box 333- 40105 Maseno, Kenya.

Sugarcane farming is a monocultural land use practice which often leads to reduction in agro-biodiversity. In Mumias division sugarcane is cultivated under small scale, large scale and nuclear estate. The study was carried out in Mumias division of western Kenya where 68% of the land is under commercial sugarcane cultivation while 32% is left for subsistence agriculture and other uses. The objectives of the study were to identify indigenous crops grown in Mumias division before the introduction of commercial sugarcane farming and to assess the effects of commercial sugarcane farming on indigenous crops. Ninety respondents were purposively selected. Data was collected using questionnaires, focus group discussion and interviews. Secondary data were obtained from documented materials. Data was analysed using means and percentages and was presented through discussions, tables and figures. With the introduction of commercial sugarcane farming in the 1970s, the land under indigenous crops declined. The research also established that sugarcane farming did not have an effect in the cultivation of groundnuts and bambara groundnuts. Our results imply that sugarcane farming is a major contributor to agro-biodiversity erosion. The results are expected to sensitize ministry of agriculture on the importance of good agricultural practices that can safeguard agro-biodiversity.

Key words: Agro- biodiversity, monoculture, indigenous crops.

INTRODUCTION

Agro-biodiversity refers to the aspects of biodiversity that affect agriculture and food production, including within-species, species and ecosystem diversity (FAO, 1999). Agro-biodiversity plays a key role in ensuring that there is increase productivity, food security, and economic returns. Monoculture is the practice of planting and cultivating crops in large tracts containing a single species. Monocultural farming involves clearing of large tracts of land to create more space for the cultivation of the single crop. Similarly in monocultural farming other subsistence crops are often abandoned with more focus and attention being given to the individual monocultural crop. At the end this is may lead to extinction of some crops that are very useful and are a source of food security to the community. Sugarcane is a monocultural crop grown in the Lake Victoria basin of Kenya and Uganda. In Kenya, sugarcane is commercially grown in Western and Nyanza provinces. Currently sugarcane occupies 107,622 ha of arable land and is grown primarily by small scale farmers followed by large-scale farmers and nucleus estates. Sixty eight percent of the land in Mumias division is put under sugarcane cultivation; this implies that a very small portion of the land (32%) in the division is left for subsistence farming. The growing of sugarcane was generally considered to alleviate poverty by expanding income generation

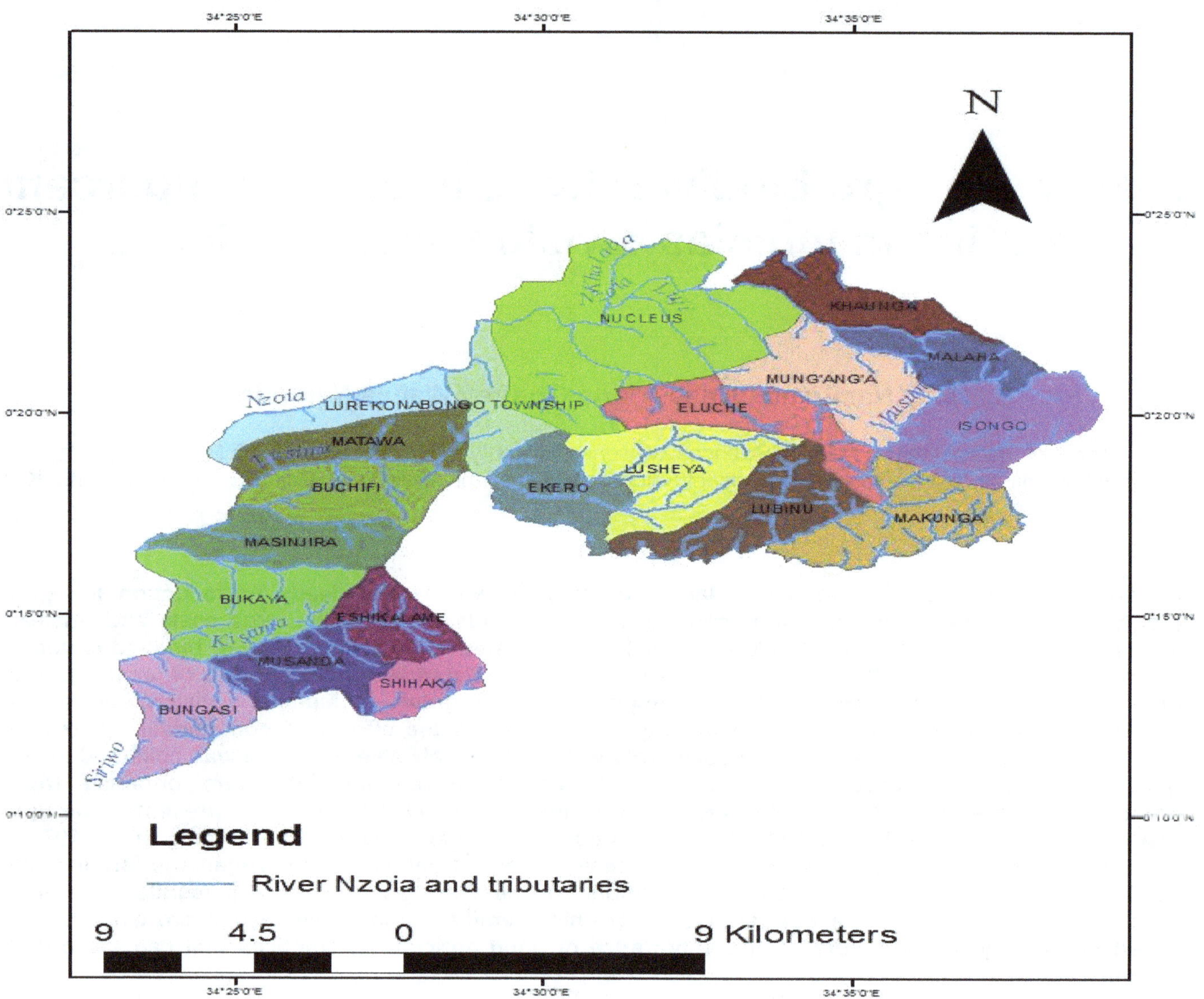

Figure 1. Map showing Mumias division.

possibilities. However statistics and observation indicate that poverty in this region remains endemic (GoK, 1999). Waswa et al. (2009) also reported that presently, sugarcane is the most widely grown commercial crop, having replaced most indigenous crops like cassava and vegetables, despite their ecological suitability and high nutritive and income value. According to World Wildlife Fund, the cultivation of sugarcane has caused a greater loss of biodiversity on planet earth than any other crop (WWF, 2004). Cheesman (2006) indicated that commercial sugarcane farming completely transformed large tracts of land especially in the coastal regions north and south of Durban, South Africa. It is widely recognized that if the remaining biodiversity is allowed to disappear as a result of socio- economic activities such as sugarcane cultivation, man's future will be at stake (Alcamo et al., 2003).

METHODOLOGY

The study was carried out in 2007 in Mumias division of Western Kenya (Figure 1). Purposive sampling techniques were used to select the respondents who included small scale, large scale farmers and key informants. The key informants included the chiefs, assistant chiefs, and District Agricultural officers. Primary data were collected using researcher administered questionnaires to 90 respondents from Mumias division and focus group discussions (FGD) involving thirty individuals who were selected with the assistance of local authorities. Both gender and age factors were put into consideration. The respondents were aged 50 years and above and were mature people who had lived in the region for more than thirty years. Secondary data on the trend in the number of farmers growing indigenous crops in the division was also acquired from Kenya Agricultural Research Institute. Data on the trend on changes in the size of land under crop species were assessed by partitioning periods into ten year intervals. Data were analyzed using descriptive statistics focusing on frequency distribution and percentages. In all cases the SPSS statistical package was used.

Table 1. Types of subsistence crops grown and acreages of land under crops in Mumias division before the introduction of commercial sugarcane farming.

Total size of land per household (acres)	Total size of land under crops (acreas)	Types of crops grown	Mean acreages per household	Percentage acreages under each crop (%)
		Sugarcane	0.16	2
		Maize	2.34	30
		Sorghum	1.34	17
		Cassava	1.22	16
8	7.82	Finger millet	0.94	12
		Ground nuts	0.64	8
		Sweet potatoes	0.56	7
		Bambara groundnuts	0.55	7
		Simsim	0.07	1

Source: Field data, 2007.

RESULTS AND DISCUSSION

Subsistence crops grown in Mumias division before the introduction of commercial sugarcane farming

In Mumias division, commercial sugarcane farming was introduced in 1972 concomitant with the introduction of Mumias Sugar Company. There are nine types of food crops that were common in Mumias division in 1960s before the introduction of commercial sugarcane farming. The field data revealed that out of the land cropped with traditional crops not all crops were given the same preference. Cereals such as finger millet, sorghum and millet together with cassava occupied a larger portion while simsim occupied the least portion. Bambara groundnuts and groundnuts were grown as intercrops with other crops such as sorghum and cassava. Subsistence crops grown in the region by the households before the introduction of commercial sugarcane farming are shown in Table 1. Maize was the dominant food crop followed by sorghum, cassava, finger millet while sugarcane was the second least grown food crop. Indigenous sugarcane was cultivated on small pieces of land of about 0.16 acres of land per household. It occupied only 2% of the total size of land under subsistence crops (Table 1). It was cultivated by about 14% of the household (Figure 2). These were the local sugarcane varieties (mikhonye cha eshinyala and mikhonye cha kampala) that were either red or green in colour and were mainly chewed raw. These varieties were commonly planted along the banks of rivers such as river Nzoia or in kitchen gardens.

Maize occupied the largest acreages per house hold with an average of 2.34 acres (Table 1). This translates to about 30% of the total size of land under subsistence crops per household in the study area. Maize, occupied the largest piece of land because ‘ugali’ is the stable food of the community. Despite maize having occupied the largest piece of land, this study established that only (51%) of the households grew it (Figure 2). The research established that this was because some households solemnly relied on other food crops such as sorghum and millet as staple food crops. The most common maize varieties grown were the unimproved landraces such as yellow maize (*shipindi)* in luhya and the white or have mixed colours (*namba nane*). Figure 2 shows the percentage number of households in Mumias division growing subsistence crops in 1960s.

Sorghum is an important crop for rural food security in Mumias division in 1960s. It occupied a mean of 1.34 acres of land per household accounting for about 17% of the land under subsistence crop and was ranked second to maize. The crop was grown by about 78% of the respondents (Figure 2). During this period, sorghum was a staple food crop in the region. Most farmers grew local land races. These varieties take 120 days to mature. Sorghum grain was utilized in preparing foods like “ugali”, porridge and for making alcoholic beverages.

Cassavas were the root crop that was given the highest priority. It occupied an area of about 1.22 acres per household (Table 1) accounting for about 16% of the total land under subsistence crops and was ranked third. It was cultivated by about 73% of the household in the division. Groundnuts were grown in Mumias division in 1960s with an average of 0.64 acres per household. This was about 8% of the total size of land under subsistence crops (Table 1). Bambara groundnut was cultivated on about 0.55 acres of land per household. This was about 7% of the total size of land under subsistence crops (Table 1). The crop was cultivated by about 68% of the household in the division in the 1960s (Figure 2). Groundnuts and bambara groundnuts were often grown as intercrops with other crops such as maize, cassava and sorghum. Sixty three percent of the household interviewed in the region grew groundnuts (Figure 2).

Sweet potatoes were an important source of food to the local Wanga community of Mumias division in 1960s. It occupied about 0.56 acres of land per household. This

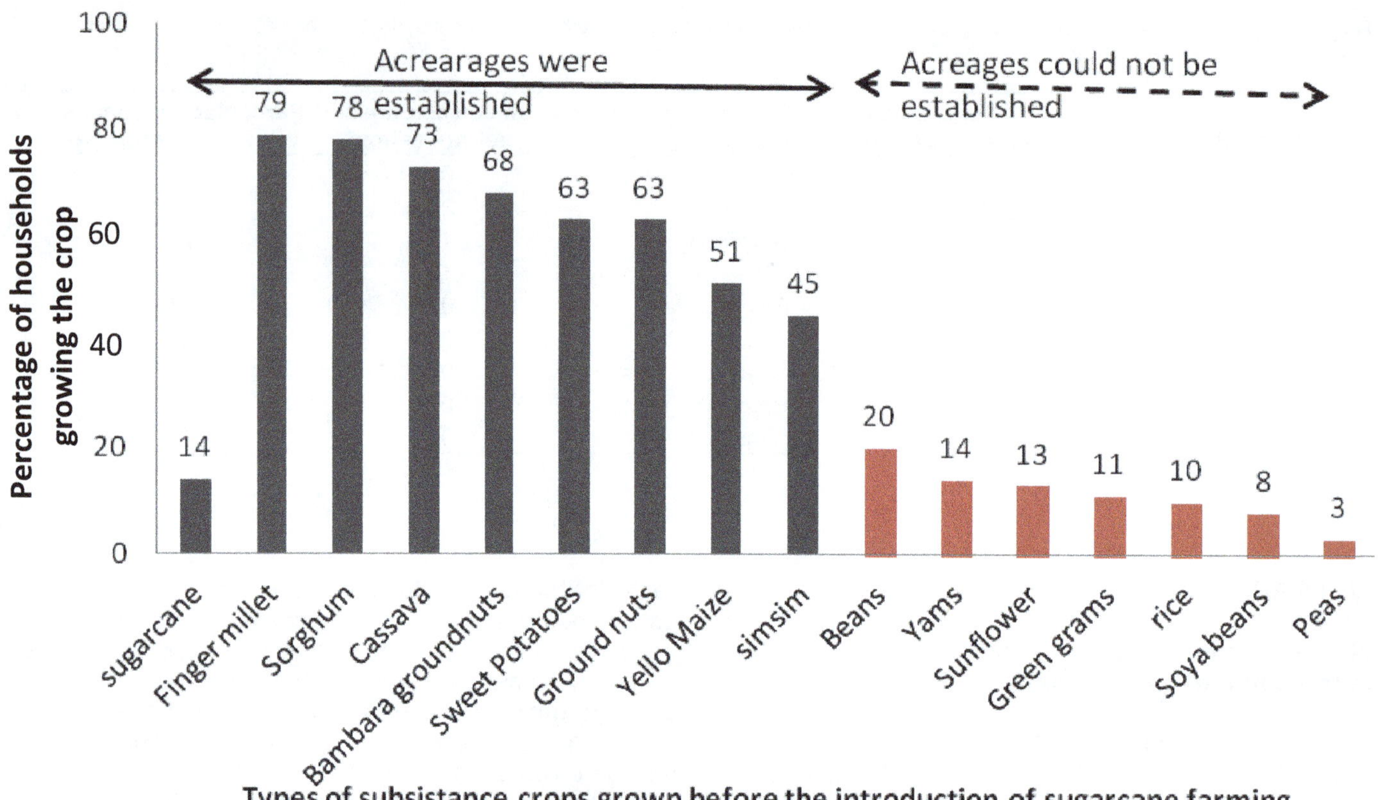

Figure 2. Percentage number of households in Mumias division growing indigenous crops in 1960s.

was about 7% of the total size of land under subsistence crops. This crop was grown by about 63% of the household in the region (Figure 2). Simsim occupied the least size of land in 1960s with 0.07 acres per household accounting for only 1% of the size of land under subsistence crops (Table 1). Despite the small size of land under simsim, it was cultivated by about 45% of the household in the division (Figure 2). Other crops that were grown in the region included, beans by 20% of the respondents interviewed, yams (14%), sunflower (13%), green grams (11%) rice (10%), soya beans (5%) and peas (3%). However, these crops occupied very small pieces of land and the acreages could not be established.

Effect of commercial sugarcane farming on subsistence crops in Mumias division in 1970s to 2000s

The study established that the average size of land under indigenous crops either decreased or remained static between 1970s and 2000s while the land under sugarcane cultivation increased over the same period. The average acreage of land per household under sugarcane increased from 1970s to 1980s when sugarcane was introduced but has tended to decline in the 1990s and 2000s. Netondo et al. (2010) reported that change in land use particularly conversion to monoculture leads to loss of agro-biodiversity.

Maize

The size of land under maize in 1970s was 1.52 (24%) acreages per household which was a 6% decline from the 2.34 (30%) acres in 1960s. The size declined to 1.12(20%) acres per household in 1980s. The size then increased to 1.3 (24%) in 1990s and increased further to 1.6 (30%) in 2000s (Figure 3). Over the same period, the size of land under sugarcane increased from 2.25 (35%) in 1970s to 2.72 (48%) in 1980s. This was an 18% increase in the size of land under sugarcane from 1970s to the 1980s. The households interviewed reported that the increase in the size of land under sugarcane was as a result of the introduction of Mumias Sugar Company which offered ready market for their sugarcane. The research findings indicated that as the size of land under sugarcane increased, the size of land under maize declined and vice versa. The respondents fifty two percent reported that in the 1970s and 1980s, much of the land that was previously under maize was transformed into sugarcane farms hence reducing the size of land available for maize cultivation.

In the 1990s and 2000s the decline in area under

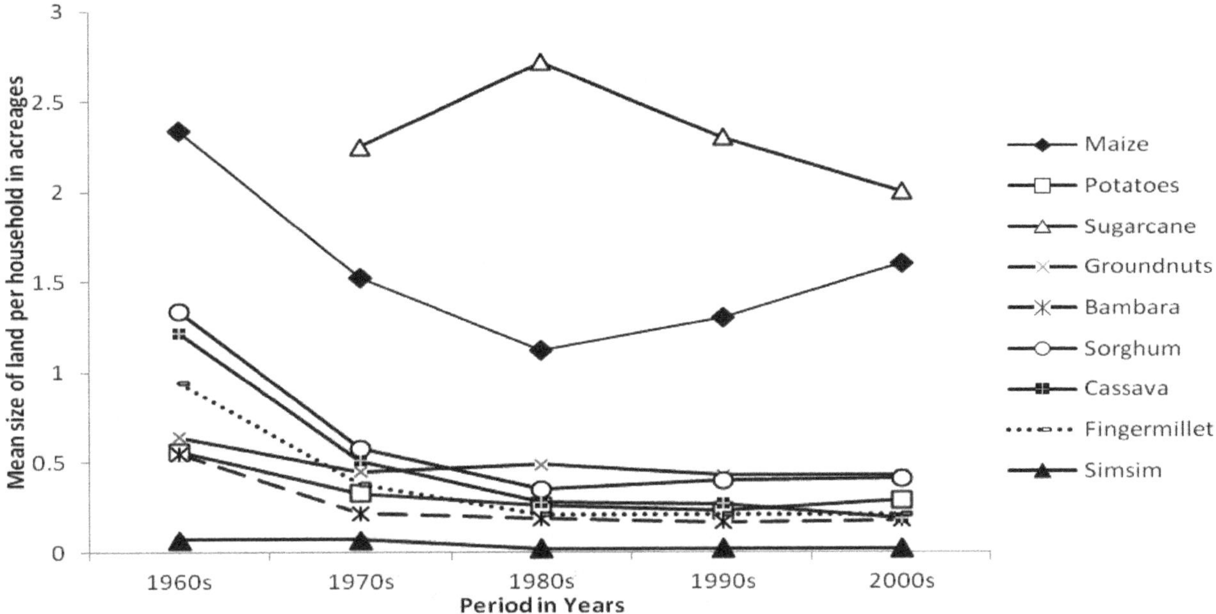

Figure 3. Size of land in Mumias division under indigenous crops between 1960s and 2000s.

individual food crops particularly maize eased. The farmers on realizing that sugarcane farming is not as profitable turned to growing some of the food crops; particularly maize which is a staple food crop. Between 1980s and 2000s, the size of land under maize increased by 10% that is from 1.12 (20%) in 1980s to 1.6 (30%) in 2000s. During the same period, the results indicated that the size of land under sugarcane declined by 0.72 (10%) acres per household that is from 2.72 (48%) in 1980s to 2 (38%) in 2000s. The research further established that in the 1970s, 1980s, 1990s and 2000s, the common maize species cultivated were the hybrid species which had replaced the local landraces that were prevalent in the 1960s. Forty eight of household interviewed indicated that land fragmentation which involves subdivision of land from father to mature sons is responsible for the decline in the size of land under maize. Previous research carried out in South Nyanza by (Eileen, 1987) revealed that as sugarcane production expands, it mainly replaces maize area. Of the plots planted in sugarcane, 95% were formerly used for maize (Eileen, 1987). Agricultural biodiversity is of fundamental importance to human survival and to the social and economic development of many countries. It supports human nutritional needs and a wide range of other crops. Pressures to develop one new species means the traditional ones are lost.

Sorghum

The land under sorghum was 0.58 (9%) acres in 1970s, it then reduced to 0.35 (6%) in 1980s. The size slightly increased to 0.4 (8%) in 1990s and 2000s (Figure 3). This implied an 8% decline in the size of land under sorghum from 1960s to 1970s, 3% decline in 1980s and a 2% increase in 1990s and 2000s. Sixty seven percent of the respondents maintained that the 7% decline in the size of land under sorghum was sugarcane farming. They reported that during this period, the size of land under sugarcane had also increased by 10% hence reducing size of land available for sorghum. Much of the land 0.99 (11%) acres per house hold that was previous under sorghum in 1960s was transformed into land for sugarcane farming in 1980s. During a focus group discussion, the respondents were asked to explain the reason for the 2% increase in the size of land under sorghum in the 1990s and 2000s. They reported that the increase experienced was as a result of the decline in the size of land under sugarcane over the same period creating more space for the cultivation of sorghum. During this period, the size of land under sugarcane had decline by 10%. The campaign encouraging people to grow more indigenous crops in Mumias was also reported to be responsible for the increased land under sorghum in the 2000s. Sorghum has starch, which makes it suitable for obese and diabetic people (Mamoudou et al., 2006). Nineteen percent of the respondents reported that the decline was as a result of maize farming whereby most farmer especially young generation aged below 40 years preferred maize over sorghum. Fourteen percent of the respondents attributed the decline to pests and diseases especially the birds that attack the crop when nearing the harvesting season. Romain (2001) reported that sorghum has lost much of its traditional area of growth in Africa to the introduced maize monoculture which is particularly vulnerable to droughts and low soil fertility. As indicated by Upreti and Ghale (2002) and FAO (2004) commercialization of agriculture such as sugarcane monoculture farming may significantly contribute to food crop diversity loss including sorghum in the study area.

Finger millet

A mean of 0.38 (6%) acres of land per household was under finger millet in 1970s. This was a 6% decline from the 0.94 (12%) acreas under the crop in the 1960s. The size of land declined further by 2 to 0.21 (4%) of acres per household in 1980s. The mean sizes increased slightly by 1% to 0.22 acres in 1990s and 2000s (Figure 3). Eighty one percent of the respondents attributed the 0.73 (8%) acres per household decline experienced in 1970s and 1980s to the conversion of land that was previously under finger millet into commercial sugarcane farming. According to fifteen percent of the respondents, the reduced finger millet on the farms may be attributed to change of taste and preference especially by the young generation between age 5 to 18 years who prefer maize meal (ugali) made from maize other than one made from, cassava, finger millet or sorghum (National Research Council, 1996). Labour intensiveness especially during weeding was cited by four percent as a major constraint to increased agricultural production of finger millet and the most difficult and time-consuming job women face in the fields.

National Research Council (1996) pointed out that finger millet is being rejected in favour of other monoculture crops such as sugarcane, maize and sorghum, which provide more income. Results from researches carried out elsewhere are in agreement with this assertion that finger millet is being abandoned in favour of monoculture. Cagley et al. (2009) reported that some indigenous crops require more labour and this difficulty is particularly pronounced in finger millet cultivation. This research finding therefore established that the causes of decline in the size of land under finger millet were, sugarcane farming, change in tastes and preference and pests and diseases. However sugarcane farming played the greatest contribution.

Cassava

The mean size of land under cassava per household has been decreasing since 1970s with 0.51(8%) acres per household in 1970s which was a 7% declines from 1.22(15%) acres per household under cassava in 1960s. There was a further decline to 0.28 (5%) acres in 1980s and 1990s, and a further decline to 0.19(3%) acres per household in 2000s (Figure 3). The respondents reported various reasons as being responsible for the 1.03 (12%) acres per household decline from 1960s to 2000s. Fifty three percent reported that households have transformed much of the land that was previously under cassavas (1.03 acres per household) to sugarcane farming. The PRA further indicated that where farmers are growing these indigenous crops the yields are generally low. This could be attributed to poor crop husbandry occasioned by the preferential treatment given to sugarcane. The respondents Twenty five percent attributed the decline in the land and cassava in 1980s to the fact that the some cassava varieties (bitter type) were termed as poisonous and had previously caused some death among the children in the region. According to twenty two percent of the respondents, the decline in the size of land under cassava was as a result of pests and diseases that have discouraged many of the farmers from growing the crop. The respondents reported that pests such as moles had led to low output hence discouraging most farmers from growing cassavas.

FAO (2004) report cautions that the replacement of local crops with large scale monocropping might lead to the simplification of agro ecosystem such as crops that are used directly or indirectly for food, fodder, fibres, fuels and pharmaceuticals. Cassava is one such crop. The potential loss of agro biodiversity presents risks of food production as well as posing a serious threat to rural livelihood and long term food security. Waswa et al. (2009) reported that presently, sugarcane is the most widely grown commercial crop, having replaced most indigenous crops like cassavas and vegetables, despite their ecological suitability and high nutritive and income value.

Sweet potatoes

A mean of 0.33 (5%) acres of land per household was under sweet potatoes in 1970s.The size of land declined to 0.26 (5%) acres per household in 1980s and 0.23 (4%) acres in 1990s. The mean sizes then increased to 0.29 (5%) acres per household in 2000s (Figure 3). Forty one percent of the respondents pointed out that the reduced size of land under sweet potatoes in 1970s, 1980s and 1990s was due to sugarcane farming. The respondents pointed out that the land that was previous allocated to the cultivation of sweet potatoes 0.33 (3%) acres per household from 1960s was transformed to land under sugarcane farming. Over the same period, the size of land under sugarcane increased by 0.47 (13%). In 2000s, there was a slight increase of 0.06 (1%) in the size of land under sweet potatoes and a 0.3 (5%) decline in the size of land under sugarcane. Similar results have been obtained in Swaziland where it is indicated in (FAO, 2008) that establishment of sugarcane plantations leads to reduction of land under sweet potatoes and other activities. In the 1990s, fifty nine percent of the respondents pointed out that pests and diseases was the main cause responsible for the further decline in size of land under sweet potatoes. Moles were mentioned as the main pest which attacks sweet potatoes. Woolfe (1992) reported that *Meloidogyne* spp. (root rot) and *Rotylenchus reniiformis* is the major known bacterial diseases of sweet potatoes in the tropics. Woolfe (1992) reported that the size of land and research under sweet potatoes and other root crops and tubers has been neglected in favour of more prestigious cereals and other export cash crops such as sugarcane. During the focus group discussion it emerged that the slight increase of 0.06 (1%) acres in the size of land under sweet potatoes was as a result of ready market for the crop in major towns such as Nairobi. FAO (2000) reported that sweet potato production has responded to strong urban demand and is a major traded commodity.

Groundnuts

The size of land under groundnuts in 1970s and 1980s were 0.45 (7%) and 0.49 (7%) acreages per household respectively, which was a 0.15 (1%) decline from the 1960s to 1980s. The size increased to 0.06 (8%) acres per household in 1990s and 2000s (Figure 3). On the other hand, the size of land under sugarcane increased by 0.47 (13%) acres per household between 1960s and 1980s. In the 1990s and 2000s, the size of land under groundnuts increased by 1% as the size of land under sugarcane declined by 0.72 (10%) per household. The research indicated that sugarcane farming had little effect on the cultivation of groundnuts. This was probably because groundnuts are often intercropped with sugarcane meaning that no special land is required to be set aside for the cultivation of groundnuts. This was reported by nine percent of the respondents. Apart from sugarcane farming, forty one percent of respondents attributed the decline in the sizes of land under groundnuts experienced in 1970s and 1980s to pests and diseases which led to poor crop yield and this discouraged most farmers from growing groundnuts. Groundnut rosette disease is a major constraint to productivity in both Kenya and Uganda (ICRISAT, 2010). Fifty percent of the respondents reported that the decline is as a result of the labour intensiveness involved in the production of groundnuts. They reported that a lot of labour is required in weeding harvesting and drying. During a focus group discussion it emerged that the main reason responsible for the increase in the size of land under groundnuts in 1990s and 2000s was changes in tastes and preferences. Most of the young generation aged below (40 years) in the region enjoys feeding on groundnuts which is served as a delicacy with tea. The crop is found to be delicious and rich in protein, minerals and edible oils. They can be eaten on their own or blended with other dishes such as finger millet, and simsim to improve taste and nutritional value. They can be roasted, boiled, pound using a mortar and pestle or grinding stone to form paste (KARI, 2000).

Bambara groundnuts

The mean size of land under bambara groundnut was 0.22 (3%) acres per household in 1970s. The size then declined to 0.19 (3%) acres per household in 1980s. In 1990s the mean size of land declined slightly to 0.17 (3%) acres per household. In 2000s, the size of land under the crop increased slightly to 0.18 (3%) acres per household (Figure 3). This research showed that the changes in the size of land under bambara groundnuts was neglegable. Thirty six percent of the respondents reported that land fragmentation was responsible for the slight decline in the size of land under bambara groundnuts in the 1970s, 1980s 1990s and 2000s. Fifteen percent indicated that the slight change in the size of land under bambara groundnuts was as a result of change in taste and preference with preference being given to groundnuts. Forty percent of the respondents attributed the slight change in the size of land under bambara groundnuts to cultural noms associated with the cultivation of bambara groundnuts. It was reported that poor yields were achieved if someone wearing shoes went into the field containing bambara groundnuts. Women having menstruation were also not allowed in the bambara groundnuts fields, since this could lead to low yields. Since many households could not successfully fulfill this cultural norms, they have abandoned the cultivation of this crop. However, despite the decline experienced in size of land under sugarcane in 1990s and 2000s, this did not have a great effect on the size of land under bambara groundnuts. Only nine percent of the respondents reported that sugarcane farming has had an effect on the size of land under bambara groundnuts. They indicated that the crops are grown on small scale and as intercrops with sugarcane and do not require special pieces of land to be set aside for their cultivation. The decline in the size of land under sugarcane in the 1990s and 2000s could have led to the decline of land under bambara groundnuts. Ngugi (1995) reported that bambara groundnuts is usually intercropped with crops such as maize, sugarcane and fingermillet. Andika et al. (2010) report indicate that cultivated area and production trend of oil crops like groundnuts, sunflower, soya beans, simsim, rapeseed, bambara groundnuts and castor have remained fairly constant in Mumias district. FAO (2008) reported that sugarcane farming has replaced many indigenous crops such as bambara groundnuts cassava and millets. Other findings by Ngugi (1995) indicated that bambara groundnut production is declining Kenya due to high cost of purchasing seedlings and cultural erosion with the young generation shifting from bambara groundnuts to groundnuts. In Nigeria Tanimu and Aliyu (1995) reported that bambara groundnuts cultivation has declined due to neglect since it is not used for industrial purposes compared to other legumes such as groundnuts. However this study established that the size of land under bambara groundnut has been declining in Mumias division from 1970s to 2000s though the decline was negligible.

Simsim

The findings of this study from 1970s to 2000s clearly indicate that in Mumias division, one of the least grown crops in the region was simsim and the mean size per household is negligible. In 1970s a mean of 0.07(1%) acres of the entire land under crops per household was under the simsim. It then declined to 0.02(1%) acres per household in 1980s and 1990s (Figure 3). Out of the respondents interviewed none of them cultivated simsim in the 2000s. Forty two (42%) of the respondents attributed the decline in the size of land under simsim to the introduction of commercial sugarcane farming. They reported that the introduction of commercial sugarcane farming may have contributed to the decline in the size of land under simsim since the land is rendered infertile. The other causes reported are being labour intensive (48%) and pests and diseases (10%). Similar results have been reported by Mishra (2008) who reported that most of the oil seeds and pulses such as simsim have been neglected through monoculture of crops such as wheat, rice and maize which occupied major areas of Indian farmlands. These crops make have nodules on their root which fix nitrogen in the soil. With the decline in the size of land under simsim, farmers have to purchase more and more nitrogenous fertilizers from the markets. This results into a pressure on the national economy. The crop is labour intensive reportedly because the drying of simsim takes a long time and requires specialized skill especially by women. Such labour is scarce.

Conclusions

Indigenous crops grown in Mumias division before the introduction of sugarcane included sorghum, fingermillet, cassava, sweet potatoes, groundnuts, bambara groundnuts, indigenous sugarcane and simsim. These subsistence crops occupied various sizes of land per household. These subsistence crops were mostly landraces which had not undergone any improvement through breeding. With the introduction of sugarcane in 1970s, the size of land under most indigenous crops since most of the land that was previously under indigenous crops was dedicated to sugarcane farming. Despite sugarcane farming being the major cause for the decline, other causes for the decline include pests and diseases, changes in tastes and preference, labour intensiveness, and lack of skill and knowledge on there preparation. Crops such as bambara groundnuts, and groundnuts are not affected by sugarcane farming because they are grown as intercrops with crops such as sugarcane, maize, sorghum and cassava. The decline in cassava is majorly as a result of pests and diseases and the fact that it contains cyanide that causes death.

ACKNOWLEDGEMENT

The authors expresses their thank to VICRES for Funding Prof Godfrey Netondo, School of Graduate Studies Maseno University and Mr Gipson Masayi for having funded the research hence making it a success. They would like to acknowledge Prof. Josephine Ngaira for having tirelessly guided me throughout the entire study.

REFERENCES

Alcamo J, Ash NJ, Butler CD (2003). Ecosystem and Human Well-being. Millennium Ecosystem Assessment. A framework for assessment. Island press, Washington.

Andika DO, Onyango AM, Onyango JC, Stutzel H (2010) Roots spatial distribution and growth in bambara groundnuts (Vigna subterranea) and nerica rice (*Oryza sativa*) intercrop system. ARPN J. Agric. Biol. Sci, 5:39-50.

Cagley JHC, Leigh AC, Marieka K (2009). Gender and Cropping: Millet in Sub-Saharan Africa. Evans School of Public affairs. University of Washington. Available at http://evans.washington.edu/files/Evans_UW_Request%2040_Gender%20and%20Cropping_Millet_06-29-2009.pdf. Accessed March 2010.

Cheesman OD (2006). Environmental impact of sugar production. Cambridge MA. Cabi Publishers Brazil.

Eileen K (1987). Effects of Sugarcane Production in South Western Kenya on Income and Nutrition.

FAO (1999). What is agro biodiversity. Available on http://www.fao.org/SD/LINKS/documents_download/FS1WhatisAgrobiodiversity.pdf. Accesed on 6th June 2011.

Food and Agriculture Organization (FAO) (2000). FAO online database. http://apps.fao.org

FAO (2004). What is agrobiodiversity. Rome: Food and Agricultural Organization of the UN. Rome. http://www.oecd.org/dataoecd/44/18/40713249.pdf. Accesed on 15 February 2009.

FAO (2008). Agriculture Outlook 2008. available at http://www.oecd.org/dataoecd/44/18/40713249.pdf. Acessed 9th July 2009.

GoK (1999). National Poverty Eradication Plan 1999-2015. Department of Development Co-ordination, Office of the President, Government Printers, Nairobi.

KARI (2000). Grow and Eat Groundnuts for More Money And Better Health. Government printers Nairobi Kenya.

Mamoudou HD, Hurry G, Alfred S, Alphons GJ, Van B (2006). Sorghum grain as human food in Africa: Relevance of content of starch and amylase activities. Afr. J. Biotechnol. 5:384-395.

Mishra MP (2008) Monoculture and Agro-chemicals: impacts on natural environment. Availabe on Ecosensorium .org. Accessed in June 2011.

National Research Council (1996). Finger millet in Lost crops of Africa: volume I: grains, National Academy of Sciences. pp. 39-57.

Netondo GW, Fuchaka W, Maina L, Naisiko T, Masayi N, Ngaira JK (2010) Agro biodiversity endangered by sugarcane farming in Mumias and Nzoia Sugar belts of Western Kenya. Afr. J. Environ. Sci. Technol. 4:437-445.

Ngugi GW (1995). Kenya Country Report in Promoting the

Conservation and Use of Underutilised and Neglected Crops. Bambara Groundnut (Vigna subterranean (L.) Verdcourt). International Plant Genetic Resources Institute (IPGRI, Rome. 9:33–44

Romain H (2001). Crop Production in Tropical Africa, Brussels publishers, Belgium.

Tanimu B, Aliyu L (1995). In Proceeding workshop on conservation and Improvement of Bambara Groundnuts(Vigna Subterranea(L) Verds) Harare Zimbambwe. Promoting the Conservation and use of underutilized and neglected crops. 45-50. Available at http://www.underutilized species.org/documents/publications/bambara_groundnut. pdf. Accessed on 21st may 2009.

Upreti RB, Ghale Y (2002). Factors leading to biodiversity loss in developing countries: The case of Nepal. Biodiversity Conserv. 11:1607-1621.

Waswa F, Netondo G, Maina L, Naisiko T, Wangamati J (2009). Potential of Corporate Social Responsibility for poverty alleviation among Contracted sugarcane farmers in the Nzoia sugarbelt, Western Kenya. J. Agric. Environ. Ethics 22:463-475.

Woolfe JA (1992). Sweet potato: An Untapped Food Resource. Cambridge University Press, Cambridge.

WWF (2004). Global Fresh Water Programme Netherlands. Available at www.panda./freshwater. Accessed 15th January 2007.

Design and implementation of a low cost computer vision system for sorting of closed-shell pistachio nuts

J. Ghezelbash[1], A. M. Borghaee[1], S. Minaei[1], S. Fazli[2] and M. Moradi[2]

[1]Department of Agricultural Machinery, Islamic Azad University, Science and Research Branch, Tehran, Iran.
[2]Department of Electrical and Computer, Zanjan University, Zanjan, Iran.

Iran is one of the most important producers and exporters of pistachio nuts in the world. Certainly, the quality of pistachio is very important for customers. Sorting of pistachio nuts is a post-harvesting process which is currently performed using a mechanical apparatus called "pin picker". Due to the possibility of kernel damage by the pin picker, mechanical methods of closed-shell pistachio removal are gradually being replaced by other appropriate systems. Low cost systems for post-harvesting processes such as computer vision-based sorting systems are experiencing rapid growth. In this paper, a low cost intelligent system is implemented for pistachio sorting using computer vision. The system uses a combination of two flat mirrors and one low cost camera to obtain suitable 3 dimensional images from pistachios which are processed to detect closed-shell nuts. The experimental results for the three varieties of pistachios nuts shows 92.7 and 86.7% average removal accuracy, respectively for open and closed shell pistachio nuts.

Key words: Pistachio, closed shell, sorting, computer vision.

INTRODUCTION

Based on FAO statistics, Iran, as the first producer of pistachio nuts, produced 35% of the world's pistachio crop in 2008 (http://faostat.fao.org). The quality of produced pistachio nuts is an important factor in consumer acceptance. In this regard, improvement of post-harvest processes is vital. One of the common methods of sorting pistachio nuts is dividing them into open (split) and closed (unsplit) shell categories. These groups are treated differently in subsequent processing. Pistachios are principally served as roasted nuts and they are usually marketed as snack food. For this end, use unsplit pistachios are considered undesirable because they are difficult to open and they may contain immature kernels. Therefore, the sorting of open and closed shell pistachio nuts is an important part of the post-harvest operations (Figure 1).

A major problem in pistachio sorting is the use of harmful mechanical equipment or expensive intelligent systems. Closed shell nuts are currently separated from open shell product by a mechanical device called "pin picker". Although pin pickers have high capacity, they may damage the kernel of open shell nuts by inserting a needle into the kernel meat. The hole created by the needle can give the appearance of an insect tunnel, and lead to rejection by the consumer (Pearson et al., 2000).

Various techniques including optical, mechanical, electrical, and acoustical methods have also been used for classification and/or sorting of pistachio nuts. Machine vision was introduced for detection of stained and early split pistachio nuts (Pearson, 1996). Ghazanfari et al.

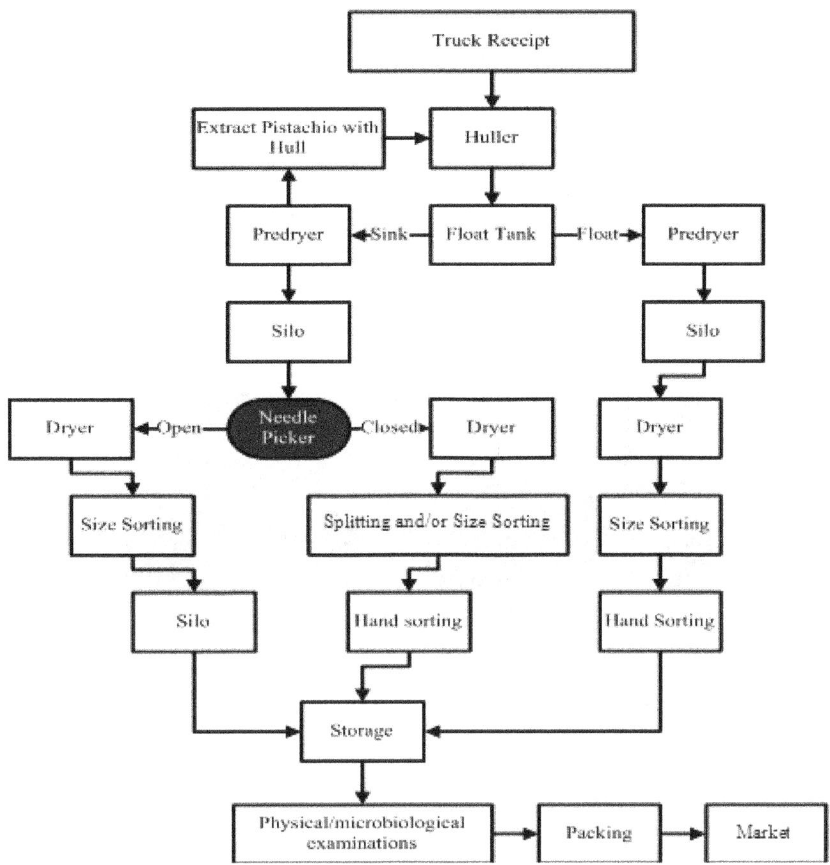

Figure 1. Post-harvest processing of Pistachio (www.viraman.com).

(1997; 1998) utilized Fourier descriptors and gray level histogram features of two-dimensional images to classify pistachio nuts into one of three USDA size grades or as having closed shells. Use of Fourier descriptors is very time-consuming and they are not suitable for real-time applications. Pearson's system demonstrates that, pistachios with either split or unsplit shells can be distinguished by using gray-scale images. Later, an automated machine vision system was developed to identify and remove pistachio nuts with closed shells from processing streams (Pearson and Toyofuku, 2000). The system includes a material handling mechanism to feed nuts through three high speed line-scan cameras without tumbling. The camera output signals are input into digital signal processing boards which extract image features characteristic of closed and open shell pistachios. The classification accuracy of this machine vision system for separating open-shell from closed-shell nuts in two passes is approximately 95%. Although the system has a throughput maximum rate of 40 nuts per second, its cost is too high for many producers in Iran.

As an alternative for vision systems, impact acoustic emission was used as the basis for a device that separates pistachio nuts with closed-shells from those with split-shells (Pearson, 2001, Cetin et al. 2004a, b). The proposed algorithm used a small number of features and achieved a classification accuracy of 91.5% on the validation dataset. Later in 2009, an intelligent pistachio nut-sorting system using a combination of acoustic emissions analysis, Principal Component Analysis (PCA) and Multilayer Feed forward Neural Network (MFNN) classifier was developed and tested (Omid et al., 2006, 2009). This system is cheaper than Pearson's vision system although it is slower.

In this study, a high performance low-cost computer vision system for automated removal of closed-shell pistachio nuts, with closed shells is designed and developed. The major contribution of this article is using two-mirrors for 3D imaging of pistachio, and implementation of a low-cost self-tuning sorter system. Ability of this system for high performance sorting is shown by the results of the experimental evaluation of the system.

MATERIALS AND METHODS

Taking a suitable image from pistachio nut is one of the major problems in the classification of pistachios using computer vision. Axial rotation of the nuts cannot be mechanically constrained as they are conveyed. This requires acquisition of three images

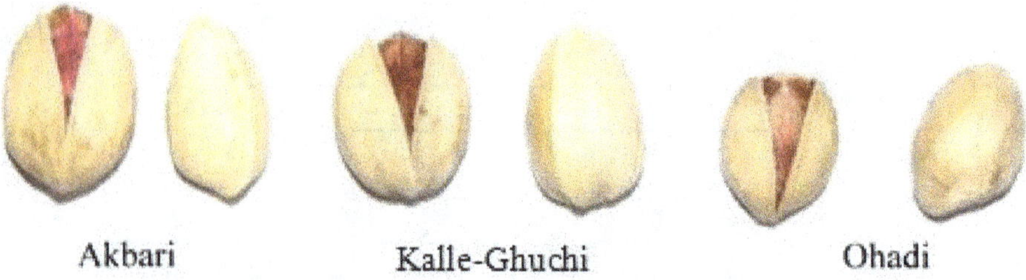

Figure 2. Common varieties of Iranian pistachio nuts (split and unsplit).

Figure 3. Arrangement of mirrors and camera.

around the perimeter of the nut in order for the whole surface to be inspected. In this study, three varieties of Iranian pistachio nuts, namely, Akbari (Ak), Kalle-Ghuchi (Ka), and O'hadi (Oh), are considered for training and analyzing. Samples of these nuts are presented in Figure 2.

For whole surface inspection, Pearson used three line-scan cameras to take several linear images from all directions of pistachio during free fall. This method of obtaining images is very expensive. To alleviate this, three cameras were replaced by one camera and two flat mirrors. This allows the system to view every side of the nut which eliminates system sensitivity to pistachio orientation. The idea is demonstrated in Figure 3. Four major possible pistachio orientations are shown in Figure 4. It can be seen that, using this method, cleavage of open shell pistachios in any orientation can be detected by only one camera.

The sorter system should be fully automatic to achieve appropriate processing. Therefore, the system should have a discrete controller with discrete events, such as pistachio arrival in the imaging corridor or sorting area. The general configuration of the designed system is shown in Figure 5.

Subsystems of the proposed system can be listed as follows:

Feeding subsystem

The feeding subsystem includes a hopper over the vibrating feeding chute. The feeding chute is vibrated by means of a spring and cam mechanism. As shown in Figure 6, rotation of the cam is obtained using an AC motor. This mechanism causes the nuts to move continuously down the chute.

Exposing subsystem

The exposing subsystem is used for correct positioning of the nuts in front of the camera. To have a constant controlled velocity, a DC motor gearbox is utilized to drive the pulley of the conveyor. The conveyor guides the pistachio nuts in a path having two inclined flat mirrors on two sides. Each pistachio is immediately detected when it crosses the path of IR rays at which time, the camera takes a

Figure 4. Main Configurations of pistachio nuts on the conveyor.

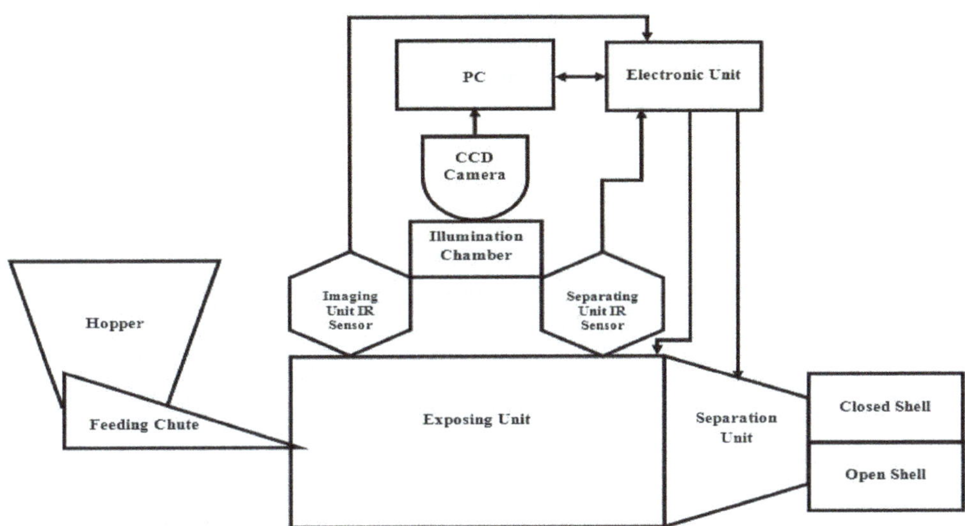

Figure 5. Block Diagram of the proposed system.

picture of the pistachio and its background after a certain delay. Therefore, no high frequency camera is required for this system.

Imaging subsystem

Sufficient illumination of the imaging chamber is achieved using two GU5.3 11W daylight lamps. This illumination is selected just for performance evaluation of the system and can be improved. A very inexpensive (POS-968A camera with SONY1/3"CCD) camera is used to acquire suitable images from nuts.

Divider subsystem

Separating subsystem includes a DC motor that provides for sideway movement of a divider vane to right or left.

Electronic subsystem

The electronic subsystem includes:

i) 12v DC power supply
ii) Two set of IR emitter and receiver sensor
iii) 12v DC conveyor driver motor-gearbox
iv) Divider vane driven by 6v DC electromotor
v) COM to USB communication cable

Low level control of the system is implemented using a microcontroller. The discrete-event based controller provides for high system performance. The communication between several subsystems and the computer is effected trough the microcontroller and USB port, which is adaptable with any personal computer. Image processing algorithm is programmed in real-time and high performance Delphi programming environment.

Sequence of system operation

The camera takes an image when a pistachio interrupts the infrared beam being detected by an IR sensor. This image is rapidly checked for split or unsplit status and the result is appended to a result queue which can consequently be used to control the sorting mechanism.

Learning and Image processing algorithm

In the real-time application, many restrictions appear and applications of many methods such as Fourier methods, spectral

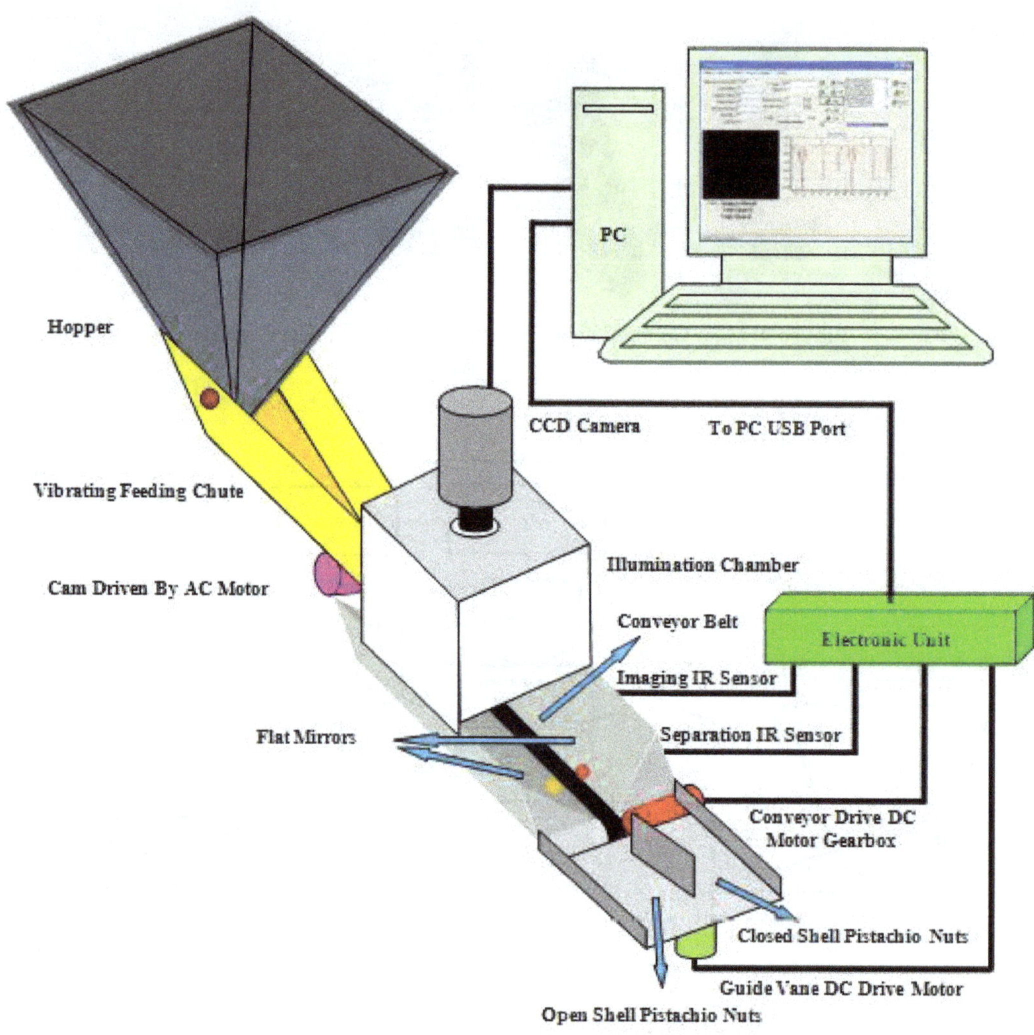

Figure 6. Schematic shape of sorting system.

methods, or active contours, are difficult to utilize. Therefore, simple methods are appropriate and should be completely analyzed and optimized off-line.

Thresholding segmentation of images is a subset of high speed methods in real-time image processing applications. In this study, these are optimized and implemented for unsplit pistachio sorting system. To the best of our knowledge, little work is published on the implementation of pistachio sorting systems using low cost and simple systems. A very inexpensive camera is employed which is bound to produce high-noise and low-resolution images. Moreover, the frame rate of the camera is confined and this restriction leads to serious problems for segmentation and evaluation of pistachio images in split or unsplit form. In order to solve these problems, multilevel thresholding is utilized for image processing.

The unknown parameters of image processing algorithm are tuned with PSO algorithm. Particle Swarm Optimization (PSO), developed by Kennedy and Eberhart, is one of the most well-known evolutionary optimization techniques (Kennedy and Eberhart 1995, 2001). PSO idea is based on social interactions such as bird flocking and fish schooling. Similar to genetic algorithms, PSO is population-based and evolutionary; with one key difference from genetic algorithms in that all chromosomes or potentials in the population, endure the entire search process. The PSO algorithm used here is explained in the following steps.

TRAINING AND TESTING OF EXPERIMENTS

Two hundred pistachio nuts were randomly selected for each variety in the training process. These include 100 open-shell and 100 closed-shell pistachios. The nuts used for training are passed through the sorting system to acquire and save the images. The system can be used under any lighting condition. Moreover, it contains mirrors which reflect the light into the camera lens. The diaphragm of the camera automatically changed in many times and this can be harmful for Multi-step threshold algorithm. Therefore, it is suitable to have background image for any image of pistachio. The training images for each variety included 400 pictures from pistachio nuts and 200 pictures from background. Thereafter, the unknown parameters of the system were tuned by PSO algorithm for each variety. Then the unknown parameters,

Table 1. Results of experiments.

Variety	Training percent	Testing percent	
		Open shell (%)	Closed shell (%)
Akbari	94	91	90.2
Ohadi	91.5	94	83.3
Kalle-Ghuchi	97	93.1	86.6

Table 2. Compare between several methods of sorting.

Parameter	Mechanical pin picker (%)	Pearson's machine vision system (two passes) (%)	Cetin's acoustics system (%)	Machine vision system designed in this study (%)
Open-shell	97	96.4	91.5	92.7
Closed-shell	95	93.2	91.5	86.7

tuned by PSO, were sent to the sorting program. Results of training and testing are presented in Table 1.

Conclusion

A low cost system for sorting of unsplit pistachio nuts was designed and implemented. Evaluation of the system in training and testing stages indicates that the system is very flexible for different varieties. The capacity of the system in initial tests is approximately 5 pistachios per second. Pearson's system is reportedly able to sort 40 pistachios per second. Since the fabrication cost of the proposed system is approximately $750 and Pearson's system is $15000, so to do the same job, we need eight of these machines ($6000). Therefore the amount of primary enterprise of this system is better than Pearson's system.

In the experiments, the system was trained for sorting Akbari and Kalle-Ghuchi varieties. But the training of the system to sort O'hadi variety is difficult. This may be because of its smallness and thin split area. The average train is 94.2% and testing is 89.7%. These results are encouraging and the system is completely implemented and analyzed. Table 2 shows a comparison between several methods for separating open shelled pistachio nuts from those with closed shells.

Future work in this field can be fulfilled by improving the whole subsystem such as feeder, exposing, separator, illumination and camera, and optimization and image processing algorithms. Multi-line systems can also increase the system capacity.

REFERENCES

Braquelaire J, Brun L, (1997). Comparison and Optimization of Methods of Color Image Quantization. IEEE Transactions on Image Processing. 6(7):1048-1052.

Cetin AE, Pearson TC, Tewfik AH, (2004). Classification of closed and open shell pistachio nuts using voice recognition technology. Transactions of the ASAE. 47(2):659–664.

Ghazanfari A, Irudayaraj J, Romaniuk M, (1997). Machine vision grading of pistachio nuts using fourier descriptors. J. Agric. Eng. Res. 68(3):247-252.

Ghazanfari A, Wulfsohn D, Irudayaraj J, (1998). Machine vision grading of pistachio nuts using gray-level histogram. Canadian Agric. Eng. 40(1):61-66.

http://faostat.fao.org.

Kennedy J, Eberhart R (1995). Particle Swarm Optimization. Proceedings of the IEEE International Conference on Neural Networks, Perth, Australia, pp. 1942-1945.

Kennedy J, Eberhart R (2001). Swarm Intelligence. Academic Press, 1st ed., San Diego, CA.

Omid M, Mahmoudi A, Omid MH (2009). An intelligent system for sorting pistachio nut vareties. Expert systems with applications 36:11528-11535.

Pearson T (1995). Machine vision system for automated detection of stained pistachio nuts. Lebensm Wiss, U Technol, 29:203-209.

Pearson T, Toyofuko N (2000). Automated sorting of pistachio nuts with closed shell. Appl. Eng. Agric. 16(1):91-94.

Pearson TC, Slaughter DC, (1996). Machine vision detection of early spilt pistachio nuts. Transactions of ASAE. 39(3):1203–1207.

22

Heterosis studies for fibre quality of upland cotton in line x tester design

Kaliyperumal Ashokkumar[1], Karuppanasamy Senthil Kumar[2] and Rajasekaran Ravikesavan[2]

[1]Centre for Plant Molecular Biology, Tamil Nadu Agricultural University, Coimbatore - 641 003, Tamil Nadu, India.
[2]Centre for Plant Breeding and Genetics, Tamil Nadu Agricultural University, Coimbatore- 641 003, Tamil Nadu, India.

Cotton fibre length, strength and fineness properties have vital influence textile industry. The developing high fibre length and strength cultivars or hybrids are essential to current modernized spinning mills. Therefore, the present study was carried out to assess the expression of *per se* performance and heterotic effect for fibre quality traits from 28 hybrids involving four adapted varieties as lines and the seven accessions as testers in Line x Tester analysis. Significant range of variability was observed all the traits in parents and their 28 hybrids. Fibre length in parents and their hybrids ranged from 23 to 32.9 mm and 25.40 to 33.3 mm, respectively. The hybrid MCU 12 x SOCC 17 and SVPR 2 x TCH 1641 for fibre elongation exhibited significant positive heterotic effects. Eleven hybrids showed significant negative heterotic effects for fibre fineness and hybrid MCU 5 × SOCC 17 displayed greater negative heterosis. Mean performance of bundle strength in parents and hybrids was 20.6 and 21.9 (g/tex), respectively. The hybrids MCU 12 x TCH 1644 and SVPR 2 x TCH 1646 for 2.5% span length, MCU 5 x SOCC 17 and Surabhi x F 776 for bundle strength exhibited significant positive heterotic effect, *per se* performance and high *sca* effects and can be utilized for direct selection.

Key words: Cotton, heterosis, fibre length and strength, micronaire, Line x Tester (L x T) analysis.

INTRODUCTION

Cotton is the leading natural fibre crop of the world. It is an important cash crop of India being used throughout the textile industry and plays a key role in the national economy. Fibre quality of a specific cotton genotype is a composite of various characteristics, including fibre length, fibre strength, fineness, and fibre elongation. These traits have their individual importance in spinning, weaving and dying units (Munro, 1987). Fibre length and strength properties mainly influence textile processing (Kohel, 1999). In addition, fibre uniformity is also of tremendous value to the textile industry. It is greatly correlated with the efficient spinning and weaving processes, which convert the fibre into fabrics. However, these traits are highly influenced by the environment with special reference to the fineness (Yuan et al., 2005; Percy et al., 2006). Ahuja. (2003) suggested that developing high fibre length and strength cultivars or hybrids is required to current modernized spinning mills. Hereafter, it is the need of the day to improve fibre quality in the dominating *hirsutum* genotypes, to fulfill the requirements of growing processing and textile industry. Current commercial cultivars of upland cotton have limited variability for these fibre quality traits. Industrial demand of cotton with the superior fibre traits is also a source of the guideline for cotton breeders. The success of these traits depends on the identification of genotypes with the ability to transmit high production potential into specific genotypic combinations (Basal and Turgut, 2005; Iqbal et al., 2003). Heterosis is a performance of F_1/F_2 genotypic combinations and is useful in determining the most appropriate parents for specific traits (Khan et al., 2010). Development of hybrids as a commercial variety is getting importance. Cotton is highly amenable for both heterosis and recombination breeding. Heterosis has substantially remained as one of the significant developments in cotton breeding programs, (Singh, 1982; Chaudhari et al., 1992; Baloch et al., 2003; Baloch, 2004; Memon et al., 2005; Ganapathy and Nadarajan, 2008; Khan et al., 2010).

Several studies have been reported on yield and yield attributing traits, but little work has been reported on the genetics and heterosis of fibre quality traits in cotton breeding. A few reports in the literature (Rahman et al., 1993; Zhang et al., 2002; Basal and Turgut 2003; Preetha and Raveendran, 2008; Karademir et al., 2009; Karademir and Gencer, 2010; Karademir et al., 2011; Bolek et al., 2011) have determined that cotton genotypes differ in fibre quality traits. The estimates of *per se* performance and heterosis provided useful information with regard to the possibilities and extent of improvement in the fibre characters of breeding material through selection. Therefore, present study objective was to estimate genetic variation for mean performance of parents and their hybrids and to estimate the effects of heterosis in F_1 cross combinations, to obtain information heterotic potential as to develop hybrid with improved fibre quality traits through Line x Tester (L x T) analysis.

MATERIALS AND METHODS

Genetic material

A field experiment was conducted to evaluate the fibre quality traits performance of four commercially cultivated varieties of upland cotton as *viz.,* MCU 12, MCU 5, SVPR 2, Surabhi and and seven *Gossypium hirsutum* genetic accessions as *viz.,* F 776, F 1861, SOCC 11, SOCC 17, TCH 1641, TCH 1644 and TCH 1646. These cultivars were cultivated in southern states of India, and collected from department of cotton, Tamil Nadu Agricultural University (TNAU), Coimbatore, Tamil Nadu, India. Genetic accessions were collected from Central Institute of Cotton Research (CICR), Coimbatore, Tamil Nadu, India.

Experimental design and field procedures

The cotton cultivars and accessions were evaluated in randomized block design (RBD) with three replications at Cotton Breeding Station, Tamil Nadu Agricultural University, Coimbatore, and Tamil Nadu in India. Each genotype was grown in a 4.5 m length row adopting a spacing of 75 cm between rows and 30 cm between the plants, to have 15 plants per row.

Sampling, traits measurements and methods

Data were recorded on five randomly selected plants per replication for all the five characters *viz.,* 2.5% span length (mm), fibre strength (g/tex), micronaire (μg/inch), uniformity ratio and fibre elongation percentage (%). Fibre quality traits were analyzed by high-volume instrument (HVI). The L x T analysis of heterosis was performed as suggested by Kempthrone (1957). Heterosis was calculated in terms of percent increase (+) or decrease (-) of the F_1 hybrids against its mid parent, better parent and standard parent value as suggested by Fehr (1987).

RESULTS AND DISCUSSION

Analysis of variance

The analysis of variance showed that line was significant for the characters viz., 2.5% span length, bundle strength and fibre elongation percentage. The testers varied significantly for the 2.5% span length, micronaire value and bundle strength. In case of the line x tester component, significant variation was observed for all the characters (Table 1). Between the parents and hybrids were observed substantial variation for all the characters. In addition, the parents versus hybrids showed significant variation for all the characters.

Mean performance

Significant variation was observed for each trait in all the genotypes. Mean expression of the fibre quality characters was recorded on the eleven genotypes and their 28 crosses. The comparison of treatment means indicated that cultivars had a significant effect on 2.5% span length (Table 2). Mean performance of parents, 2.5% span length had a minimum expression of 23 (TCH 1641) to 32.9 mm (MCU 5). Previous studies reported that fiber length could vary widely with plant variety and growing conditions. Ehsan et al. (2008), Copur (2006) and Khan et al. (1989) reported similar results for fiber length. The hybrids SVPR 2 x TCH 1641 and MCU 5 x TCH 1646 registered minimum (25.40 mm) and maximum (33.3 mm) length. The mean fibre length of hybrids was 29.83 mm. Niagun and Khadi (2001) observed that mean fibre length for *Gossypium barbadense* crosses was 35. 9 mm and this is the greater value compare to our results, and is in conformation with *G. barbadanse* which is higher in fibre length than *G. hirsutum.*

Fibre fineness or micronaire and fibre strength are very important characteristic of the fiber quality of cotton and are extremely useful for textile industry. In parents, bundle strength was ranged from 18.9 to 22.9 (g/tex). Maximum and minimum values recorded in hybrids were 23.80 g/tex in MCU 5 x SOCC 17 and 19.60 g/tex in SVPR 2 x TCH 1644. These results were supported by earlier studies (Khan, 2002; Karademir et al., 2011). The average bundle strength of parents and hybrids was 20.6 and 21.9 g/tex, respectively. Niagun and Khadi (2001) observed mean fibre strength for *G. hirsutum* crosses

Table 1. Analysis of variance showing mean square for fibre quality traits.

Source of variation	df	2.5 % Span length (mm)	Uniformity Ratio (%)	Micronaire (µg/inch)	Bundle strength (g/tex)	Elongation Percentage
Replication	2	117.64	23.05	41.97	13.20	18.49
Parents	10	108.48**	29.30**	5.42**	22.74**	25.06**
Hybrids	27	77.15**	9.59**	4.19**	23.27**	11.73**
Parents Vs Hybrids	1	322.09**	6.36*	16.70**	105.71**	4.35*
Error	76	0.18	0.55	0.08	0.19	0.14
Replication	2	78.11	19.00	43.17	8.41	10.28
Lines	3	6.59**	1.45	0.71	8.74**	6.76**
Tester	6	3.80*	1.58	3.81*	3.14*	1.02
line x tester	18	32.45**	8.74*	5.02**	12.90**	6.38**
Error	54	0.21	0.63	0.07	0.18	0.15

Table 2. Ranges for mean expression of parents in fibre quality characters.

S/N	Character	Minimum value	Maximum value	Mean	Parents recording	
					Lowest	Highest
1	2.5% Span length (mm)	23.00	32.90	28.15	SOCC 11	Surabhi
2	Bundle strength(g/tex)	18.93	22.90	20.66	TCH 1641	Surabhi
3	Micronaire (µg/inch)	3.40	4.60	4.03	Surabhi and F 1861	SOCC17
4	Uniformity ratio (%)	44.00	51.00	48.36	Surabhi & TCH 1641	MCU 5
5	Fibre elongation (%)	5.80	9.60	7.53	TCH 1641	SOCC11

was 26.6 g/tex, and this is greater than the value of present results.

The tester SOCC 17 recorded the greatest micronaire value of 4.6 µg/inch and lowest of 3.4 µg/inch in F 1861. Differences between the cultivars with respect to fiber fineness were also found significant by Copur (2006) and Ehsan et al. (2008). Of the 28 hybrids, MCU 12 x F 776 recorded the lowest value of 3.40 µg/inch. The highest value (4.80 µg/inch) was recorded by MCU 12 x TCH 1646. The average value of micronaire value in hybrids was 3.79 µg/inch (Table 3). Niagun and Khadi (2001) observed mean micronaire for *G. hirsutum and G. barbadense* crosses was 2.95 µg/inch, and this showed that the present study significantly exploited the hybrids than earlier studies.

The range for uniformity ratio was from 44 to 51(%), the genotypes exhibiting the values being Surabhi and MCU 5 respectively. The mean value of uniformity ratio for tester was 48.42. Among the hybrids, Surabhi x TCH 1641 recorded the lowest value of 45% and highest value (50%) recorded the SVPR 2 x TCH 1644 and SVPR 2 x SOCC 17 (Table 3). The hybrids had a mean uniformity ratio of 47.98 (%) which was intermediate between their parental group values. Niagun and Khadi (2001) observed mean uniformity ratio for *G. barbadense* crosses was 44.3, and this is lower than the present results. Fibre elongation percentage in testers had a range of 5.8 in TCH 1641 to 9.6 in SOCC 11. MCU 12 and SVPR 2 among the lines recorded the lowest and highest elongation of 6.8 and 8.5, respectively. The mean values of fibre elongation percentage for lines and testers were 7.27 and 7.67, correspondingly. The hybrids varied from 6.3 (Surabhi x TCH 1641) to 9.50 (SVPR 2 x TCH 1641) in respect of elongation percentage and displayed a mean of 7.37.

Expression of Heterosis for fibre quality traits

Estimation of heterotic effects is necessary to identify the new cross combinations that are suitable for direct exploitation. Therefore, heterosis over mid parent, better parent and standard parent was estimated in the entire cross combinations under study. For fibre length or 2.5% span length, eighteen hybrids displayed significant positive relative heterosis over mid parental value, with a range of 2.44% in MCU 12 x TCH 1646 to 17.35% in MCU 5 x SOCC 11 (Table 4). The results of heterosis are in conformity with the reports of Tuteja et al. (2005), Iraddi and Kajjidoni (2009), Karademir and Gencer (2010) and Karademir et al. (2011). Significant positive heterobeltiotic effect was discernible in thirteen hybrids, and ranged from 2.36 to 14.43%. Standard heterosis ranged from 2.75% in MCU 12 x F 1861 to 14.43% in

Table 3. Mean performance of F_1 crosses for fibre quality traits.

Hybrids	2.5% span length (mm)	Bundle strength (g/tex)	Micronaire (µg/inch)	Uniformity ratio (%)	Fibre elongation (%)
MCU 5 x F 776	32.0	22.4	3.6	46	7.1
MCU 5 x F 1861	29.2	23.6	4.1	49	7.5
MCU 5 x SOCC 11	30.1	23.0	3.7	49	7.0
MCU 5 x SOCC 17	30.5	23.8	3.6	48	6.8
MCU 5 x TCH 1641	30.5	21.1	3.5	48	7.4
MCU 5 x TCH 1644	31.1	21.1	3.7	46	7.1
MCU 5 X TCH 1646	33.3	22.1	3.5	48	6.6
MCU 12 x F 776	33.1	22.6	3.4	48	7.7
MCU 12 x F 1861	29.9	22.1	3.5	48	7.4
MCU 12 x SOCC 11	30.3	23.0	3.9	46	6.5
MCU 12 x SOCC 17	27.0	20.3	4.4	48	8.5
MCU 12 x TCH 1641	30.4	22.4	3.6	47	7.1
MCU 12 x TCH 1644	31.8	22.3	4.0	47	6.8
MCU 12 x TCH 1646	30.1	21.9	4.8	49	6.8
SURABHI x F 776	32.5	23.6	3.5	47	7.1
SURABHI x F 1861	31.0	20.9	4.5	49	6.5
SURABHI x SOCC 11	30.2	21.8	3.5	47	6.9
SURABHI x SOCC 17	29.2	20.5	3.5	49	7.2
SURABHI x TCH 1641	32.1	21.6	4.1	45	6.3
SURABHI x TCH 1644	26.7	20.9	3.6	50	8.0
SURABHI x TCH 1646	30.8	22.1	3.6	47	6.5
SVPR 2 x F 776	30.0	19.8	3.6	49	8.3
SVPR 2 x F 1861	27.1	20.7	3.8	48	8.3
SVPR 2 x SOCC 11	25.8	20.9	3.6	48	8.2
SVPR 2 x SOCC 17	26.6	20.0	3.9	50	8.0
SVPR 2 x TCH 1641	25.4	19.8	3.9	49	9.5
SVPR 2 x TCH 1644	27.4	19.6	3.7	50	7.8
SVPR 2 x TCH 1646	31.1	20.5	3.9	47	7.3
Mean of parent	28.1	20.6	4.0	48.3	7.5
Mean of cross	30.0	21.9	3.7	49.9	7.6
SE	0.24	0.25	0.16	0.43	0.21

MCU 5 x TCH 1646 exhibited by 22 hybrids for this character. The present findings was substantiate with Tuteja et al. (2005), Iraddi and Kajjidoni (2009) and Tuteja and Banga (2011).

The heterotic expression of relative heterosis for uniformity was significant and positive for four hybrids and negative for ten hybrids (Table 4). Niagun and Khadi (2001) and Karademir et al. (2009) noticed significant negative and positive heterosis for uniformity ratio, and this is in conformity to our results. All the hybrids showed heterobeltiosis in the negative direction. Twenty seven hybrids recorded significant heterosis ranging from 4.55 to 13.64% over standard parental value in positive direction.

The hybrid MCU 12 x TCH 1646 and Surabhi x F 1861 displayed significant positive relative heterosis over the mid parent and better parent value for micronaire. Eleven hybrids expressed significant negative heterosis over mid parent for micronaire. Which indicates that the greater the micronaire value, the lower the fineness. These results are in the agreement with earlier research findings of Carvalho et al. (1994), Soomro (2000), Rauf et al. (2005), Karademir and Gencer (2010) and Karademir et al. (2011) who reported varying degree of heterosis and heterobeltiosis for micronaire. Significant positive standard heterosis of 14.29% was evident in the hybrid MCU 12 x TCH 1646 for microanire value.

For bundle or fibre strength, fourteen hybrids displayed significant positive relative heterosis over mid parental value, with a range of 3.27% in Surabhi x TCH 1641 to 18.70% in MCU 5 x SOCC 17. Hybrid vigour was also observed by Hassan et al. (1999), Soomro (2000), Rauf et al. (2005) and Karademir et al. (2011). Significant positive heterobeltiosis over the better parents was discernible in seven hybrids, and it ranged from 3.60 to

Table 4. Expression of heterosis in hybrids (%) for 2.5% span length, uniformity ratio, and micronaire.

Hybrids	2.5 % Span length			Uniformity Ratio			Micronaire		
	di	dii	diii	di	dii	diii	di	dii	Diii
MCU 5 x F 776	14.70 **	13.07 **	9.97 **	-8.91 **	-9.80 **	4.55 **	-10.00	-16.28 **	-14.29 *
MCU 5 x F 1861	3.00 **	2.82 *	0.34	-1.01	-3.92 **	11.36 **	6.49	-4.65	-2.38
MCU 5 x SOCC 11	17.35 **	6.36 **	3.44 **	-2.97 **	-3.92 **	11.36 **	-13.95 **	-13.95 *	-11.90 *
MCU 5 x SOCC 17	15.46 **	7.77 **	4.81 **	-4.95 **	-5.88 **	9.09 **	-19.10 **	-21.74 **	-14.29 *
MCU 5 x TCH 1641	6.27 **	4.81 **	4.81 **	1.05	-5.88 **	9.09 **	-17.65 **	-18.60 **	-16.67 **
MCU 5 x TCH 1644	9.89 **	9.89 **	6.87 **	-8.00 **	-9.80 **	4.55 **	-10.84 *	-13.95 *	-11.90 *
MCU 5 x TCH 1646	16.03 **	14.43 **	14.43 **	-3.03 **	-5.88 **	9.09 **	-14.86 **	-17.83 **	-15.87 **
MCU 12 x F 776	15.94 **	11.82 **	13.75 **	-3.03 **	-4.00 **	9.09 **	-12.82 *	-17.07 **	-19.05 **
MCU 12 x F 1861	3.10 **	1.01	2.75 *	-1.03	-2.04	9.09 **	-6.67	-14.63 *	-16.67 **
MCU 12 x SOCC 11	15.21 **	2.36 *	4.12 **	-7.07 **	-8.00 **	4.55 **	-7.14	-9.30	-7.14
MCU 12 x SOCC 17	-0.25	-8.78 **	-7.22 **	-3.03 **	-4.00 **	9.09 **	1.15	-4.35	4.76
MCU 12 x TCH 1641	3.58 **	2.70 *	4.47 **	1.08	-4.08 **	6.82 **	-13.25 **	-14.29 *	-14.29 *
MCU 12 x TCH 1644	9.84 **	7.43 **	9.28 **	-4.08 **	-4.08 **	6.82 **	-1.23	-2.44	-4.76
MCU 12 x TCH 1646	2.44 *	1.58	3.32 **	1.03	0.00	11.36 **	18.52 **	17.07 **	14.29 *
SURABHI x F 776	7.62 **	-1.22	11.68 **	0.00	-6.00 **	6.82 **	-1.41	-5.41	-16.67 **
SURABHI x F 1861	1.14	-5.78 **	6.53 **	6.52 **	2.08	11.36 **	32.35 **	32.35 **	7.14
SURABHI x SOCC 11	8.29 **	-8.00 **	4.01 **	0.00	-6.00 **	6.82 **	-9.09	-18.60 **	-16.67 **
SURABHI x SOCC 17	1.68	-11.25 **	0.34	4.26 **	-2.00	11.36 **	-12.50 *	-23.91 **	-16.67 **
SURABHI x TCH 1641	3.55 **	-2.43 *	10.31 **	2.27	2.27	2.27	7.89	-2.38	-2.38
SURABHI x TCH 1644	-12.75 **	-18.84 **	-8.25 **	7.53 **	2.04	13.64 **	-2.70	-10.00	-14.29 *
SURABHI x TCH 1646	-0.54	-6.28 **	5.96 **	2.17	-2.08	6.82 **	-2.70	-10.00	-14.29 *
SVPR 2 x F 776	6.32 **	3.69 **	3.09 *	-1.01	-2.00	11.36 **	-10.00	-16.28 **	-14.29 *
SVPR 2 x F 1861	-5.35 **	-6.22 **	-6.76 **	-1.03	-2.04	9.09 **	-1.30	-11.63 *	-9.52
SVPR 2 x SOCC 11	-0.64	-10.83 **	-11.34 **	-1.68	-2.67 *	10.61 **	-16.28 **	-16.28 **	-14.29 *
SVPR 2 x SOCC 17	-0.50	-8.06 **	-8.59 **	1.01	0.00	13.64 **	-12.36 **	-15.22 **	-7.14
SVPR 2 x TCH 1641	-12.46 **	-12.71 **	-12.71 **	6.81 **	1.36	12.88 **	-8.24	-9.30	-7.14
SVPR 2 x TCH 1644	-4.25 **	-5.30 **	-5.84 **	2.04	2.04	13.64 **	-10.84 *	-13.95 *	-11.90 *
SVPR 2 x TCH 1646	7.18 **	6.87 **	6.87 **	-3.09 **	-4.08 **	6.82 **	-6.02	-9.30	-7.14

*, **, Significant at 5% and 1% level respectively; di, Relative heterosis; dii, heterobeltiosis; diii, standard heterosis.

18.41%. Carvalho et al. (1994) while studying heterosis in cotton also reported similar results for fibre strength. Standard heterosis was positive in twenty-seven hybrids, and the range was from 4.58% in SVPR 2 x TCH 1641, SVPR 2 x F 776 to 25.70% in MCU 5 x SOCC 17 for fibre strength. Tuteja and Banga (2011) observed positive standard heterosis in all the conventional hybrids, and few of male sterility based hybrids for fibre strength, and these are conformed in the present study results.

Five hybrids exhibited significant positive relative heterosis over mid parental value for fibre elongation percentage (Table 5). The estimates ranged from 10.34% in Surabhi x TCH 1644 to 32.87% in SVPR 2 x TCH 1641, and none of the hybrids were found to be significant over the better parent. Karademir et al. (2009) and Karademir et al. (2011) observed significant both positive and negative heterosis for fibre elongation percentage in cotton. Standard heterosis was positive in eighteen hybrids, and the range was exhibited from 12.07% in Surabhi x TCH 1646, Surabhi x F 1861, MCU 12 x SOCC 11 to 63.79% in SVPR 2 x TCH 1641 for elongation percentage.

Conclusion

Fibre quality parameters of cotton, fibre length and fineness have a vital influence on the yarn strength. The increasing fibre length results in improved yarn strength because a long fibre generates a greater frictional resistance to an external force. High fibre length and the tensile strength of the fibres becomes the controlling factor of yarn strength. The developing high fibre length and strength cultivars or hybrids are essential to current modernized spinning mills. Therefore, the present study was carried out for improving fibre quality traits from upland cotton by line x tester design. The results showed that hybrids are superior to the parents for all the fibre

Table 5. Expression of heterosis in hybrids (%) for bundle strength and elongation percentage.

Hybrids	Bundle strength			Elongation percentage		
	di	dii	diii	di	dii	diii
MCU 5 x F 776	6.92 **	2.75	18.31 **	-5.33	-12.35 **	22.41 **
MCU 5 x F 1861	14.66 **	12.03 **	24.65 **	4.90	1.35	29.31 **
MCU 5 x SOCC 11	12.75 **	11.11 **	21.48 **	-15.15 **	-27.08 **	20.69 **
MCU 5 x SOCC 17	18.70 **	18.41 **	25.70 **	-12.26 **	-20.93 **	17.24 **
MCU 5 x TCH 1641	8.11 **	4.98 **	11.44 **	16.54 **	7.25	27.59 **
MCU 5 x TCH 1644	6.37 **	5.31 **	11.80 **	-2.07	-6.58	22.41 **
MCU 5 x TCH 1646	9.95 **	9.95 **	16.73 **	-2.22	-4.35	13.79 *
MCU 12 x F 776	2.73	1.80	19.37 **	3.36	-4.94	32.76 **
MCU 12 x F 1861	2.16	-0.45	16.73 **	4.23	0.00	27.59 **
MCU 12 x SOCC 11	7.23 **	3.60 *	21.48 **	-20.73 **	-32.29 **	12.07 *
MCU 12 x SOCC 17	-3.79 *	-8.56 **	7.22 **	10.39 **	-1.16	46.55 **
MCU 12 x TCH 1641	8.91 **	0.90	18.31 **	12.70 **	4.41	22.41 **
MCU 12 x TCH 1644	6.44 **	0.45	17.78 **	-5.56	-10.53 *	17.24 **
MCU 12 x TCH 1646	3.55 *	-1.35	15.67 **	1.49	-0.00	17.24 **
Surabhi x F 776	5.59 **	3.06	24.65 **	-4.44	-11.52 **	23.56 **
Surabhi x F 1861	-4.93 **	-8.73 **	10.39 **	-9.09 *	-12.16 **	12.07 *
Surabhi x SOCC 11	0.00	-4.80 **	15.14 **	-16.36 **	-28.13 **	18.97 **
Surabhi x SOCC 17	-4.43 **	-10.48 **	8.27 **	-7.10 *	-16.28 **	24.14 **
Surabhi x TCH 1641	3.27 *	-5.68 **	14.08 **	-1.31	-9.18 *	8.05
Surabhi x TCH 1644	-1.88	-8.73 **	10.39 **	10.34 **	5.26	37.93 **
Surabhi x TCH 1646	2.79	-3.49 *	16.73 **	-3.70	-5.80	12.07 *
SVPR 2 x F 776	-4.81 **	-9.17 **	4.58 *	0.00	-2.35	43.10 **
SVPR 2 x F 1861	1.31	-1.74	9.33 **	4.40	-2.35	43.10 **
SVPR 2 x SOCC 11	3.37 *	1.13	10.56 **	-9.39 **	-14.58 **	41.38 **
SVPR 2 x SOCC 17	0.50	0.00	5.63 **	-6.43 *	-6.98	37.93 **
SVPR 2 x TCH 1641	2.24	0.00	4.58 *	32.87 **	11.76 **	63.79 **
SVPR 2 x TCH 1644	-0.76	-1.01	3.52	-3.11	-8.24 *	34.48 **
SVPR 2 x TCH 1646	2.76	1.99	8.27 **	-3.31	-14.12 **	25.86 **

*, **, Significant at 5 and 1% level,respectively; di, relative heterosis; dii, heterobeltiosis; diii, standard heterosis.

quality traits except micronaire value. Of the 28 hybrids, MCU 12 x TCH 1644 and SVPR 2 x TCH 1646 for 2.5% span length, MCU 5 x SOCC 17 and Surabhi x F 776 for bundle strength exhibited highest in itself performance, and heterotic effect was found to be utilized for directional choice. Increasing the fibre quality traits are a valuable addition to cotton cultivars or hybrids, and it will be useful for textile industries.

REFERNECES

Baloch MJ (2004). Genetic variability and heritability estimates of some polygenic traits in upland cotton. Pak. J. Sci. Indust. Res. 42(6):451-454.

Baloch MZ, Ansari BA, Memon N (2003). Performance and selection of intraspecific hybrids of spring wheat (*Triticum aestivum* L.). Pak. J. Agric. Agril. Engg. Vet. Sci. 19(1):28-31.

Basal H, Turgut I (2003). Heterosis and combining ability for yield components and fiber quality parameters in a half diallel cotton (*G. hirsutum*L.) population. Turk. J. Agric. For. 27(2):207-212.

Basal H, Turgut I (2005). Genetic analysis of yield components and fibre strength in upland cotton (Gossypium hirsutum L.). Asian J Plant Sci 27(4):207-212.

Bolek Y, Cokkizgin H, Bardak A (2011). Combining ability and heterosis for fiber quality traits in cotton. Plant Breed. Seed Sci. 62(1):3-16.

Carvalho LD, Moraes CF, Cruz CD (1994). Combining ability and heterosis in upland cotton. Revista. Ceres. 41(3):514–527.

Chaudhari PN, Borole DN, Tendulkar AV, Narkhede BN (1992). Heterosis for economic traits in intra-specific crosses of desi cotton. J. Maharashtra Agric. Univ. 17(2):273-276.

Copur O (2006). Determination of yield and yield components of some cotton cultivars in semi-arid conditions. Pak. J Biol. Sci. 9(14):2572-2578.

Ehsan F, Nadeem A, Tahir MA, Majeed A (2008). Comparitive yield performance of new cultivars of cotton (*Gossypium hirsutum* L.). Pak. J. Life Soc. Sci. 6(1):1-3.

Fehr WR (1987). Principles of cultivar development. Theory and technique. Macmillan Pub. Co. Inc. New York. USA. pp. 115-119.

Ganapathy S, Nadarajan N (2008). Heterosis studies for oil content, seed cotton yield and other economic traits in cotton *(G. hirsutum* L.). Madras Agric. J. 95(7-12):306-310.

Iqbal M, Chang M, Iqbal MZ (2003) Breeding behavior effects for yield, its components and fibre quality in intraspecific crosses of cotton (G. *hirsutum* L.). OnLine. J. Biol. Sci. 4:451-459.

Iraddi V, Kajjidoni ST (2009). A Comparative study on heterosis for productivity and fibre quality traits in intra-herbaceum and interspecific (*G. herbaceum* x *G. arboreum*) crosses of diploid cotton. J. Res. Angrau. 37(3-4):35-43.

Karademir C, Karademir E, Ekinci R, Gencer O (2009). Combining Ability Estimates and Heterosis for Yield and Fiber Quality of Cotton in Line x Tester Design. Not. Bot. Hort. Agrobot. Cluj. 37(2):228-233.

Karademir C, Karademir E, Gencer O (2011). Yield and Fiber Quality of F 1 and F 2 Generations of Cotton (Gossypium hirsutum L.) Under Drought Stress Conditions. Bulg. J. Agric. Sci. 17(6):795–805.

Karademir E, Gencer O (2010). Combining Ability and Heterosis for Yield and Fiber Quality Properties in Cotton (G. hirsutum L.) obtained by Half Diallel Mating Design. Not. Bot. Hort. Agrobot. Cluj. 38(1):222-227.

Kempthrone O (1957). An introduction to genetic statistics. John Wiley and Sons, Ist edition, New York, USA. 456-471.

Khan UQ (2002). Studies of heterosis in fibre quality traits of cotton. Asian J. Plant Sci. 1(5):593-595.

Khan WS, Khan AA, Naz AS, Ali S (1989). Performance of six Punjab commercial varieties of Gossypium hirsutum L. under Faisalabad conditions. The Pak. Cottons. 33(2):60-65.

Khan N, Basal H, Hassan G (2010). Cottonseed oil and yield assessment via economic heterosis and heritability in intraspecific cotton populations. Afr. J. Biotechnol. 9(44):7418–7428.

Kohel RJ (1999). Cotton germplasm resources and the potential for improved fibre production and quality. Cotton Fibres, The Haworth Press, Inc., NY, USA. pp. 167-182.

Munro JM (1987) Cotton: Tropical Agriculture series, 2nd edition. Longman Scientific and technical JohnWilley Sons Inc, New York, USA, P. 161.

Niagun HG, Khadi BM (2001). Progeny analysis of fibre characteristics of DCH 32 an interspecific cotton hybrid. J . Genet. Breed. 55(2):209-216.

Percy RG, Cantrell RG, Zhang J (2006). Genetic variation for agronomic and fiber properties in an introgressed recombinant inbred population of cotton. Crop Sci. 46(3):1311-1317.

Preetha S, Raveendran TS (2008). Combining ability and heterosis for yield and fibre quality traits in line x tester crosses of Upland cotton (*G. hirsutum*. L). Int. J. Plant Breed. Gene. 2(2):64-74.

Rahman S, Khan MA, Ayub M, Khan MA (1993). Heterosis and heterobeltiosis for lint percentage, seed and lint indices and stable length in different cross combination of five cotton cultivars. Pakistan J. Agric. Res. 14 (2-3):115-120.

Rauf S, Khan TM, Nazir S (2005). Combining ability and heterosis in (*G. hirsutum* L.) Int. J. Agric. Biol. 7(1):109-113.

Singh P (1982). Note on useful heterosis in upland cotton. Indian J. Agric. Sci. 52(1):29-31.

Soomro AR (2000). Assessment of useful heterosis in glandless *Gossypium hirsutum* cotton strains through their performance in hybrid combination. Pakistan J. Bot. 32(1):5–68.

Tuteja OP, Banga M (2011). Effects of cytoplasm on heterosis for agronomic traits inupland cotton (*Gossypium hirsutum).*Indian J. Agric. Sci. 81(11):23-29.

Tuteja OP, Kumar S, Hasan H, Singh M (2005). Heterosis and interrelationship between seed cotton yield and qualitative characters in upland cotton (*Gossypium hirsutum*). Indian J. Agric. Sci. 75(3):167-171.

Yuan YL, Zhang TZ, Guo WZ, Pan JJ, Kohel RJ (2005). Diallel analysis of superior fibre quality properties in selected upland cottons. Acta Gene. Sinica. 32(1):79-85.

Zhang ZS, Li XB, Liu DJ, Huang SL, Zhang FX (2002). Study on heterosis utilization of upland cotton (*G. hirsutum* L.) lines with high fibre quality. Cotton Sci. 14(1):264-268.

23

Productivity enhancement of sesame (*Sesamum indicum* L.) through improved production technologies

R. S. Raikwar and P. Srivastva

Jawaharlal Nehru Krishi Vishwa Vidyalaya, College of Agriculture Tikamgarh (M.P.) India.

Sesame (*Sesamum indicum* L.) is most important oil seed crop in Madhya Pradesh. One of the major constraints of its low productivity is non-adoption of improved technologies. Front line demonstration were conducted at 65 farmers field, to demonstrate production potential and economic benefit of improved technologies comprising short duration, phillody (mycoplasma) resistant varieties, line sowing, integrated nutrient management and timely weed removal (TKG-55,TKG-306 and JTS-8), line sowing (45 × 10 cm), integrated nutrient management (60:30:15:40, NPKS kg/ha). The seeds were treated with phosphate-solubilzing bacteria each at 20 g/kg of seeds. Pre-emergence application of weedicide Pendimethalin at 1 kg a.i /ha used for effective control of the weeds during *Kharif* season of 2007 to 2008 to 2011 to 2012 in rainfed condition. The improved technology recorded a mean yield of 5.34 q/ha which was 34% higher than that obtained with farmers practice yield of 3.45 q/ha. The improved technologies resulted higher mean net income of Rs.12913.80/ha with a benefit cost ratio of 2.49 as compared to local practice (7740/ha, 2.20).

Key words: Sesame, frontline demonstration, improved technologies, net return, productivity.

INTRODUCTION

Sesame or gingelly (*Sesamum indicum*) commonly known as til (Hindi) is an ancient oilseed crop grown in India, and perhaps the oldest oilseed crop in the world. The crop is now grown in a wide range of environments, extending from semi-arid tropics and subtropics to temperate regions. Consequently, the crop has a large diversity in cultivars and cultural systems. India is the largest producer of sesame in the world. It also ranks first in the world in terms of sesame-growing area (24%). Figure 1 shows that the increase in sesame productivity is about 2% for Ethiopia and India and 2.8% for China in the period of 1990 to 2007 (FAO, 2008). Perhaps the productivity increase should better be interpreted as a linear trend with and increase of 7 kg/ha per year in India, 13 kg/ha per year in Ethiopia and 22 kg/ha per year in China. Clearly, the level and rate of increase of yield per hectare of sesame in China is more than 50% higher than in Ethiopia. This probably indicates a great opportunity for a prolonged and higher increase in sesame productivity in India. In order to realise this opportunity, an analysis is needed of the major current constraints limiting sesame productivity in India. Due to the increased production per hectare and the increase in acreage, India has become the first producing and exporting country (Figure 2).

The yield increase is due both to development and use of improved varieties and improved agronomy practices and crop protection. The potential yield of sesame still is much higher than actual yield, as still much damage occurs by pests and diseases, insufficient weed control, to high levels of monocropping, lack of mechanisation

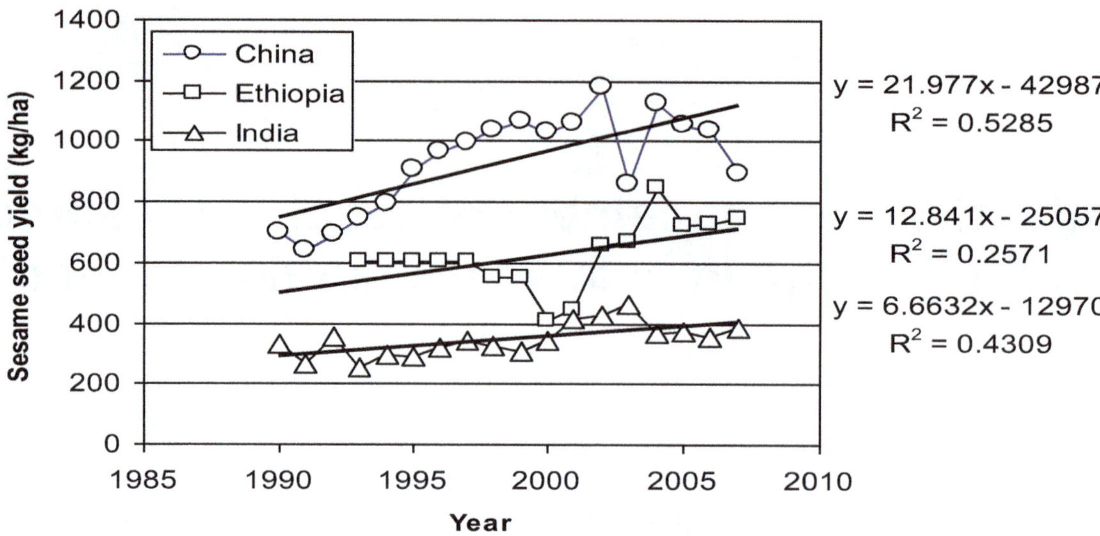

Figure 1. Sesame productivity in India, China and Ethiopia from 1990 to 2007 (FAO, 2008).

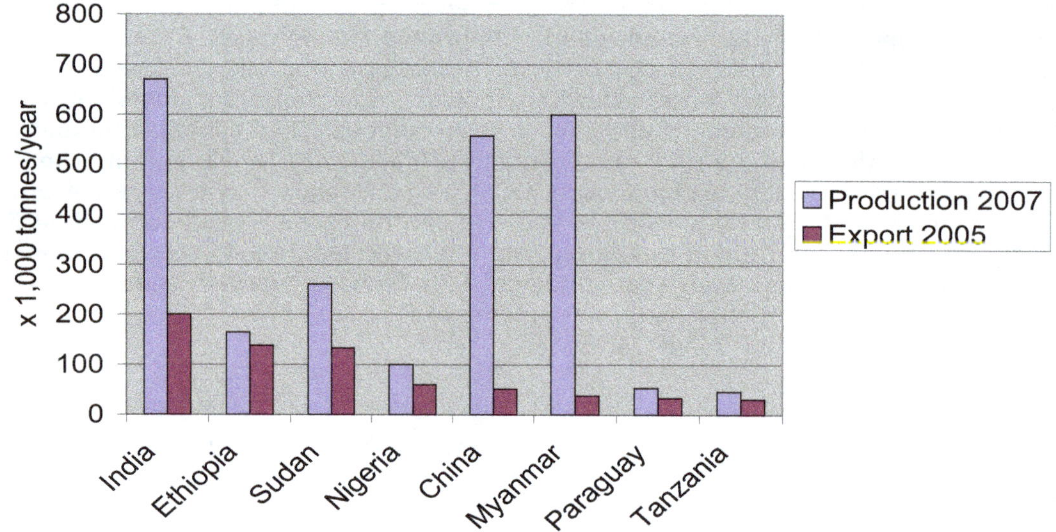

Figure 2. Comparison of total production and export of sesame seed of India compared with the major producing and exporting countries (data of 2007, FAOSTAT, 2008).

(amongst others causing seed shattering when not enough labour is available during harvest) and unrealised genetic potential. Potential yields are probably as high as 2000 kg/ha (Mkamilo and Bedigian, 2007).

Its seeds may be eaten fried, mixed with sugar or in the form of sweat meats. Sesame oil is used as a cooking oil in southern India. It is also used for anointing the body, for manufacturing perfumed oils and for medicinal purposes. Sesame cake is a rich source of protein, carbohydrates and minerals, such as calcium and phosphorus. The cake is edible and is eaten widely by working classes. It is also a valuable and nutritious feed for milch cattle. The oil is highly resistant to oxidative rancidity and exerts synergistic affect on the action of certain insecticides like pyrethrins and rotenone. Ryu et al. (1972) reported that sesame oil contains sesamolin and sesamine which is used as synergist for insecticides.

Sesame is grown in an area of 7.54 million hectares with a production of 3.34 million tonnes in the world with a productivity of 443 kg/ha (FAI, 2011). China, Myanmar and Sudan account for 40% of the world's sesame production. In India, sesame is grown in about 1.8 million hectares with a total production of 0.76 million tonnes and productivity of 422 kg/ha (FAI, 2011). West Bengal alone accounts for 25% of the total sesame production in India. The other major sesame-producing states are Gujrat, Madhya Pradesh, Tamil Nadu, Maharashtra, Karnataka, Rajasthan and Uttar Pradesh. The effect of plant population on yield and yield components have been reported by several workers. For example, seed yield per unit area increases with increases population density from 80,000 to 160,000 plant/ha and beyond this density in becomes counterproductive (Delgado and Yermanos, 1975). Also increased number of seed per capsule, number of capsule per plant, and dry mater production increased when the intra-row spacing increased from 30 to 90 cm (Weiss, 1983; Olowe and Busari, 1994).

In general, average productivity of sesame continues to be lower (144 to 234 kg/ha) than expected from agricultural technology for the last 20 years, mainly due to its cultivation on marginal lands, under poor management and without inputs except seed. The major constraint responsible for lower yield are inappropriate production technologies viz; broadcast method of sowing, no use of fertilizer and untimely weed management (45 DAS), (Khaleque and Begum, 1991). The yield of sesame can be increased by 21 to 53% with adoption of improved technologies such as improved variety, recommended dose of fertilizer, weed management and plant protection. Keeping this in view, frontline demonstrations on sesame were conducted to demonstrate the production potential and economic benefits of latest improved technologies on farmer's fields.

MATERIALS AND METHODS

Front line demonstrations were conducted on 65 farmers fields of five adopted villages viz; Sukwa, Barath, Doriya, Pannapura and Chokhada of Chhatarpur, District in Bundelkhand region of Madhya Pradesh during Kharif seasons of 2007 to 2008 to 2011 to 2012 in rainfed condition, on light to medium soil with low to medium fertility status under sesame-gram production system. Each demonstration was conducted on an area of 0.4 ha and the same area adjacent to the demonstration plot was kept as farmers practices. For many of the diseases and pests resistance occurs in sesame, e.g. resistance to phyllody (Singh et al., 2007), resistance to powdery mildew in India and phyllody (Gopal et al., 2005). The package of improved technologies included phillody (mycoplasma) resistant varieties, line sowing, integrated nutrient management and timely weed removal. The varieties of sesame TKG-55 in 2007 to 2008, TKG-306 2008 to 2009 and JTS-8 in 2009 to 2010 to 2011 to 2012 were included in demonstrations. The spacing was 45 cm between rows and 10 cm between plants in the row. Thinning should be done scrupulously to ensure recommended plant spacing within a row. The first thinning is done invariably 14 days after sowing and the second thinning 21 days after sowing. Excess population adversely affects growth and yield of crop. Seed was treated with Thiram at 2.5 g/kg seed for prevention of seed-borne diseases. Seed sowing was done between July-8-July 28 in 2007 to 2008, July 8-July 30 in 2008 to 2009, July 8-August 3 in 2009 to 2010, July 5-July 30 in 2010 to 2011 and July 5-July 28 in 2011 to 2012 with a seed rate of 5 kg/ha. Entire dose of N and P through diammonium phosphate and K through muriate of potash and sulphur at 60:30:15:40, respectively, was applied as basal before sowing. In India, highest net returns were found with 60 kg N/ha, 30 kg P/ha and 15 kg K/ha (Tripathi and Rajput, 2007). 40 kg S/ha productivity increased from 700 to 800 kg/ha with also a 3% higher oil content (increase from approx. 47 to 50%) (Maragatham et al., 2006). Hand weeding was done once at 25 days after sowing. Weeds were also controlled effectively by using of proper herbicides. In 2007 to 2008 the weedicide was used Diuron at 400-600 g/ha in 2008 to 2009 Basalin was used at 1 kg a.i/ ha, in 2009 to 2010 to 2011 to 2012 Pendimethalin was used at 1 kg a.i/ha as pre-emergence treatment for effective control of weed. The crop was harvested during September 30 to October 15 after the leaves turn yellow and start dropping while the capsules are still greenish-yellow.

RESULTS AND DISCUSSON

A total rainfall of 348, 623, 799, 653 and 556 mm was recorded in 32, 40, 47,45 and 42 rainy days during the crop season of 2007 to 2008, 2008 to 2009, 2009 to 2010, 2010 to 2011 and 2011 to 2012, respectively however, heavy rainfall (779 mm) was received in the first week of July (2009 to 2010). This caused an unusual delay in sowing during 2009 to 2010 and lowered productivity of sesame. The late planted crop finds relatively less time for plant growth and development. The sesame crop received 213.8 mm rains in 2008 to 2009 at maturity stage during first week of October. This caused seed sprouting and shattering in the capsule itself in standing crop conditions which lowered the productivity.

The yield attributing characters of number of capsule per plant under improved technology were 135.60, 115.20, 76.40, 95.00 and 95.60 as against local check (farmers practice), 100.2, 85.6, 45.8, 75.6 and 80.6 (Table 1) during the year 2007 to 2008, 2008 to 2009, 2009 to 2010, 2010 to 2011 and 20011 to 2012, respectively. There were 26.10, 25.69, 40.05, 20.37 and 15.69% increase in number of capsules under demonstration of improved technology over and above local check (farmers practice). The average number of capsules per plants were 103.56 under improved technology and 77.57 under local check, thus there were 25.58% more capsules per plant under improved technology demonstration as compared to local check. The average number of seeds per plant observed in improved technology was 71.06 as compared to 66.32 in local check. In the year 2007 to 2008, 2008 to 2009, 2009 to 2010, 2010 to 2011 and 2011 to 2012 the number of seeds per plants under improved technology and local check were 76.60 and 70.10, 72.50 and 69.50, 65.20 and 60.60, 70.2 and 63.20 and 70.80 and 68.20, respectively. The percentage increase in seeds per plants during these years were 8.48, 4.13, 7.05, 9.97 and 3.67 respectively with and overall average 6.66 seeds per plants. As regards test weight (g/100 seed) the observation showed that during the years 2007 to 2008, 2008 to 2009, 2009 to 2010, 2010 to 2011 and 2011 to 2012 the test weight under improved technology and local check were 3.00

Table 1. Yield attributing characters of Sesame.

Year	Rainfall during crop season (mm)	Rainy days during crop season (no.)	Yield attributing characters								
			No. of capsules/plant			No. of seeds/Capsule			Test weight(gm)		
			Improved technology	Local check	% increased	Improved technology	Local check	% increased	Improved technology	Local check	% increased
2007-2008	348	32	135.60	100.20	26.10	76.60	70.10	8.48	3.00	1.80	40.00
2008-2009	623	40	115.20	85.60	25.69	72.50	69.50	4.13	2.12	1.38	34.91
2009-2010	779	47	76.40	45.80	40.05	65.20	60.60	7.05	2.00	1.20	40.00
2010-2011	653	45	95.00	75.65	20.37	70.20	63.20	9.97	2.25	1.45	35.56
2011-2012	556	42	95.60	80.60	15.69	70.80	68.20	3.67	2.50	1.50	40.00
Average	591.8	41.2	103.60	77.60	25.60	71.10	66.30	6.70	2.40	1.50	38.10

Table 2. Seed yield of Sesame as affected by improved and local practices in farmers fields.

Year	Area (ha)	Demonstration (No.)	Yield (q/ha)				Additional yield (q/ha) over local check	% increased in yield over local check
			Improved technology			Local check		
			Maximum	Minimum	Average			
2007-2008	5.0	13	6.84	5.23	6.13	4.40	1.73	28.00
2008-2009	5.0	13	6.56	4.91	5.20	3.58	1.62	21.00
2009-2010	5.0	13	5.60	3.85	4.72	2.10	2.62	56.00
2010-2011	5.0	13	5.75	4.25	5.26	3.53	1.73	33.00
2011-2012	5.0	13	6.50	4.80	5.40	3.65	1.75	32.00
Average	5.0	13	6.25	4.61	5.34	3.45	1.89	34.00

and 1.80, 2.12 and 1.38, 2.00 and 1.20, 2.25 and 1.45 and 2.50 and 1.50 respectively with an average test weight 2.37 under improved technology and 1.47 under local check. The per cent increase in test weight during above year was found to be 40.00, 34.91, 40.00, 35.56 and 40.00 with an average of 38.09.

The productivity of sesame in Chhatarpur District of Madhya Pradesh in India under improved production technology ranged between 385 and 684 kg/ha with mean yield of 625 kg/ha. The productivity under improved technology varied from 523 to 684, 491 to 656, 385 to 560, 425 to 575 and 480 to 650 kg/ha with a mean yield of 613, 520, 472, 526 and 540 kg/ha during 2007 to 2008, 2008 to 2009, 2009 to 2010, 2010 to 2011 and 2011 to 2012, respectively (Table 2) as against a yield range between 210 and 440 kg/ha with a mean of 345 kg/ha under farmer's practices (local check). The additional yield under improved technologies over farmers practice ranged from 162 to 262 kg/ha with a mean of 189 kg/ha. In comparison to farmer's practice, there was an increase of 28, 21, 56, 33 and 32% in productivity of sesame under improved technologies in 2007 to 2008, 2008 to 2009, 2009 to 2010, 2010 to 2011 and 2011 to 2012. Respectively the increased grain yield with improved technologies was mainly because of line sowing, use of nutrient management and timely

Table 3. Cost of cultivation (Rs/ha), net return (Rs/ha) and benefit: cost-ratio of Sesame as affected by improved and local practices.

Year	Total cast of cultivation		Net return (Rs/ha)		B:C ratio		Additional cost of cultivation (Rs/ha)	Additional net return (Rs/ha)
	Improved technology	Local check	Improved technology	Local check	Improved technology	Local check		
2007-2008	6920	6348	11470	6852	2.66	2.08	572	4618
2008-2009	7040	5460	8560	4980	2.21	1.91	1580	3580
2009-2010	8075	5560	8919	5384	2.10	1.97	2515	3535
2010-2011	9620	8225	14366	7872	2.00	1.96	1395	6494
2011-2012	8500	6500	21254	13612	3.50	3.09	2000	7642
Average	8031.00	6418.60	12913.80	7740.00	2.49	2.20	1612.40	4457.80

Sale rate of Sesame 2007-2008 Rs. 3000/q, 2008-2009 Rs.3000/q, 2009-2010 Rs.3600/q, 2010-2011 Rs.4560/q 2011-2012 Rs.5510/q.

weed management.

Fertiliser response has been widely studied in other countries and the extent of the response depends on many factors: with high yielding varieties higher fertiliser rates are needed and also in cases of lower soil fertility. (Tripathi and Rajput, 2007). Sometimes micronutrients and improvement of cation exchange capacity proved helpful, for example by use of humix (humic acid preparation) (Abo-El-Wafa and Abd-El-Lattief, 2006) Hegde DM.1998 reported that integrated nutrient management increased productivity by 36% as compared to local variety of sesame. Kinman and Stark (1954) reported that adoption of improved variety increased productivity by 32% as compared to local variety of sesame. Improved technology produced higher grain yield in 2007 to 2008 to 2011 to 2012 as compared to local check. The reason for this could be the inter plant competition for the moisture and nutrients which could be more severe under local check demonstration (Farmers practice). Also, the higher weed infestation under the local check as evident from the higher weed cover and reduced the amount of nutrients and water available to the local check. This agrees with the findings of Weiss (1971), Imoloame et al. (2007) and Stonebridge (1963) who reported the superiority of row planting over broad casting to control weed and that this factor resulted in considerable yield increased and also grain yield increased significantly.

Phytopthora and Phyllody resistant variety, integrated The economic viability of improved technologies over traditional farmer's practices was calculated depending on prevailing prices of inputs and output costs (Table 3). It was found that cost of production of sesame under improved technologies varied from Rs 6920 to 9620/ha with an average of Rs.8031/ha as against Rs. 5460 to 8225/ha with an average of Rs.6418.60/ha under farmers practice (local check). The improved production technologies registered an additional cost of production ranging from Rs.572 to 2515/ha with an average of Rs. 1612/ha over local check. The additional cost increased in the improved technologies was mainly due to more cost involved in balanced fertilizer, improved seed and weed management practices. Cultivation of sesame under improved technologies gave higher net return which ranged from Rs 8560 to 21254/ha, with a mean of Rs.12913.80/ha as compared to farmers practices which recorded Rs. 4980 to 13612/ha with mean of Rs. 7740/ha. Similar results also have been reported by Khan et al. (2009). There was an additional net return of 4618 in 2007to 2008, 3580 in 2008 to 2009, 3535 in 2009 to 2010, 6494 in 2010 to 2011 and 7642 in 2011 to 2012 under demonstration plots. The improved technologies also gave higher benefit cost ratio, 2.26 2.21, 2.10, 2.00 and 3.50 compared to 2.08, 1.91,1.97, 1.96 and 3.09 under local check in the corresponding season the results from the current study clearly brought out the potential of improved production technologies in rainfed condition of Madhya Pradesh in India. To get maximum yield of sesame recommended package of practices should be followed. By not following any one management practice yield may be reduced severely and it was also observed that delay in sowing, unbalanced does of fertilizer, untimely weed management and plant protection drastically reduced the grain yield of sesame.

REFERENCES

Abo-El-Wafa AM, Abd-El-Lattief EA (2006). Response of some sesame (*Sesamum indicum L.*) cultivars to fertilization treatments by micronutrients, biofertilizer and humix. Assiut J. Agric. Sci. 37:55-65.

Delgado M, Yermanos DM (1975). Yield component of sesame (*Sesamum indicum* L.) under different population densities. Econ. Bot. 29(1):68-78.

FAO (2008). FAO Agricultural Production Statistics, http://faostat.fao.org/ as accessed on 30 November 2008.

FAI (2011). Fertiliser Statistics, Fertiliser Association of India, New Delhi.

Gopal K, Jagadeswar R, Babu GP (2005). Evaluation of sesame (*Sesamum indicum*) genotypes for their reactions to powdery mildew and phyllody diseases. Plant Dis. Res Ludhiana 20:126-130.

Imoloame EO, Gworgwor NA, Joshua SD (2007). Sesame (*Sesamum indicum* L.) weed infestation, yield and yield components as influenced by sowing method and seed rait in Sudan Savanna agro-ecology of Nigeria. Afr. J. Agric. 2(10):528-533.

Khaleque MA, Begum D (1991). Area and production of Oilseed crops, 1988-90. In fifteen years of oilseed research and development in Bangladesh. AST/CIDA 28:190.

Khan MAH, Sultana NA, Islam MN, Hasanuzzaman M (2009). Yield and yield contributing of Sesame as affected by different management practices. Am.-Eur. J. Sci. Res. 4(3):195-197.

Kinman ML, Stark SM (1954). Yield and composition of sesame (*Sesamum indicum* L.) as affected by variety and location. J.A.O. Chem. Soc. 31(3):104-108.197.

Mkamilo GS, Bedigian D (2007). In *PROTA* (Plant Resources of Tropical Africa / Ressources végétales de l'Afrique tropicale), Wageningen, Netherlands. < http://database.prota.org/search.htm>. Accessed 30 November, 2008. (Van der Vossen, H. A. M. & Mkamilo, G. S., eds.).

Olowe VIO, Busari LD (1994). Appropriate plant population and spacing for sesame (*Sesamum indicum L.)* in the Southern Guinea Savanna of Nigeria. Trop. Oil Seeds J. 2:18-27.

Singh PK, Mohammad A, Madhu V, Srivastava RL, Kumud K, Ram N (2007). Screening and development of resistant sesame varieties against phytoplasma. Bull. Insectol. 60:303-304.

Stonebridge WC (1963). Benniseed variety and sowing method trails. Technical Report of Institute for Agricultural Research, Northern Nigeria 28:1-9.

Tripathi ML, Rajput RL (2007). Response of sesame (*Sesamum indicum*) genotypes to levels of fertilizers. Adv. Plant Sci. 20:521-522.

Experimental and modelling study of thin layer drying kinetics of pellets millet flour

P. T. Bassene[1], S. Gaye[1], A. Talla[2], V. Sambou[1] and D. Azilinon[1]

[1]Laboratoire d'Energétique Appliquée (LEA), Ecole Supérieure Polytechnique (ESP), Université Cheikh Anta Diop (UCAD) BP 5085 Dakar-Fann, Dakar, Sénégal.
[2]Laboratoire d'Energétique d'Eau et de l'Environnement, Ecole Nationale Supérieure Polytechnique (ENSP), BP 8390, Yaoundé, Cameroun.

The thin layer drying kinetics of pellets millet flour was studied at drying air temperatures of 40, 50 and 60°C using a constant air velocity of 1.0 m/s for pellets size 2.5 mm < d_1 < 3.15 mm and 5 mm < d_2 < 6.30 mm. Ten thin layer drying models were evaluated by fitting the experimental moisture data. The goodness of fit of each model was evaluated using the coefficient of determination (R^2) the reduced chi-square (χ^2) and root means square error (RMSE). Among these models, the modified Henderson and Pabis model is the best one to describe the behaviour in the drying of pellets millet. The effective diffusivity has been found to be varying between 4.4676 × 10^{-9} and 31.94 x 10^{-9} m^2/s and activation energy was 33.048 and 42.658 kJ/mol for pellets size of d_1 and d_2 respectively.

Key words: Thin layer drying, mathematical modelling, effective diffusivity, activation energy.

INTRODUCTION

Cereals are the main food in Sahel countries. There is a strong demand for grain products in both rural and urban areas. This sudden and unexpected interest for local cereals, once abandoned, is interesting in agriculture and agribusiness. It will finally allow finding enough opportunities for these local grains and thus reviving the cultivation of cereals including millet.

Drying process is one of the oldest methods of agricultural products preservation (Meziane, 2011). It is the most important process to preserve grains, crops and foods of all varieties. The removal of moisture prevents the growth and reproduction of microorganisms causing decay and minimises many of the moisture-mediated deterioration reactions. It brings about substantial reduction in weight and volume, minimising packing, storage and transportation costs and enables storability of the product under ambient temperatures (Doymaz, 2008).

In order to study the effects of the drying on the pellets millet it is necessary to have a deep knowledge of the mass transfer parameters and the drying kinetics. These characteristics are considered to be important for the design, simulation and optimization of the drying process by using mathematical modelling. The mathematical models have proved to be very useful in the new design and/or improvement of drying systems and analysis of mass transfer phenomena involved during drying (Akpinar et al., 2008).

Thus it is necessary to have an accurate model able to predict water removal rates and describe the drying performance of each product under the common conditions used in normal commercial relevant facilities (Flores et al., 2012). Therefore several researchers have investigated the drying kinetics of various agricultural products in order to evaluate different mathematical models for describing drying characteristics. Belghit and Bennis (2009) made experimental analysis of the drying kinetics of cork. Doymaz (2008) investigated the drying kinetics of strawberry. Drying kinetics of nopal *(Opuntia ficus-indica)* using three different methods and their effect on their mechanical properties were studies by Torres et al. (2008); Lertworasirikul (2008) investigated drying kinetics of semi-finished cassava crackers. Roberts et al.(2008) studied drying of grape seeds representing waste products from white wine processing *(Riesling)*, red wine processing *(Cab Franc)*, and juice processing *(Concord)*. Chong et al. (2008) investigated drying kinetics and product quality of dried chempedak. Recently, Folres et al. (2012) studied the drying kinetics of castor oil seeds during fluidized bed drying and Mezaine (2011) investigated drying kinetics of olive pomace in a fluidized dryer. Arumuganathan et al. (2009) studied the drying kinetics of milky mushroom slices in a fluidized bed dryer.

Thin layer drying equations describe the drying phenomena in a unified way, regardless the controlling mechanism. They have been used to estimate drying times of several products and to generalize drying curves. In the development of thin layer drying models for agricultural products, generally the moisture content of the material at any time after it has been subjected to a constant relative humidity and temperature condition is measured and correlated to the drying parameters (Akpinar et al., 2008). Many studies have recently focused on thin layer drying of agricultural products, fruits, and food materials. Thin layer drying characteristics of the finger millet *(Eluesine coracane)* were studied by Radhika et al. (2011) and in their work drying data were fitted to nine thin layer models. The logarithmic model was found to satisfactorily describe the drying kinetics of the finger millet. Shen et al. (2011) studied thin-layer drying kinetics and quality changes of sweet sorghum stalk for ethanol production as affected by drying temperature. The drying kinetics and the effects of drying temperature on the qualities of sweet sorghum stalk were investigated in their work. The results showed that the drying process could be well simulated by Wang and Singh's model. Mathematical modelling of the thin layer drying kinetics of *G. tournefortii* was investigated for both the microwave and open sun drying conditions by Evin (2011). The experimental moisture loss data were fitted to the 14 thin layer drying models and the logarithmic model was found to best describe the open sun drying kinetics of *G. tournefortii.*

Characteristics of Amelie and Brooks mangoes were experimentally determined using a solar dryer made up of four trays and used under weather conditions of fruit harvest period by Dissa et al. (2011). Direct solar drying curves were established, fitted using 10 mathematical models and simulated with a direct solar drying model.

Other investigators have successfully used thin layer equations to explain drying of several agricultural products. For example, Roselle (I) (Saeed et al, 2008); green pepper (Akpinar et al., 2008). However, in our knowledge, no information is available about thin layer drying of pellets millet. Therefore, in the present study, thin layer drying kinetic of pellets millets was investigated.

The main objectives of the present work were:

1. To investigate the drying kinetics of pellets millet
2. To discuss the influence of temperature and size pellets on the drying kinetics
3. To fit the drying kinetics with ten mathematical models and select, the best model to represent accurately the drying kinetics of pellets millets
4. To evaluate the effective moisture diffusivity coefficients and activation energy.

MATERIALS AND METHODS

Sample preparation

Pellets are prepared from the flour of millet. The manufacture of the product consists of humidifying the flour then to roll it to the hand in order to obtain pellets the dimensions of which depend on the quality of the required product. Pellets sizes are obtained by using a sieves system and the initial moisture content is 0.602 kg water / kg dry matter for pellets $2.5 \text{ mm} < d_1 < 3.15 \text{ mm}$ and 0.628 kg water / kg dry matter for pellets $5 \text{ mm} < d_2 < 6.30 \text{ mm}$.

Experimental apparatus

The experimental device used is a blower with air called vein of drying. It includes:

1. An axial fan driven by an engine;
2. A comprising low tension battery of resistance which makes it possible to heat the air;
3. A test vein of section 175 × 175 mm^2
4. A convergent at the entry of the sheath for better channelling the flow of air;
5. A power station of data acquisition connected to a microcomputer equipped with the software for the management of the power station, storage and the data processing;
6. A site to put the product to be dried;
7. An electronic balance of precision ± 0.1 g for monitoring the evolution of the mass in the course of drying;
8. An digital anemometer for the control rate of air flow in the test vein.

Experimental procedure

Initially, the desired conditions of drying, namely the temperature T, the relative humidity H_R and the air velocity have to be fixed. By varying fan tension, the air flow and temperature of drying was

Table 1. Thin layer mathematical models used to describe the drying kinetics of pellets.

N°	Model name	Model equation	Reference
1	Newton	$X_R = \exp(-kt)$	Shen et al. (2011), Evin (2011), Meziane (2011), Akpinar et al.(2008)
2	Page	$X_R = \exp(-kt^n)$	Flores et al. (2012) ,Radhika et al.(2011), Shen at al.(2011), Evin (2011), Meziane (2011) , Doymaz (2008) , Akpinar et al.(2008)
3	Henderson and Pabis	$X_R = a\exp(-kt)$	Flores et al. (2012) , Radhika et al. (2011), Shen et al. (2011), Evin (2011), Meziane (2011) , Doymaz (2008) , Akpinar et al. (2008)
4	Logarithmic	$X_R = a\exp(-kt) + c$	Radhika et al.(2011), Shen at al.(2011), Evin (2011), Meziane (2011), Doymaz (2008) , Akpinar et al.(2008)
5	Two-term	$X_R = a\exp(-k_0 t) + b\exp(-k_1 t)$	Flores et al.(2012) , Shen at al.(2011), Evin (2011), Meziane (2011), Akpinar et al.(2008)
6	Diffusion approach	$X_R = a\exp(-kt) + (1-a)\exp(-kbt)$	Flores et al. (2012) , Meziane (2011), Akpinar et al. (2008)
7	Modified Henderson and Pabis	$X_R = a\exp(-kt) + b\exp(-gt) + c\exp(-ht)$	Evin (2011), Meziane (2011), Akpinar et al. (2008)
8	Two term exponential	$X_R = a\exp(-kt) + (1-a)\exp(-kat)$	Shen at al.(2011), Evin (2011)
9	Wang and Sing	$X_R = 1 + at + bt^2$	Flores et al. (2012) , Radhika et al(2011), Shen et al. (2011), Meziane (2011), Doymaz (2008)
10	Midilli et al	$X_R = a\exp(-kt^n) + bt$	Evin (2011), Meziane (2011)

controlled during the whole operation. The fan makes the air pass through electrical resistances where it is heated. Once the experimental conditions are stable, the product is put on an aluminium plate placed in parallel to the hot air flow. To follow the product weight losses during drying, weight measurements were taken every five minutes. Drying procedure goes on until there is no weight change after three successive readings.

Mathematical modelling of the drying curves

Ten simplified drying models given in Table 1 have been used to describe the drying kinetics of pellets millet. In these models, X_R represents moisture ratio expressed by the following equation:

$$X_R = \frac{X_t - X_e}{X_0 - X_e} \qquad (1)$$

Where Xt is the moisture content at t time, X_0 and X_e are respectively the initial and equilibrium moisture contents on dry basis. Xe is relatively small compared to X_t and X_0 (Meziane, 2011; Radhika et al., 2011; Evin, 2011; Dissa et al., 2011; Arumuganathan et al., 2009; Akpinar et al., 2008; Doymaz, 2007;). Thus, X_R can be simplified by $X_R = X_t / X_0$.

A non-linear regression analysis was performed using the Microsoft excel 2010 solver to fit the experimental data with chosen mathematical models. The statistical validity of the models was evaluated and compared by means of the coefficient of determination (R^2) the reduced chi-square (χ^2) and root means square error (RMSE). The best fit is defined by the highest value of R^2, and the lowest values of χ^2 and RMSE. These parameters can be calculated as follows:

$$RMSE = \left[\frac{1}{N} \sum_{1=1}^{N} \left(X_{R\exp,i} - X_{R\mathrm{pre},i} \right)^2 \right]^{1/2} \qquad (2)$$

$$\chi^2 = \frac{\sum_{i=1}^{N} \left(X_{R\exp,i} - X_{R\mathrm{pre},i} \right)^2}{N-p} \qquad (3)$$

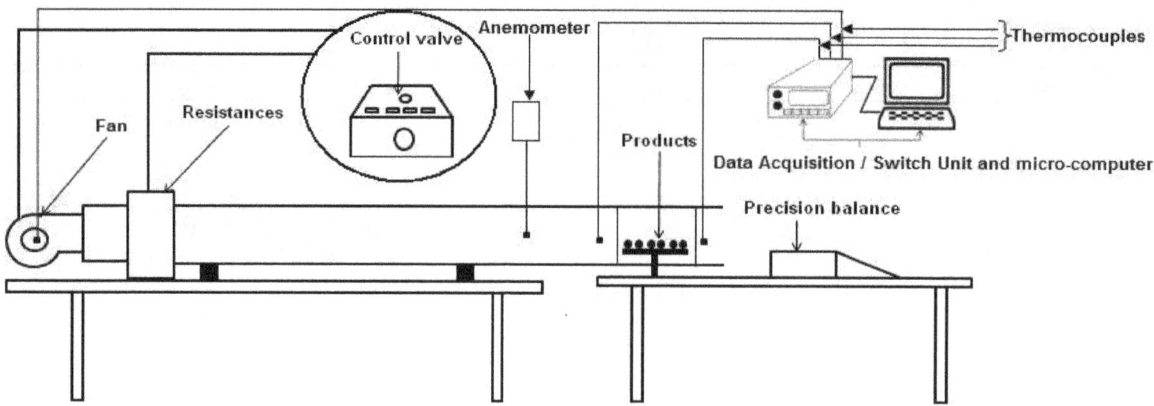

Figure 1. Schema of the thin layer drying apparatus used in the experiments.

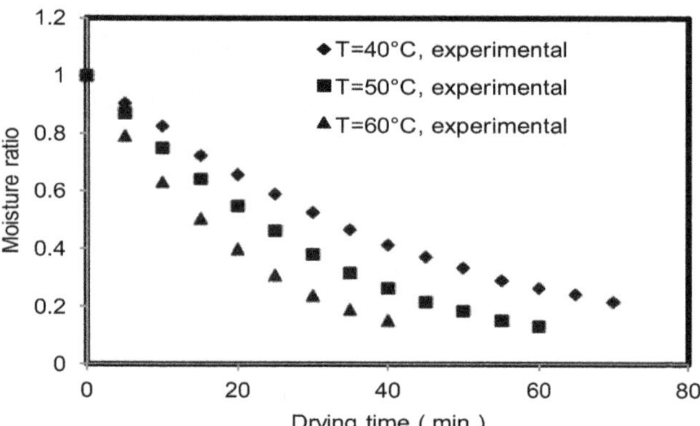

Figure 2. Effect of drying air temperature on moisture ratio for pellets size d_1.

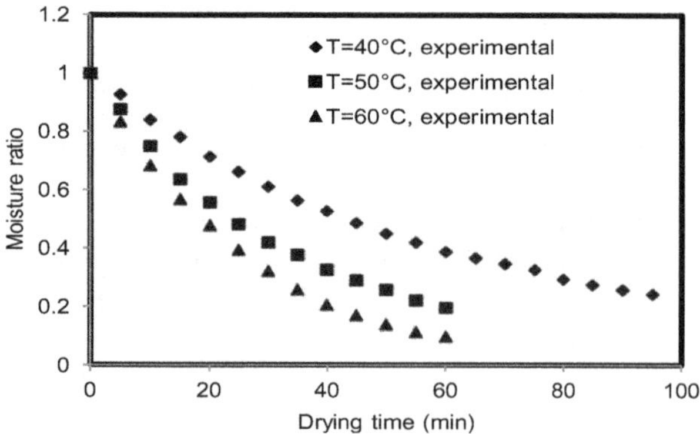

Figure 3. Effect of drying air temperature on moisture ratio for samples pellets size d_2.

Where $X_{Rexp,i}$ is the *i*th experimental moisture ratio, $X_{Rpre,i}$ is the *i*th predicted moisture ratio, *N* is the number of experimental data points, and *p* is the number of parameters in model (Dissa et al., 2011 ; Roberts et al., 2008; Doymaz , 2008).

RESULTS AND DISCUSSION

Effect of drying air temperature on moisture ratio of pellets millet

Pellets were dried in thin layer at the drying air temperature of 40, 50 and 60°C in a convective hot air dryer using a constant air velocity of 1.0 m/s (Figure 1). Figures 2 and 3 present the variations of moisture ratio depending on time for various used air temperatures and diameters 2.5 mm < d_1 < 3.15 mm and 5 mm < d_2 < 6.30 mm respectively.

Moisture ratio decreases continuously with drying time and a constant-rate period is not observed at none of the experiments. All the drying process occurred during the falling rate-drying period. This indicates that the process describing the drying behaviour of the pellets millet is governed by diffusion. In fact, an increase in drying air temperature not only modifies water activity but also influences the diffusion coefficient and to a lesser extends the vaporization enthalpy. Similar results related to behaviour or drying rate curves have also been reported in several drying studies of biological materials by Flores et al. (2012) for castor oil seeds *(Ricinus communis)*, Radhika et al. (2011) for Finger Millet *(Eluesine coracana)*, Shen et al. (2011) for sweet sorghum; Meziane (2011) for olive pomace, and Doymaz (2008) for strawberry.

Effect of the pellets size

The influence of the pellets size is highlighted by changing pellets size from one test to another. Figure 4 shows the effect of the pellets size on drying kinetics at various drying air temperatures. Decreasing pellets size increases moisture ratio and consequently drying time

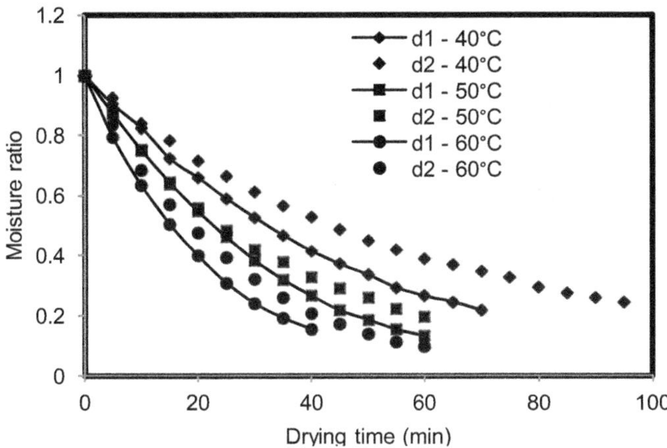

Figure 4. Effect of the pellets size on the kinetics of drying at various drying air temperatures.

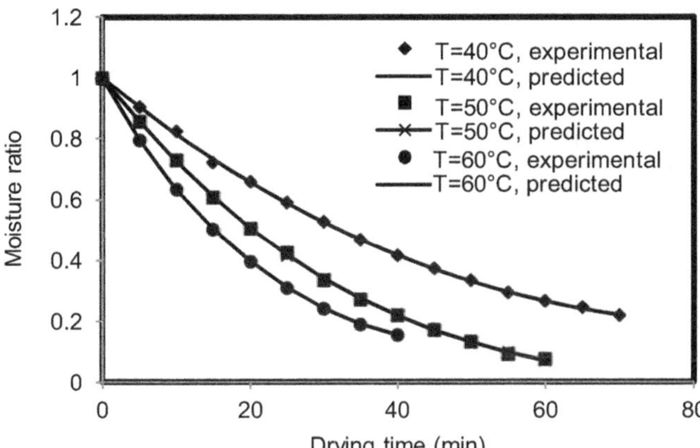

Figure 5. Variations of experimental and predicted moisture ratios by the Henderson and Pabis modified drying model with drying time for pellets size d_1 and V= 1 $m.s^{-1}$.

decreases. Decreasing pellets size accelerates the migration of water from inside to the surface of the product and a decrease of surface exchange increases heat transfer between the air and the product. Similar result has been noted by Madamba et al. (1995) for garlic slices.

Modelling of drying curves

The moisture ratio data obtained from the drying experiments were fitted to the 10 thin layer models listed in Table 1. The values of the coefficient of determination (R^2), the reduced chi-square (χ^2) and the root mean square error (RMSE) for different temperatures and pellets size determined by non-linear regression analysis

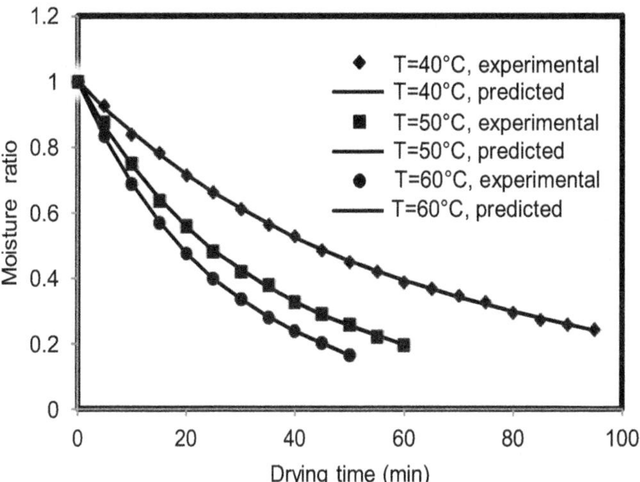

Figure 6. variations of experimental and predicted moisture ratios by the Henderson and Pabis modified drying model with drying time for pellets size d_2 and V= 1 $m.s^{-1}$.

are presented in Table 2. In all cases, the R^2 values for the models were greater than the acceptable R^2 value of 0.90, indicating a good fit (Doymaz, 2008). For all cases, Henderson and Pabis' modified model gives the best fit of experimental values since highest value of R^2 and lowest values of χ^2 and RMSE were obtained. Accordingly, the Henderson and Pabis' modified model can be selected as a suitable model to represent the thin layer drying behaviour of pellets millets. Coefficients and constants of the best drying model for different drying condition are given in Table 3. Figures 5 and 6 represent the variations of experimental moisture ratio and predicted moisture ratio by the Henderson and Pabis modified drying model for various drying air temperatures and for pellets size d_1.and d_2 respectively.

Estimation of effective diffusivities

It has been accepted that the drying characteristics of biological products in falling rate period could be described by Fick's diffusion equation (Flores et al., 2012; Shen et al., 2011; Doymaz, 2008). General series solution of Fick's second law in spherical coordinates, with the assumptions of moisture migration being by diffusion, negligible shrinkage, constant diffusion coefficients and temperature are given as follows (Crank, 1975; Doymaz, 2008; Flores et al., 2012):

$$X_R = \frac{6}{\pi^2} \sum_{n=1}^{\infty} \frac{1}{n^2} exp \left[- n^2 \frac{\pi^2 D_{eff} t}{r^2} \right] \tag{4}$$

Where X_R is the moisture ratio, D_{eff} is the effective

Table 2. Statistical results obtained from 10 thin layer drying models.

Model	T°C	Pellets size					
		2.5 mm < d_1 < 3.5 mm			5 mm < d_2 < 6.30 mm		
		R^2	χ^2	RMSE	R^2	χ^2	RMSE
Newton	40	0.99969	0.000063	0.007642	0.99931	0.000121	0.010723
	50	0.99915	0.000292	0.016406	0.99932	0.000109	0.010049
	60	0.99987	0.000027	0.004872	0.99979	0.000053	0.006980
Page	40	0.99980	0.000027	0.004867	0.99975	0.000030	0.005180
	50	0.99992	0.000014	0.003477	0.99957	0.000065	0.007392
	60	0.99991	0.000018	0.003690	0.99988	0.000023	0.004371
Henderson and Pabis	40	0.99967	0.000045	0.006241	0.99915	0.000099	0.009428
	50	0.99891	0.000191	0.013279	0.99927	0.000111	0.009682
	60	0.99986	0.000028	0.004693	0.99976	0.000048	0.006372
Logarithmic	40	0.99973	0.000041	0.005707	0.99982	0.000021	0.004220
	50	0.99965	0.000069	0.007304	0.99976	0.000039	0.005453
	60	0.99991	0.000019	0.003603	0.99988	0.000024	0.004288
Two term	40	0.99974	0.000042	0.005573	0.99986	0.000018	0.003773
	50	0.99993	0.000020	0.003320	0.99977	0.000042	0.005369
	60	0.99992	0.000022	0.003475	0.99976	0.000059	0.006372
Diffusion approach	40	0.99983	0.000026	0.004522	0.99929	0.000146	0.011152
	50	0.99915	0.000292	0.015273	0.99975	0.000044	0.005806
	60	0.99987	0.000027	0.004872	0.99979	0.000063	0.006980
Modified Henderson and Pabis	40	0.99986	0.000027	0.004010	0.99987	0.000019	0.003647
	50	0.99993	0.000020	0.003288	0.99977	0.000052	0.005285
	60	0.99996	0.000016	0.002333	0.99994	0.000018	0.003133
Two term exponential	40	0.99976	0.000033	0.005362	0.99929	0.000138	0.011152
	50	0.99915	0.000318	0.016406	0.99974	0.000041	0.005866
	60	0.99987	0.000031	0.004872	0.99979	0.000058	0.006980
Wang and Singh	40	0.99982	0.000025	0.004671	0.99881	0.000167	0.012276
	50	0.99994	0.000011	0.003088	0.99854	0.000240	0.014254
	60	0.99945	0.000140	0.010433	0.99882	0.000367	0.015949
Midilli et al.	40	0.99985	0.000024	0.004188	0.99977	0.000027	0.004808
	50	0.99992	0.000016	0.003378	0.99975	0.000045	0.005588
	60	0.99991	0.000023	0.003585	0.99989	0.000025	0.004153

moisture diffusivity (m^2/s), r is the equivalent radius (m) and t is the time (s). For long drying time, Equation (4) could be further simplified to retain only the first term (Flores et al., 2012; Radhika et al., 2011). It could be rewritten in a logarithmic as shown in Equation (5). The effective moisture diffusivity was calculated from the slope of a straight line, plotting experimental data in terms of ln (X_R) versus drying time (Flores et al., 2012; Doymaz, 2008):

$$\ln X_R = \ln\frac{6}{\pi^2} - \left(\frac{\pi}{r}\right)^2 D_{eff}\,t \quad (5)$$

The effective moisture diffusivity was estimated by using

Table 3. Coefficients and constants of the Modified Henderson and Pabis model for different drying condition.

Pellets size	Model constants						
	T°C	k	a	b	c	g	h
$2.5\ mm < d_1 < 3.5\ mm$	40	0.0101	2.7666	-2.1394	0.3741	0.0022	-0.0089
	50	0.0449	3.1382	1.9919	-0.1454	0.0540	0.0455
	60	0.0441	1.0049	-0.3054	0.2983	0.0802	0.0807
$5\ mm < d_2 < 6.30\ mm$	40	0.0188	0.9399	0.758	-0.0131	-0.0107	-0.0237
	50	0.0349	0.8863	0.3135	-0.1935	-0.0075	-0.0123
	60	0.0372	1.0019	0.0025	-0.0034	-0.0965	-0.0916

Table 4. Effective diffusivity values of pellets millet.

Diameter	d_1			d_2		
T°C	40	50	60	40	50	60
$D_{eff} \times 10^{-9}$ ($m^2.s^{-1}$)	4.468	6.974	9.562	11.967	21.832	31.940

the method of slopes. From Equation (5), a plot of ln X_R versus time gives a straight line with a slope of:

$$\text{slope} = \frac{\pi^2 D_{eff}}{r^2} \tag{6}$$

The effective diffusivity values for different temperature and pellets size are presented in Table 4. The effective diffusivities vary from 4.468×10^{-9} to 6.974×10^{-9} m^2/s for pellets size d_1 and from 11.967×10^{-9} to 31.940×10^{-9} m^2/s for pellets size d_2. Generally in the biological products, this value varies between 10^{-9} and 10^{-11} m^2/s (Doymaz, 2008; Flores et al., 2012; Madamba et al., 1995). The effective diffusivity increases when drying air temperature s and pellets size increase. These results are in agreement with those of Meziane (2011) for olive pomace.

Estimation of activation energies

The temperature dependence of effective moisture diffusivity can be described by an Arrhenius type relationship:

$$\text{Deff} = D_0 \exp\left(-\frac{E_a}{RT}\right) \tag{7}$$

Where D_{eff} is the effective moisture diffusivity (m^2/s), D_0 is the pre-exponential factor equivalent to the diffusivity at a high temperature (m^2/s), E_a is the activation energy (kJ/mol), R is the universal gas constant (8,314 J/mol K), and T is absolute temperature (K). The activation energy E_a and the constant D_0 could be determined by plotting ln (D_{eff}) versus 1/T after linearization for Equation (7). The values of E_a calculated from the slope and intercept of each plot are given in Table 5. As it can be seen, the activated energy increases when pellets size increase. This observation is in agreement with those reported by Meziane (2011).

Similar activation energy values have been found by several authors, from 35.37 kJ/mol for finger millet (Radhika et al., 2011); in the range of 34.05 to 38.10 kJ for olive pomace (Meziane, 2011), and 41.41 kJ/mol for castor oil seeds *(Ricinus cumminis)* (Flores et al., 2012).

Conclusions

In this study, thin layer drying kinetics of pellets millet flour was studied at drying air temperatures of 40, 50 and 60°C.using a constant air velocity of 1.0 m/s for diameters $2.5\ mm < d_1 < 3.15\ mm$ and $5\ mm < d_2 < 6.30\ mm$. The drying process takes place only in falling rate period. Drying air temperature and the pellets size are influencing factors to drying kinetics. Statistical results of ten mathematical models at different drying conditions showed that the modified Henderson and Pabis is the best model to describe the behaviour of pellets millet drying. The effective diffusivity varies from 4.468×10^{-9} to 9.562×10^{-9} m^2/s for d_1 and from 11.967×10^{-9} to 31.940×10^{-9} m^2/s for d_2. It increases when drying air

Table 5. Arrhenius parameters for the drying kinetics of granulated of flour of millet.

D(mm)	d_1	d_2
E_a (kJ.mol^{-1})	33.05	42.65
D_0 x10^{-4} (m^2.s^{-1})	1.024	5.332

temperatures and pellets size increase. The activation energy increases with the increasing of pellets size and its values were 33.05 kJ/mol for d_1 and 42.65 kJ/mol for d_2.

Nomenclature: a,b,c,n, Coefficients in models; **k,k$_0$,k$_1$,g ,h,** Constants in models (s^{-1}); **D$_{eff}$,** effective moisture diffusivity (m^2/s); **D$_0$,** effective moisture diffusivity for an infinite temperature (m^2/s) **E$_a$,** activation energy (kJ/mol); **X$_e$,** equilibrium moisture content (kg water/kg dry matter); **X$_0$,** initial moisture content (kg water/kg dry matter); **X$_R$,** moisture ratio; **X$_t$,** moisture content at T time (kg water/kg dry matter); **X$_{Rexp,l}$,** experimental moisture ratio; **X$_{Rpre,l}$,** predicted moisture ratio; **N,** number of observation; **p,** number of parameters in a model; **r,** radius; **RMSE,** root mean square error; **R,** universal gas constant (J/mol.K); ***R*^{2},** coefficient of determination; ***T*,** drying temperature (°K); ***t*,** drying time (s).

Greek letter: χ^2, Chi-square.

REFERENCES

Akpinar E, Kavak, Bicer Y (2008).Mathematical modelling of thin layer drying process of long green pepper in solar dryer and tinder open sun. Energy Convers. Manage. 49(6):1367-1375.

Arumuganathan T, Manikantan M R, Rai RD , Anandakumar S, Khare V (2009). Mathematical modelling of drying kinetics of milky mushroom in a fluidized bed dryer, Int. Agrophys. 23:1-7.

Belghit A, Bennis A (2009). Experimental analysis of drying kinetics of cork. Energ. Convers. Manage. 50:618-625.

Chong CH, Law CL, Cloke M, Hii C L, Abdullah LC, Daud WRW (2008). Drying kinetics and product quality of dried Chempedak. J. Food Eng. 88:522–527.

Crank J (1975). The mathematics of diffusion. Clarendon Press, Oxford, ISBN 0-19-853344-6, England.

Dissa AO, Bathielo DJ, Desmorieux H, Coulibaly O, Koilidiati J (2011). Experimental characterisation and modelin of thin direct solar drying of amelie and brooks mangoes. Energy 36:2517-2527.

Doymaz I (2008). Convective drying kinetics of strawberry. Chem. Eng. Process. 47:914-919.

Evin D (2011). Thin layer drying kinetics of gundelia tournefortii L. Food. Bioprod.Process. doi:10.1016/j.fbp.2011.07.002.

Flores MJP, Febles VG, Pérez JJC, Dominguez G, Mendez JVM, Gonzalez EP, Lopez GFG (2012). Mathematical modelling of castor oil seeds *(Ricinus Communis)* drying kinetics in fluidized bed at high temperatures. Ind. Crop. Prod. 38:64-71.

Lertworasirikul S (2008). Drying kinetics of semi-finished cassava crackers: A comparative study. LWT 41:1360–1371.

Madamba SP, Driscoll HR, Buckle AK (1995). The thin-layer drying characteristics of garlic slices. J. Food Eng. 29:75-97.

Meziane S. (2011). Drying kinetics of olive pomace in a fluidized bed dryer. Ind. Crop. Prod. 52:1644-1649.

Radhika GB, Satyanarayana SV, Rao DG (2011). Mathematical model on thin Layer drying of Finger Millet *(Eluesine coracana)*. Adv. J. Food Sci. Technol. 3(2):127-131.

Roberts JS, Kidd RD, Zakour OP (2008). Drying kinetics of grape seeds. J. Food Eng. 89:460-465.

Saeed IE, Sopian K, Zainol AZ (2008). Thin Layer Drying of Rosell (I): Mathematical Modeling and Drying Experiments. Agr. Eng. Int: the CIGR Ejournal. Manuscritpt FP 08 015.

Shen F, Peng L, Zhang Y, Wu J, Zhang X, Yang G, Peng H, Qi Hui, Deng S (2011). Thin-layer drying kinetics and quality changes of sweet sorghum stalk for ethanol production as affected by drying temperature. Ind. Crop. Prod. 34:1588-1594.

Torres ML, Infante JAG, Laredo RF, Guzman NER (2008). Drying kinetics of nopal *(Opuntia ficus-indica)* using three different methods and their effect on their mechanical properties. LWT 41:1183–1188.

Activity of rice bran proteic extracts against *Fusarium graminearum*

Fernanda A. P. , Cristiana C. B., Sílvia L. R. M., Jaqueline G. B. and Eliana B. F.

Rio Grande Federal University, Food Science and Engineering Graduate Program, street Engineer Alfredo Huch, 475, P. O. box 474, CEP 96201-900, RS, Brazil.

The application of natural antifungal substances is motivated by the need for alternatives to existing methods that are not always applicable, efficient, or that do not pose risk to consumers or the environment. Furthermore, studies on the behaviour of toxigenic species in the presence of natural fungicides have enabled their safe application in the food chain. This study aimed to identify the fraction of the rice grain with greater inhibitory activity of amylase and related to its antifungal and antimycotoxigenic potential against *Fusarium graminearum* CQ 244 biomass. The greatest inhibitory effect was observed in extracts of bran, which inhibited by 90% the fungal amylase activity. The primary fractionation of the rice bran extract was more efficient when ethanolic extracts was precipitated by acetone, resulting in a specific inhibition estimated at 20 $\mu g_{hydrolysed\ starch}$ $min^{-1}mg\ protein^{1}$, PF 45 and recovery 61%. The rice bran protein extracts showed fungistatic activity against *F. graminearum*, with MIC_{50} of 419 $\mu g\ ml^{-1}$ and 168 mg ml^{-1} estimated from glucosamine and amylase inhibition, respectively, which cause 63% biomass inhibition and 40% of the nivalenol (NIV) production.

Key words: Rice, *Fusarium graminearum*, glucosamine, amylase, nivalenol (NIV).

INTRODUCTION

The presence of alpha-amylase inhibitors in beans, corn, rice, rye and other grains is related to the mechanism of germination cycle regulation and defense against contamination by fungi or other pests (Mosolov et al., 2001; Figueira et al., 2003; Marsaro-Júnior et al., 2005; Mosca et al., 2008). These amylase inhibitors present proteic character (Pagnussatt et al., 2011; Pagnussatt et al., 2012) and its extraction and purification based on proteic properties like salt solubility, adsorption by solvents, organic polymers and pH variation associated to separation operations, the make primary purification possible and allowed the study of the specific effect against fungal growth related to amylase activity (Iulek et al., 2001).

Reports of grain losses by contamination by toxigenic fungal species such as *Fusarium graminearum* are frequent in different production regions (Del Ponte et al., 2012, Kokkonen and Laitila, 2012; Abia et al., 2013; Morcia et al., 20013), despite the plants having natural defenses. Unlike other grains, rice is not considered a preferential target of this specie, because crop damage are not very frequent although the mycotoxins they produce have been found even with the use of fungicides (Dors et al., 2011; Heidtmann-Bemvenuti et al., 2012).

The fungal contamination may occur in any part of the plant, but the grains, are the most susceptibilities part (Usha et al., 1993). In rice, the external grain portion, compound by lignocelulosic and proteic material, is related to the fungal resistance associated to physical barriers and enzymatic inhibition. The mycotoxin production is a stress consequence promoted by the resistant cultivars and fungicide sprays during flowering, but losses in both the yield and quality of grain cannot be prevented under environmental conditions favorable for epidemics (Keller et al., 2013). During rice milling, the endosperm, bran and husk are separated with the bran accounting for 8 to 10% of the grain (Liu et al., 2009). The endosperm is widely consumed in the world diet, but the bran and husk are underutilized, this has demanded the search for innovative solutions to use these portions efficiently in a nutritional, functional and economical manner (Singh, et al., 2000). Thus, we seek the presence of enzyme inhibitors in fractions of rice milling tocontribute as an alternative to the valuation of co-products and also to reduce the mycotoxin contamination, since the inhibitors are natural substances found in cereals and do not cause stress to the fungus.

This study aimed to identify the fraction of the rice grain with greater inhibitory activity of amylase and related to their properties against on *F. graminearum*.

MATERIALS AND METHODS

Samples

BR-IRGA 417 rice (*Oryza sativa* L.) was grown in experimental fields of the Riograndense Rice Institute, (IRGA-Brazil), 2010 crop. After harvesting, the grains were milled in a Suzuki mini laboratory mill, separating the husk, bran and endosperm fractions. These were ground in slicer (Tecnal, model TE-631) and sieved to obtain uniform particle size (32 mesh). The bran was defatted with petroleum ether, on cold by stirring.

Identification of rice fraction with inhibitory activity

The enzyme inhibitors were extracted under orbital shaking (200 rpm) with 95% ethanol, a ratio of 1:7 (w/v) for 7 h and the crude extract was separated by centrifugation and filtration (Pagnussatt et al., 2011). Crude extracts of rice fractions were assessed for their ability to inhibit the action of commercial fungal amylase from *Aspergillus oryzae* (Fungamyl®), provided by Novozymes®, Brazil, containing 0.05 mg protein mL^{-1}. Enzyme activity was determined by the iodometric method (Baraj et al., 2010, Pagnussatt et al., 2012) in experiments and in control.

The extracts containing the amylase inhibitor were incubated with commercial fungal α--amylase and sodium acetate buffer (pH 7.0) for 30 min at 25°C. After a soluble starch solution 0.5% (w/v) was added maintaining the reaction for 30 min at 25°C. The reaction was interrupted by hydrochloric acid 0.1 M addition. The residual starch was determined by iodometry, and the absorbance of the formed complex measured at 620 nm (quadruplicate). The amylase activity was expressed as µg starch mL^{-1} min^{-1}. One unit of amylase was defined as the amount of enzyme required to hydrolyze 0.06 mg starch per min (Pagnussatt et al., 2011). The starch hydrolyzed by the enzyme in the control experiment (without inhibition extraction) was considered the maximum velocity reaction (v).

Primary purification of the extract inhibitor

The rice bran composition was determined by the AOAC (2000) methods: Humidity (No. 935.29), ash (No. 923.03), protein (No. 920.87), lipids (No. 920.85) and crude fiber (No. 991.43). Carbohydrates were estimated by difference. The crude extract containing the bran enzyme inhibitors was obtained with 95% ethanol as described above and also with water in the ratio 1:3 (w/v) (Pereira et al., 2010). The protein extracted was precipitate by decreasing extract dielectric constant with organic solvents and a change in pH was also tested about their efficiency to protein recuperation. The total protein in the purified extracts was carried out by the Lowry (1951) method. The specific inhibitory activity (µg hydrolyzed starch min^{-1} mg protein^{-1}) was estimated through the inhibition of enzyme activity per mg of soluble protein.

Protein precipitation by adding organic solvents

In the ethanol extract (F1) of the proteins precipitation was accomplished adding acetone, 1:3 (v / v) standing 12 h at 4°C and after being centrifuged at 2250 x g, 20 min at 4°C, the precipitate was allowed to stand for 30 min at room temperature and then resuspended with 5 ml of water (F2). In the aqueous extract (F1'), the precipitation was performed with acetone or ethanol in proportion 1:3 (v / v) standing 12 h at 4°C and after, centrifuging at 2250 *x g*, 20 min at 4°C the precipitate was allowed to stand for 30 min and the precipitated from acetone (F2'a) or ethanol (F2'b) were resuspended with 5 mL of water (Figure 1).

Protein precipitation by pH change

In the ethanol extract (F1), the fractioning was performed by acidification. The protein contained in this extract were precipitated with HCl 6M (F2a), in a 1:1 ratio (v/v) and allowed to stand and centrifuged as described above. The precipitate was resuspended in water to protein and enzyme inhibition activity determination. In the aqueous extract (F1') with initial pH 6.0 (F2'd) two precipitations were performed, one with adjustment to pH 5 (F2'c) and the other to pH 7 (F2'e) using HCl 1M and NaOH 2.5 M, followed by standing for 12 h at 4°C and centrifugation at 2250 *x g*, 20 min at the same temperature. In this case, the precipitated proteins were resuspended with 5 mL of 0.02 M NaOH (Figure 2).

The efficiency of the extraction process was evaluated by calculating the purification factor (PF), comparing the specific inhibition of alpha-amylase in the fraction from the purification with the specific inhibition of alpha-amylase in the crude extract, FP = (final specific enzyme inhibition / initial specific enzyme inhibition). Protein inhibitor recovery (PR) was calculated by the equation: PR = (final enzyme inhibition) / initial enzyme inhibition) x 100.

Antifungal activity of protein extracts

The toxigenic species of *F. graminearum* CQ 244 given by the Laboratory of Plant Epidemiology of the Federal University of Rio Grande do Sul (UFRGS) was maintained on potato dextrose agar (PDA) during 14 days and the spores recovered from the agar surface with sterile solution Tween 80 (1%, v/v) following the direct enumeration in a Neubauer chamber. Fungal growth was tested by the agar dilution method, in PDA medium. The proteic compounds (100 mg of soluble protein, corresponding to 4 ml of extract in 26 ml of medium) was added to the culture medium at a temperature of 35 to 40°C and poured into Petri dishes. After solidification of the media, a spore solution containing 4×10^6 spores *F. graminearum* ml^{-1} was added at the Petri dishes (Pagnussatt et al., 2013). The experiment was conducted at 30°C for 35 days in dark condition, with samples taken from the culture every 7 days. The inhibition percentage was obtained by the equation (%I) = (C - T) / C × 100, where C is the development indicator (glucosamine or amylase) in the control (absence of the rice bran extract, sterile water was added in place of the extract) and T was the treated with rice bran extract (RBE) (Souza et al., 2011). The median inhibitory

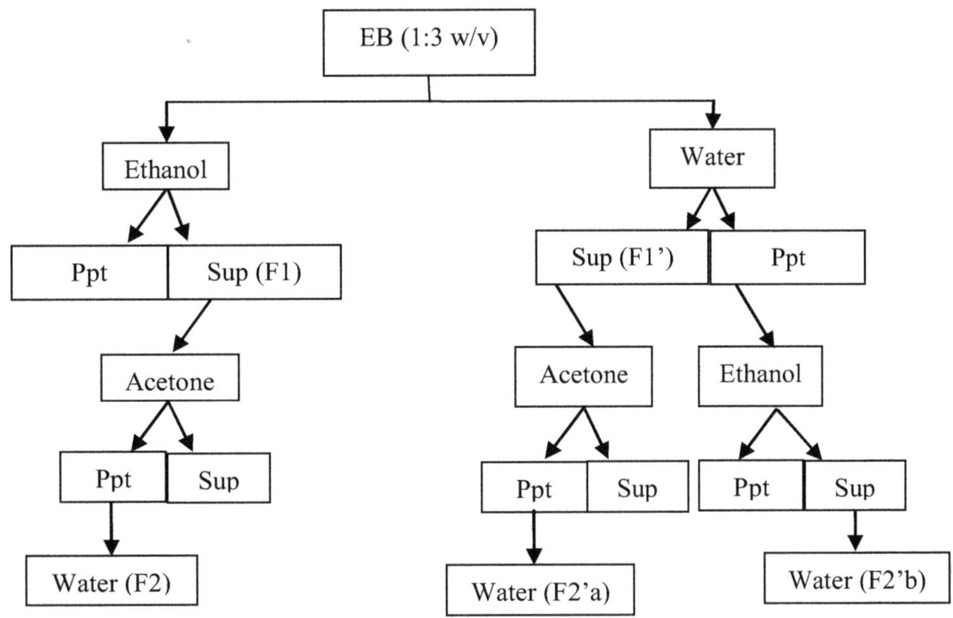

Figure 1. Protein fractionation by adding organic solvents. Sup, Supernatant; Ppt, precipitate.

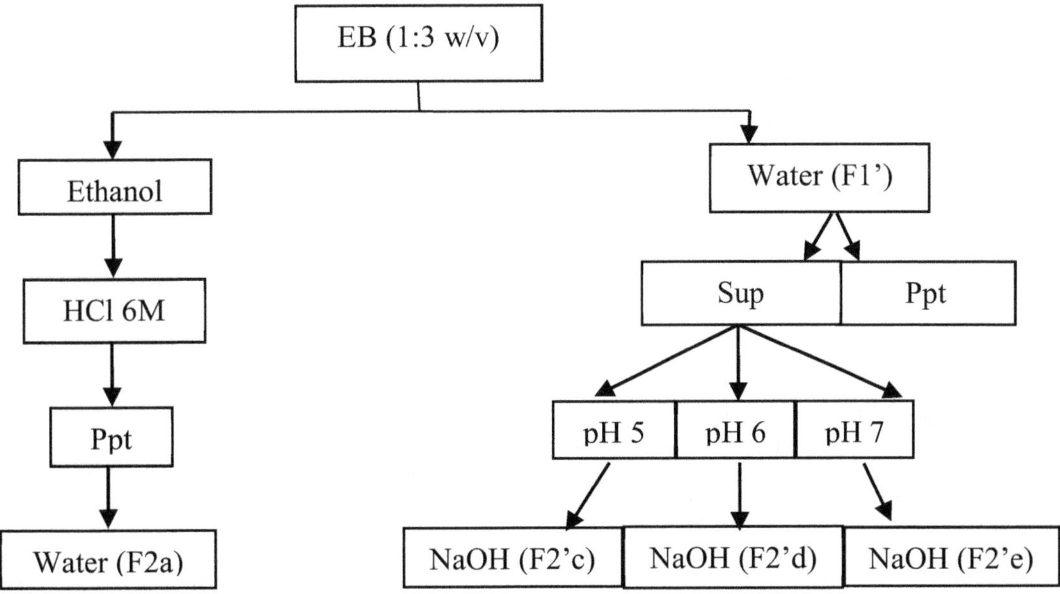

Figure 2. Protein fractionation by pH variation. Sup, Supernatant; Ppt, precipitate.

concentration (MIC_{50}) was considered as the RBE concentration that resulted in 50% inhibition of the fungal growth when compared to the control groups.

Determination of glucosamine production and amylase activity in the *F. graminearum*

For the quantification of the glucosamine content, each 1 g of biomass was dried at 60°C for 4 h, 5 ml of 6 M HCl 6 mol L-1 was added and the mixture was autoclaved at 121°C for 20 min. The hydrolyzed material was neutralized with NaOH 3 M, and reverse titration was carried out with $KHSO_4$ (1 g 100 ml^{-1}). Finally, the colorimetric method was used for the determination of glucosamine (Souza et al., 2011). The absorbance units were obtained by spectrophotometry (Varian, Cary 100, California, EUA) at 530 nm, and the concentrations were established using a standard curve for glucosamine (0.01 to 0.2 g L^{-1}). The measurements were carried out in triplicate, and the results were expressed as glucosamine per mg sample.

The enzyme extract was obtained from the fungal biomass with 20 mL of NaCl 0.9% in an ultrasonic bath (Unique, USC-800A, São Paulo, Brasil) for 40 min, centrifuged (Cientec, CT-5000R, São Paulo, Brasil), and filtered. The α-amylase activity was determined

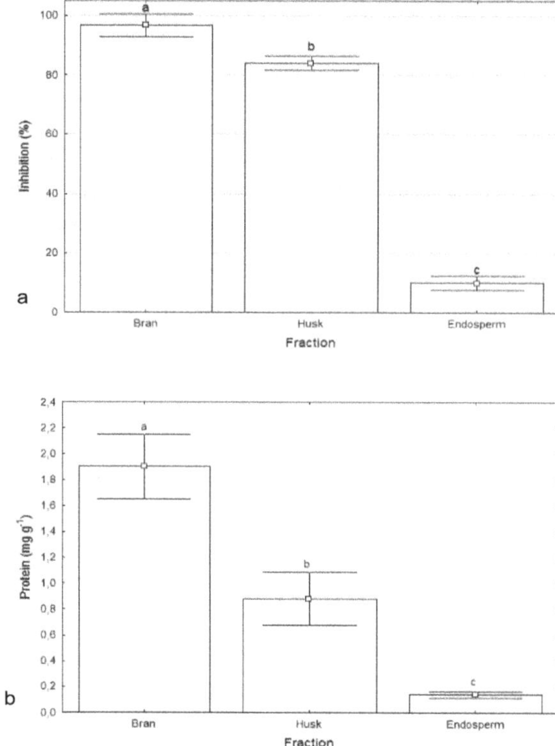

Figure 3. Inhibition (a) and soluble protein (b) of commercial fungal amylase in the presence of extracts of rice bran, husk and endosperm.* Same lower case letters mean non-significant differences between the means at 0.05 probability by Tukey test.

by starch degradation estimated quantitatively by iodometric titration.

Determination of mycotoxins in biomass

The chromatographic patterns for the determination of deoxynivalenol (DON), nivalenol (NIV) and zearalenone (ZEA) purchased from Sigma-Aldrich ® were used to prepare stock solutions. From the stock solutions dissolved in benzene: acetonitrile (95:5, v/v) working solutions were prepared, which was quantified spectrophotometrically (AOAC, 2000).

Mycotoxins were extracted using the method adapted from Tanaka (2000). The extracts were resuspended in benzene (200 µL) and applied to plates for high performance thin layer chromatography plates (HPTLC) (nano-adamant, specific surface area ~ $500 m^2 g^{-1}$, average pore diameter of 60A, specific pore volume of 0.75 mL g^{-1} particles 2-10µm Macherey-Nangel®, Germany) for their detection and quantification in the resuspended extracts. The mycotoxins were eluated with toluene: ethyl acetate: formic acid in the proportions (6:4:1, v/v /v) and quantified by comparison with fluorescence intensity of standards after derivatization with aluminum chloride solution 15% (w/v). The detection limits, quantification and retardation factor (R_f) of the method (Anvisa, 2003) were evaluated.

Statistical analysis

The significance of each treatment was determined by analysis of variance (ANOVA) and the means were compared with each other by Tukey's test at the 5% level considering the amylase and fungal growth.

RESULTS AND DISCUSSION

Inhibitory activity of rice derivatives

In our previous studies, the soluble proteins of rice grain had been characterized and the class of prolamine (18.5%, v/v) was the predominant one (Pagnussatt et al., 2012). In view of this, it was decided to verify the presence of prolamines in different portions of the grain of rice separated during processing. Moreover, this class of proteins is that whose peptide chains possess inhibitory activity of amylolytic enzymes, that is, part of the grain defense mechanisms against the attack by pests and germination control (Figueira et al., 2003). The inhibitory activities of fractions of rice milling had their crude extracts tested against the commercial alpha-amylase activity (Figure 3), a variation directly proportional to the increase in protein content present in them being shown, confirming the presence of inhibitors and their proteic character.

The total soluble protein content and inhibitory effect

Table 1. Specific inhibition profile of alpha-amylase by fractionated extracts *

Fraction†	Activity inhibited‡	Proteins§	Specific inhibition#	PF††	RP‡‡ (%)
EB‡‡‡ ET (F1)	13.47	30.50	0.44 ± 0.00f	1	
ET: AC: AG (F2)	16.35	0.83	19.90 ± 0.03 a	44.99	61
ET: HCl 6M: AG (F2a)	3.54	1.53	2.32 ± 0.55 e.f	5.26	26
EB‡‡‡AG (F1')	11.63	32.37	0.36 ± 0.15 f	1	
AG: AC: AG (F2'a)	20.59	3.15	6.54 ± 0.91 c	18.17	89
AG: ET:AG(F2'b)	25.18	3.13	8.05 ± 0.44 b.c	22.36	108
AG: pH 5:NaOH (F2'c)	20.45	34.50	0.59 ± 0.03 e.f	1.65	176
AG: pH 6:NaOH (F2'd)	8.53	3.46	2.47 ± 0.08 d.e	6.86	37
AG: pH 7:NaOH (F2'e)	33.50	11.90	2.82 ± 0.69 d	7.83	173

*Significant at 0.05 probability by Tukey test. †The proteins were extracted, precipitated and resuspended in: ET: Ethanol; AG: water; AC: acetone. ‡ (mg starch hydrolyzate g sample min^{-1}); § (mg g^{-1}); # (starch hydrolyzate mg min^{-1} mg $protein^{-1}$). ††PF, purification factor, within a column differences in specific inhibition were significant at p< 0.05 on the bases of Tukey's Test . ‡‡RP, recovery; ‡‡‡EB, crude extract.

were higher in the bran extracts than the extracts from rice husk and endosperm rice bran inhibitory was around 90% (w/v) against fungal alpha-amylase, while the husk inhibited by 80% (w/v). These results demonstrated that the inhibitors are distributed in the outer portions of the grain, which is consistent with the fact that these fractions are most affected by physical damage during cultivation and processing, therefore where the risk of fungal contamination is greater and the defense mechanisms are necessary (Sospedra et al., 2010). From this evidence and depending on the availability in the supply chain, the rice bran was selected as the source of enzyme inhibitors for use in fractionation and in *in vitro* tests against the *F. graminearum* growth and the production of mycotoxins.

Purification primary of rice bran inhibitors

The rice bran composition is influence by the origin of raw materials and milling process conditions. The levels recommended by industrial grain applications are (w/v): 16% minimum lipids, 13% minimum protein, 9% maximum fiber, 12% maximum humidity and maximum ash of 10%[11]. Brazilian law does not set specific values for the rice bran composition. The rice bran employed is this work presented 21.3 ± 0.3% lipids, 10% ± 0.1 humidity, 6.9 ± 0.8% crude fiber, 10.5 ± 0.4% ash and 11.6 ± 2.0% protein (w/v), similar to that frequently found by other authors (Silveira et al., 2007; Kupski et al., 2012).

The alpha-amylase inhibitor parcial purification by ethanol and acetone precipitation were adopted considering that these solvents, when decreasing the extracts dielectric constant, favor proteins separation from other compounds, and allows the concentration of the fraction containing the inhibitory activity (Pereira et al., 2010). The fungal amylase inhibition by the crude extracts and subjection to primary purification with the organic solvents did not cause the loss of inhibitory activity, which remained around 56% (w/v). The highest specific inhibition values were obtained after precipitation using ethanol and acetone (19.9 $\mu g_{\text{hydrolysed starch}}$ min^{-1} mg_{protein}^{-1}), among other purification conditions (Table 1).

The lower amylase inhibitory effects of the extracts obtained by pH precipitation suggest that this properties determined the molecules hydrophobic regions which exposure was determined by the pH variation. The highest PF was obtained when using ethanol in extraction step and acetone in the precipitation, with a value of 45, reinforce this and also showing a specific inhibition higher than the crude extract. The protein recovery under this condition was 61%. Higher protein recovery (PR) was observed in the tests with water extraction and ethanol precipitation (108%, w/v), indicating that this condition allowed other proteins extraction whose inhibitor character was lower than the further.

The best conditions to obtain the fungal alpha-amylase inhibition from bran rice, demonstrated by the higher purification factor were adopted to follow the study of effect against *F. graminearum* CQ 244. Since precipitation is considered a low resolution stage, depending on the type of application. In this work, the interest was to verify the antifungal properties in the rice and understanding better the effect on *F. graminearum* before to continue the inhibitor purification.

Antifungal effect

The choice of *F. graminearum* for fungal inhibition model is justified because it is a contaminant of grains with the highest susceptibilities between the flowering -to-early stages of grain (Scotti et al., 2001), when there is greater susceptibility of the plant structure, instead of it the low occurrence in rice. The inhibition of growth by measuring

Table 2. Minim Inhibitory concentration (IC_{50}) and inhibition of fungal growth (% I) of the rice bran against *Fusarium graminearum*[†].

Days	IC_{50} mg mL^{-1}†		I^2 (%)	
	Glucosamine*	Amylase*	Glucosamine	Amylase
7°	1652±12,3^{f}	64±0,88^{A}	2	79
14°	166±8,70^{e}	336±64,1^{E}	84	85
21°	95±3,76^{c}	178±16,5^{D}	62	72
28°	63±0,66^{ab}	84±5,64^{AB}	44	39
35°	119±20,1^{d}	180±6,51^{D}	88	82
Average over time	419	168	56	72

*Significant at 0.05 probability by Tukey's test. [†]Results represent the mean of three determinations ± standard deviation. [2]Dados standardized. C (control) = 100%.

halos was not considered appropriate to monitor fungal growth, an uneven distribution of colonies in culture media was observed, especially those containing extract inhibitors that modified morphological hyphae characteristics. In this work, it was possible to visualize changes in the morphological characteristics of the colonies which did not present cottony hyphae and color characteristic of the species. In view of this, levels of glucosamine and amylase activity in fungal biomass were adopted as indicators of *F. graminearum* growth, since the inhibition of amylase had already been verified during the protein inhibitors screening.

Through the content of glucosamine, maximal inhibition of 88% (w/v) was observed in dry biomass in the presence of the proteic extracts, on the 35th day of the experiment. Enzyme inhibitors from cereals inhibited 80% (w/v) average content of glucosamine against the same fungi studied in this work (Pagnussatt et al., 2012). Indeed very promising, because the cell wall is a good indication of the viability of fungal cultures (Scotti et al., 2001).

The determination of reduced amylase activity is another important variable when monitoring fungal inhibition, since these micro-organisms produce extracellular enzymes capable of hydrolyzing the starch medium. For amylase, MIC_{50} was 168 mg mL^{-1} culture medium (Table 2). On the 35th day the extracts showed of 82% (w/v) inhibition in the activity of amylase from fungal biomass compared to the control. Considering previously performed tests with commercial enzyme (Fungamyl®) (Pagnussatt et al., 2011), the results are consistent and confirm that inhibition of fungal growth occurs mainly by decreasing the availability of nutrients from carbohydrates, which is also reflected in the decreased production of glucosamine biomass.

The antifungal effect of the extracts was tested at the concentration required to inhibit 50% of glucosamine production and enzyme activity (MIC_{50}). The fungal cultivation in media containing rice bran extract showed MIC_{50} of 419 mg mL^{-1} culture medium. Considering the content of glucosamine, this value was higher than the MIC_{50} 199 and 207 μg ml^{-1} in extracts of wheat applied to *F. culmorum* and *F. graminearum*, respectively (Chilosi et al., 2000). The protein efficiency of prolamine fraction from rice as antifungal agent were also demonstrated before against *F. oxysporum*; *Fusaruim solani*; *Aspergillus flavus*, *Aspergillus fumigatus* and *Aspergillus parasiticus* with MIC_{50} from 0.6 to 20 μg mL^{-1} (Lee et al., 2007).

The fungal growth indicators studied in this research reinforce the idea that there are natural fungicides in controlling microbial growth in rice grain seeds which is located in the bran, since the extracts studied showed fungistatic effect. They still suggest that the recovery of these compounds from rice bran protein may be more feasible to establish a strategy for the control of fungal contamination losses. After the demonstration of the action of protein extracts of rice bran as a natural antifungal agent, its effect on the inhibition of the *Fusarium* toxins production was also evaluated. HPTLC was adopted due to the speed, low cost and good analytical like: limit of detection (LOD) = 30 ng, 20 ng and 60 ng, limit of quantification (LOQ) = 2.4 μg kg^{-1}, 0.07 μg kg^{-1} and 0.12 μg kg^{-1}; retardation factor Rf = 0.1 cm, 0.3 cm and 0.8 cm for NIV, DON and ZEA, respectively. These parameters were very similar to others reported by gas chromatography methods (Garda-Buffon and Badiale-Furlong, 2004).

F. graminearum produced nivalenol, detected at day 28, whereas in the control culture produced 3.2 μg g^{-1} the treated by the inhibition extracts, the production was 1.92 μg g^{-1} (40% inhibition, w/v). deoxynivalenol (DON) and zearalenone (ZEA) were not produced by the fungus under the conditions of the study. The synthesis of the trichothecenes comprises of cyclized sesquiterpene ring, catalyzed by the tricodiene synthase enzyme, followed by four oxygenations and eight esterifications. This sequence leads to the formation of basic structures such as DON, NIV and its acetylates (Garda-Buffon et al., 2010). Each metabolic pathway requires the expression of a carrier protein and a network of regulatory genes and each chemotype encodes specific proteins. The strain used in this work, classified as nivalenol chemotype (Astolfi et al., 2010) was susceptible to the effect of the

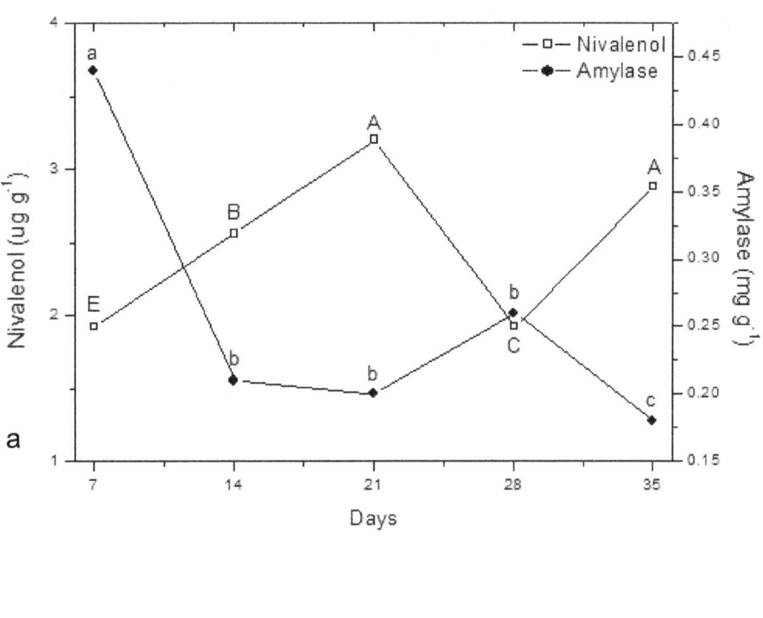

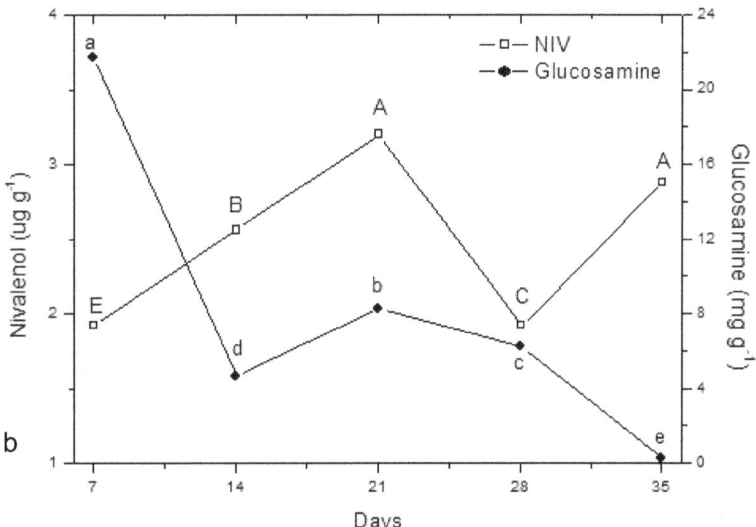

Figure 4. Production of nivalenol (µg g^{-1}), glucosamine (mg g^{-1}) and amylase (mg g^{-1}) in the presence of rice bran extracts. *Same lower case letters mean non-significant differences between the means at 0.05 probability by Tukey test. Same capital letters mean non-significant differences between the means at 0.05 probability by Tukey test.

inhibitors in some point of the route, possibly by complexing the carrier protein by the fungal multiplication was mostly affected by the bran extracts.

The lower production of fungal biomass in the presence of protein inhibitor extract of rice bran was on day 28 and 35, coincident with the inhibition of the production of NIV, relating the decreased amounts of glucosamine with the lowest levels of NIV detected (Figure 4b). At the point of greatest inhibition of amylase activity, higher production of NIV (Figure 4a) was also observed, indicating that the micro-organism could be trying to compensate for the stress caused by lack of nutrients.

However, these results were also promising and suggest that the proteic extract from bran rice might be efficient to be applied inhibiting others toxigenic species multiplication without the understanding the natural plant resistance mechanism are interesting to the economic, environmental and health point of view because it can prevent the indiscriminate use of pesticides. Moreover, it has shown that these inhibitory substances could be extracted from underused industrial fractions and reapplied during cultivation of cereals naturally during different stages of growth. The identification of inhibitors and their genetic coding can also be interesting for genetic improvement of others cereal species not possessing these defenses, increasing their natural resistance to pathogen attack.

Abbreviations: PF, Purification factor; **PR,** protein inhibitor recovery; **T,** treated with rice bran extract; **RBE,** rice bran extract; **MIC_{50},** median inhibitory concentration; **DON,** deoxynivalenol; **NIV,** nivalenol; **ZEA,** zearalenone; **HPTLC,** high performance thin layer chromatography plates; **LOD,** limit of detection; **LOQ,** limit of quantification; Rf, retardation factor; **PDA,** potato dextrose agar.

REFERENCES

Abia W A, Warth B, Sulyk M, Krska R, Tchana AN, Njobeh PB, Dutton MF, Moundipa PF (2013). Determination of multi-mycotoxin occurrence in cereals, nuts and their products in Cameroon by liquid chromatography tandem mass spectrometry (LC-MS/MS). Food Cont. 31:438-453.

Agência Nacional de Vigilância Sanitária (Anvisa) (2003) Resolução RE nº 899 de 29/05/2003: *Guia para validação de métodos analíticos e bioanalíticos*, Ministério da Saúde: Brasil.

Association Official Analytical Chemists (AOAC) (2000). Official Methods of Analysis. 17 th Ed. Method 935.29, humidity; method 923.03, ash; method 920.87, protein; method 991.43, crude fiber. Horwitz, W. Arlington (ed.), USA, 2:49.

Astolfi P, J Santos, P Spolti, DJ Tessmann, EM Del Ponte (2010). Complexo *Fusarium graminearum*: Taxonomia, potencial toxigênico e genética populacional na era molecular. Rev Anual Patol Plant. 18:78-119.

Baraj E, Garda-Buffon J, Badiale-Furlong E (2010) Influence of the trichothecenes DON and T-2 toxin in malt aminolitic enzymes activity. Braz. J. Food. Technol. 53:14.

Chilosi G, Caruso C, Caporale C, Leonard L, Buzio MA, Nobile Magro P, Buonoccore, V (2000) Antifungal Activity of a Bowman-Birk type Trypsin Inhibitor from Wheat Kernel. J. Phytopathol. 148:477-481.

Del Ponte EM, Garda-Buffon J, Badiale-Furlong E (2012). Deoxynivalenol and nivalenol in commercial wheat grain related to *Fusarium* head blight epidemics. Food Chem. 132:1087-1091.

Dors GC, Primel EG, Fagundes CA, Mariot C H P, Badiale-Furlong E (2011). Distribution of pesticides in rice grain and rice bran. J. Braz. Chem. Soc. 22:1921-1930.

Figueira ELZ, Hirooka EY, Mendiola-Olaya E, Blanco-Labra, A (2003). Characterization of a Hydrophobic Amylase Inhibitor from Corn (*Zea mays*) Seeds with Activity Against Amylase from *Fusarium verticillioides.* Biochem Cell Biol. 93:917-922.

Garda-Buffon J, Baraj E, Badiale- Furlong E B (2010) Effect of the trichothecenes DON and T-2 toxin in malt aminolitic enzymes activity. Braz Arch Biol Technol. 53: 505-511.

Garda-Buffon J, Badiale-Furlong E (2004) Determinação de tricotecenos em cerveja e avaliação da incidência no produto comercializado no Rio Grande do Sul. Ciênc Tecnol Aliment. 24:657-663.

Iulek J, Franco OL, Silva M, Slivinski CT, Block C, Rigden DJ, Grossi-de-Sá MF (2001) Purification, biochemical characterization and partial primary structure of a new a-amylase inhibitor from *Secale cereale* (rye). Int J Biochem Cell Biol. 32:1195–1204.

Kupski L, Cipolatti E, Rocha M, Oliveira MS, Souza-Soares L, Badiale-Furlong E (2012) Solid-State Fermentation for the enrichment and extraction of proteins and antioxidant compounds in rice bran by *Rhizopus oryzae*. Braz. Arch. Biol. Technol. 55(6):939-944.

Lee JR, Seong-Cheol P, K Mi-Hyun, J Ji-Hyun, S Mi-Rim, D Ho-Lee, M Gyeong-Cheon, P Yoonkyung, H Kyung-Soo, L Sang-Yeol (2007). Antifungal activity of rice Pex5p, a receptor for peroxisomal matrix proteins. Biochem Biophys Res Comm. 359:941–946.

Lowry OH, NJ Rosebrough, AL Farr, RJ Randall (1951) Protein measurement with the folin-phenol reagent. J Biol Chem. 193:265-275.

Keller LAM, González Pereyra ML, Keller KM, Alonso VA, Oliveira AA, Almeida TX, Barbosa TS, Nunes LMT, Cavaglieri LR, Rosa CAR (2013) Fungal and mycotoxins contamination in corn silage: Monitoring risk before and after fermentation. J Store Products Res. 52:42-47.

Kokkonen M, Laitila JA (2012) Mycotoxin production of Fusarium langsethiae and Fusarium sporotrichioides on cereal-based substrates. Mycotoxin Res. 28:25–35.

Liu K, Cao X, Bai Q, Wen H, Gu Z (2009) Relationships between physical properties of brown rice and degree of milling and loss of selenium. J Food Eng. 94:69–74.

Marsaro-Júnior AL, Lazzari SMN, Figueira ELZ, Hirooka EY (2005) Inibidores de amilase em híbridos de milho como fator de resistência a *Sitophilus zeamais* (Coleoptera: Curculionidae). Neotrop Entomol. 34 (3):443-50.

Morcia C, Ratotti E, Stanca MA, Tumino G, Rossi V, Ravaglia S, Germeirer CU, Hermann M, Polisenska I, Terzi V (2013). Fusarium genetic traceability: Role for mycotoxin control in small grain cereals. J Cer Sci. 57:175-182.

Mosca M, Boniglia C, Carratu B, Giammarioli S, Nera V, Sanzin E (2008) Determination of alpha-amylase inhibitor activity of phaseolamin from kidney bean (*Phaseolus vulgaris*) in dietary supplements by HPAEC-PAD. Anal Chim Acta. 617:192-195.

Mosolov VV, Grigor'eva LI, Valueva TA (2001). Involvement of Proteolytic Enzymes and Their Inhibitors in Plant Protection (Review). Appl Biochem Microbiol. 37(2):115-123.

Pagnussatt FA, Bretanha CC, Garda-Buffon J, Badiale-Furlong E (2011) Extraction of fungal amylase inhibitors from cereal using response surface methodology. Int Res J Agric Sci Soil Sci. 1:428-434.

Pagnussatt FA, Meza SLR, Garda-Buffon J, Badiale-Furlong E (2012) Procedure to determine enzyme inhibitors activity in cereal seeds. J. Agric. Sci. 4:85-92.

Pagnussatt FA, Bretanha CC, Kupski C, Garda-Buffon J, Badiale-Furlong E (2013) Promising Antifungal Effect of Rice (*Oryza sativa* L.), Oat (*Avena sativa* L.) and Wheat (*Triticum aestivum* L.) Extracts. J. Appl. Biotechnol. 1:37-44.

Pereira LLS, Santos CD, Pereira CA, Marques TR, Sátiro LC (2010) Precipitação do inibidor de α-amilase de feijão branco: avaliação dos métodos. Aliment Nutr. 21:15-20.

Silveira CM, Badiale-Furlong E (2007) Caracterização de compostos nitrogenados presentes em farelos fermentados em estado sólido. Cienc Tecnol Alim. 27 (5):805-811.

Scotti C, Vergoignam T, Feron C, Durand G (2001) Glucosamine measurement as indirect method for biomass estimation of *Cunninghamella elegans* grown in solid state cultivation conditions. Biochem Eng. J. 7:1–5.

Singh S, Srivastava R, Choudhary S (2010) Antifungal and HPLC analysis of the crude extracts of *Acorus calamus, Tinospora cordifolia* and *Celestrus paniculatus*. J Agric Technol. 6(1):149-158.

Sospedra J, Blesa J, Soriano JM, Mãnes J (2010) Use of the modified quick easy cheap effective rugged and safe sample preparation approach for the simultaneous analysis of type A- and B-trichothecenes in wheat flour. J Chromatogr A. 1217:1437–1440.

Souza MM, Prieto L, Ribeiro AC, Souza TD, Badiale-Furlong E (2011) Assesment of the antifungal activity of *Spirulina platensis* phenolic extract against *Aspergillus flavus*. Cienc Agrotec. 35:1050-1058.

Tanaka T, Yoneda A, Inoue S, Sugiura Y, Ueno Y (2000) Simultaneous determination of trichothecene mycotoxins and zearalenone in cereals by gas chromatography–mass spectrometry. J. Chromatogr A. 882:23-28.

Usha CM, Patkar KLP, Shett S, Kennedy R, Lace J (1993) Fungal colonization and mycotoxin contamination of developing rice grain. Mycol. Res. 97(7):795-798.

Permissions

All chapters in this book were first published in AJAR, by Academic Journals; hereby published with permission under the Creative Commons Attribution License or equivalent. Every chapter published in this book has been scrutinized by our experts. Their significance has been extensively debated. The topics covered herein carry significant findings which will fuel the growth of the discipline. They may even be implemented as practical applications or may be referred to as a beginning point for another development.

The contributors of this book come from diverse backgrounds, making this book a truly international effort. This book will bring forth new frontiers with its revolutionizing research information and detailed analysis of the nascent developments around the world.

We would like to thank all the contributing authors for lending their expertise to make the book truly unique. They have played a crucial role in the development of this book. Without their invaluable contributions this book wouldn't have been possible. They have made vital efforts to compile up to date information on the varied aspects of this subject to make this book a valuable addition to the collection of many professionals and students.

This book was conceptualized with the vision of imparting up-to-date information and advanced data in this field. To ensure the same, a matchless editorial board was set up. Every individual on the board went through rigorous rounds of assessment to prove their worth. After which they invested a large part of their time researching and compiling the most relevant data for our readers.

The editorial board has been involved in producing this book since its inception. They have spent rigorous hours researching and exploring the diverse topics which have resulted in the successful publishing of this book. They have passed on their knowledge of decades through this book. To expedite this challenging task, the publisher supported the team at every step. A small team of assistant editors was also appointed to further simplify the editing procedure and attain best results for the readers.

Apart from the editorial board, the designing team has also invested a significant amount of their time in understanding the subject and creating the most relevant covers. They scrutinized every image to scout for the most suitable representation of the subject and create an appropriate cover for the book.

The publishing team has been an ardent support to the editorial, designing and production team. Their endless efforts to recruit the best for this project, has resulted in the accomplishment of this book. They are a veteran in the field of academics and their pool of knowledge is as vast as their experience in printing. Their expertise and guidance has proved useful at every step. Their uncompromising quality standards have made this book an exceptional effort. Their encouragement from time to time has been an inspiration for everyone.

The publisher and the editorial board hope that this book will prove to be a valuable piece of knowledge for researchers, students, practitioners and scholars across the globe.

List of Contributors

Chen Hongzhang
National Key Laboratory of Biochemical Engineering, Institute of Process Engineering, Chinese Academy of Sciences, Beijing 100190, China

Zhao Junying
National Key Laboratory of Biochemical Engineering, Institute of Process Engineering, Chinese Academy of Sciences, Beijing 100190, China
Graduate University of Chinese Academy of Sciences, Beijing 100049, China

A. Kassim
School of Engineering, Bioresources Engineering, University of KwaZulu-Natal, Pietermaritzburg, Private Bag X01, Scottsville, 3209, South Africa

T. S. Workneh
School of Engineering, Bioresources Engineering, University of KwaZulu-Natal, Pietermaritzburg, Private Bag X01, Scottsville, 3209, South Africa

C. N. Bezuidenhout
School of Engineering, Bioresources Engineering, University of KwaZulu-Natal, Pietermaritzburg, Private Bag X01, Scottsville, 3209, South Africa

J. M. Ithiru
Coffee Research Foundation, P.O. Box 4 - 00232, Ruiru, Kenya

E. K. Gichuru
Coffee Research Foundation, P.O. Box 4 - 00232, Ruiru, Kenya

P. N. Gitonga
Kenya Methodist University, P.O. Box 267 - 60200, Meru, Kenya

J. J. Cheserek
Coffee Research Foundation, P.O. Box 4 - 00232, Ruiru, Kenya

B. M. Gichimu
Coffee Research Foundation, P.O. Box 4 - 00232, Ruiru, Kenya

Cemile Temur
Department of Plant Protection, Faculty of Seyrani Agriculture, Erciyes University, Kayseri, 38039, Turkey

Osman Tiryaki
Department of Plant Protection, Faculty of Seyrani Agriculture, Erciyes University, Kayseri, 38039, Turkey

P. L. Saran
Indian Agricultural Research Institute, Regional Station, Pusa, Samastipur (Bihar)-848 125, India

Ravish Choudhary
Indian Agricultural Research Institute, Regional Station, Pusa, Samastipur (Bihar)-848 125, India

Hayrunnisa Nadaroglu
Department of Food Technology, Erzurum Vocational Training School, Ataturk University, 25240 Erzurum, Turkey

Neslihan Celebi
Department of Food Technology, Erzurum Vocational Training School, Ataturk University, 25240 Erzurum, Turkey

Nazan Demir
Department of Chemistry, Science Faculty, Mugla University, 48000 Mugla, Turkey

Yasar Demir
Department of Chemistry, Science Faculty, Mugla University, 48000 Mugla, Turkey

Radheshyam Sharma
Institute of Agri-Biotechnology (IABT), College of Agriculture, University of Agricultural Sciences, AC, Dharwad-580 005 Karnataka, India

Sumangala Bhat
Institute of Agri-Biotechnology (IABT), College of Agriculture, University of Agricultural Sciences, AC, Dharwad-580 005 Karnataka, India

Tie Manman
College of Horticulture, Sichuan Agricultural University, Ya'an 625014, Sichuan, China

Luo Qian
College of Horticulture, Sichuan Agricultural University, Ya'an 625014, Sichuan, China

Tan Huaqiang
College of Horticulture, Sichuan Agricultural University, Ya'an 625014, Sichuan, China

Zhu Yongpeng
College of Horticulture, Sichuan Agricultural University, Ya'an 625014, Sichuan, China

Lai Jia
College of Horticulture, Sichuan Agricultural University, Ya'an 625014, Sichuan, China

Li Huanxiu
College of Horticulture, Sichuan Agricultural University, Ya'an 625014, Sichuan, China

Alireza Yousefi
Department of Food Science and Technology, Shiraz University, Shiraz, Iran

Mehrdad Niakousari
Department of Food Science and Technology, Shiraz University, Shiraz, Iran

Mehdi Moradi
Department of Food Science and Technology, Shiraz University, Shiraz, Iran

B. O. Mbah
Department of Home Science, Nutrition and Dietetics, University of Nigeria Nsukka, Enugu State, Nigeria

P. E. Eme
Department of Home Science, Nutrition and Dietetics, University of Nigeria Nsukka, Enugu State, Nigeria

C. N. Eze
Department of Home Science, Nutrition and Dietetics, University of Nigeria Nsukka, Enugu State, Nigeria

Abid Ali Lone
Division of Post Harvest Technology, Sher-e-kashmir University of Agricultural Sciences and Technology of Kashmir, Shalimar, Srinagar 191121, India

Qazi Nissar Ahmed
Division of Post Harvest Technology, Sher-e-kashmir University of Agricultural Sciences and Technology of Kashmir, Shalimar, Srinagar 191121, India

Shaiq A. Ganai
Division of Post Harvest Technology, Sher-e-kashmir University of Agricultural Sciences and Technology of Kashmir, Shalimar, Srinagar 191121, India

Imtiyaz A Wani
Division of Pomology, Sher-e-kashmir University of Agricultural Sciences and Technology of Kashmir, Shalimar, Srinagar 191121, India

Rayees A. Ahanger
Division of Plant Pathology, Sher-e-kashmir University of Agricultural Sciences and Technology of Kashmir, Shalimar, Srinagar 191121, India

Hilal A. Bhat
Division of Plant Pathology, Sher-e-kashmir University of Agricultural Sciences and Technology of Kashmir, Shalimar, Srinagar 191121, India

Tauseef A. Bhat
Division of Agronomy, Sher-e-kashmir University of Agricultural Sciences and Technology of Kashmir, Shalimar, Srinagar 191121, India

D. D. Nangare
Central Institute of Post Harvest Engineering and Technology, (CIPHET), Abohar, Punjab, India

K. G. Singh
Department of Soil and Water Engineering, PAU, Ludhiana, Punjab, India

Satyendra Kumar
Central Soil Salinity Research Institute, Karnal, Haryana, India

Nyembezi Marasha
Department of Agriculture and Animal Health, College of Agriculture and Environmental Sciences, University of South Africa, Private Bag X6, Florida, 1710, South Africa

Irvine Kwaramba Mariga
Department of Soil Science, Plant Production and Agricultural Engineering, University of Limpopo, Private Bag X1106, Sovenga, 0727, South Africa

Wonder Ngezimana
Department of Agriculture and Animal Health, College of Agriculture and Environmental Sciences, University of South Africa, Private Bag X6, Florida, 1710, South Africa

Fhatuwani Nixwell Mudau
Department of Agriculture and Animal Health, College of Agriculture and Environmental Sciences, University of South Africa, Private Bag X6, Florida, 1710, South Africa

Muhammad Asim Shabbir
National Institute of Food Science and Technology, University of Agriculture, Faisalabad, Pakistan

Faqir Muhammad Anjum
National Institute of Food Science and Technology, University of Agriculture, Faisalabad, Pakistan

Moazzam Rafiq Khan
National Institute of Food Science and Technology, University of Agriculture, Faisalabad, Pakistan

Muhammad Nadeem
National Institute of Food Science and Technology, University of Agriculture, Faisalabad, Pakistan

Muhammad Saeed
National Institute of Food Science and Technology, University of Agriculture, Faisalabad, Pakistan

S. Muthu Kumar
Horticultural College and Research Institute, Tamil Nadu Agricultural University, Periyakulam, Theni District-625604, Tamil Nadu, India

V. Ponnuswami
Horticultural College and Research Institute, Tamil Nadu Agricultural University, Periyakulam, Theni District-625604, Tamil Nadu, India

Niharendu Saha
AICRP on Soil Test Crop Response Correlation, Directorate of Research, Bidhan Chandra Krishi Viswavidyalaya Kalayni-741235, Nadia, West Bengal, India

Stephan Wirth
Centre for Agriculture and Land use Research (ZALF), Institute of Microbial Ecology, MÜncheberg, Germany

Andreas Ulrich
Centre for Agriculture and Land use Research (ZALF), Institute of Microbial Ecology, MÜncheberg, Germany

S. Daudu
Department of Agricultural Extension and Communication, University of Agriculture, Makurdi, Benue State, Nigeria

M. C. Madukwe
Department of Agricultural Extension, University of Nigeria, Nsukka, Enugu State, Nigeria

Lawali Abubakar
Department of Crop Science, Faculty of Agriculture, Usmanu Danfodiyo University, P. M. B. 2346, Sokoto State, Nigeria

T. S. Bubuche
Department of Agricultural Education, College of Education, P. M. B. 1012, Argungu, Kebbi State, Nigeria

V. Arumugam
Tamil Nadu Agricultural University Coimabtore 641003, Tamil Nadu, India

E. Vadivel
Tamil Nadu Agricultural University Coimabtore 641003, Tamil Nadu, India

Nelly Nambande Masayi
School of Environment and Earth Sciences, Maseno University, P. O. Box 333- 40105 Maseno, Kenya

Godfrey Wafula Netondo
Department of Botany and Horticulture, Faculty of Science, Maseno University, P. O. Box 333- 40105 Maseno, Kenya

J. Ghezelbash
Department of Agricultural Machinery, Islamic Azad University, Science and Research Branch, Tehran, Iran

A. M. Borghaee
Department of Agricultural Machinery, Islamic Azad University, Science and Research Branch, Tehran, Iran

S. Minaei
Department of Agricultural Machinery, Islamic Azad University, Science and Research Branch, Tehran, Iran

S. Fazli
Department of Electrical and Computer, Zanjan University, Zanjan, Iran

M. Moradi
Department of Electrical and Computer, Zanjan University, Zanjan, Iran

Kaliyperumal Ashokkumar
Centre for Plant Molecular Biology, Tamil Nadu Agricultural University, Coimbatore - 641 003, Tamil Nadu, India

Karuppanasamy Senthil Kumar
Centre for Plant Breeding and Genetics, Tamil Nadu Agricultural University, Coimbatore- 641 003, Tamil Nadu, India

Rajasekaran Ravikesavan
Centre for Plant Breeding and Genetics, Tamil Nadu Agricultural University, Coimbatore- 641 003, Tamil Nadu, India

R. S. Raikwar
Jawaharlal Nehru Krishi Vishwa Vidyalaya, College of Agriculture Tikamgarh (M.P.) India

P. Srivastva
Jawaharlal Nehru Krishi Vishwa Vidyalaya, College of Agriculture Tikamgarh (M.P.) India

P. T. Bassene
Laboratoire d'Energétique Appliquée (LEA), Ecole Supérieure Polytechnique (ESP), Université Cheikh Anta Diop (UCAD) BP 5085 Dakar-Fann, Dakar, Sénégal

S. Gaye
Laboratoire d'Energétique Appliquée (LEA), Ecole Supérieure Polytechnique (ESP), Université Cheikh Anta Diop (UCAD) BP 5085 Dakar-Fann, Dakar, Sénégal

A. Talla
Laboratoire d'Energétique d'Eau et de l'Environnement, Ecole Nationale Supérieure Polytechnique (ENSP), BP 8390, Yaoundé, Cameroun

V. Sambou
Laboratoire d'Energétique Appliquée (LEA), Ecole Supérieure Polytechnique (ESP), Université Cheikh Anta Diop (UCAD) BP 5085 Dakar-Fann, Dakar, Sénégal

D. Azilinon
Laboratoire d'Energétique Appliquée (LEA), Ecole Supérieure Polytechnique (ESP), Université Cheikh Anta Diop (UCAD) BP 5085 Dakar-Fann, Dakar, Sénégal

A. P. Fernanda
Rio Grande Federal University, Food Science and Engineering Graduate Program, street Engineer Alfredo Huch, 475, P. O. box 474, CEP 96201-900, RS, Brazil

C. B. Cristiana
Rio Grande Federal University, Food Science and Engineering Graduate Program, street Engineer Alfredo Huch, 475, P. O. box 474, CEP 96201-900, RS, Brazil

L. R. M. Sílvia
Rio Grande Federal University, Food Science and Engineering Graduate Program, street Engineer Alfredo Huch, 475, P. O. box 474, CEP 96201-900, RS, Brazil

G. B. Jaqueline
Rio Grande Federal University, Food Science and Engineering Graduate Program, street Engineer Alfredo Huch, 475, P. O. box 474, CEP 96201-900, RS, Brazil

B. F. Eliana
Rio Grande Federal University, Food Science and Engineering Graduate Program, street Engineer Alfredo Huch, 475, P. O. box 474, CEP 96201-900, RS, Brazil

www.ingramcontent.com/pod-product-compliance
Lightning Source LLC
Chambersburg PA
CBHW080646200326
41458CB00013B/4747
* 9 7 8 1 6 8 2 8 6 1 4 3 1 *